CONTROL SYSTEM SYNTHESIS

The MIT Press Series in Signal Processing, Optimization, and Control
Alan S. Willsky, editor

1. *Location on Networks: Theory and Algorithms*, Gabriel Y. Handler and Pitu B. Mirchindani, 1979

2. *Digital Signal Processing and Control and Estimation Theory: Points of Tangency, Areas of Intersection, and Parallel Directions*, Alan S. Willsky, 1979

3. *Stability and Robustness of Multivariable Feedback Systems*, Michael George Safanov, 1980

4. *Issues in the Implementation of Digital Feedback Compensators*, Paul Moroney, 1983

5. *Theory and Practice of Recursive Identification*, Lennart Ljung and Torsten Söderström, 1983

6. *Approximation and Weak Convergence Methods for Random Processes, with Applications to Stochastic Systems Theory*, Harold J. Kushner, 1984

7. *Control System Synthesis: A Factorization Approach*, M. Vidyasagar, 1985

CONTROL SYSTEM SYNTHESIS: A FACTORIZATION APPROACH

M. Vidyasagar

The MIT Press
Cambridge, Massachusetts
London, England

Printed and bound in the United States of America

Library of Congress Cataloging in Publication Data

Vidyasagar, M. (Mathukumalli), 1947-
Control system synthesis.

(MIT Press series in signal processing, optimization, and control; 7)
Bibliography: p.
Includes index.
1. Control theory. 2. Factorization (Mathematics)
I. Title. II. Series.
QA402.3.V53
1985 629.8'312 84-14411
ISBN 0-262-22027-X

To the memory of my grandfather
Mathukumalli Narasimharao (1881–1965)

Contents

Series Foreword

The fields of signal processing, optimization, and control stand as well-developed disciplines with solid theoretical and methodological foundations. While the development of each of these fields is of great importance, many future problems will require the combined efforts of researchers in all of the disciplines. Among these challenges are the analysis, design, and optimization of large and complex systems; the effective utilization of the capabilities provided by recent developments in digital technology for the design of high-performance control and signal processing systems; and the application of systems concepts to a variety of applications, such as transportation systems, seismic signal processing, and data communication networks.

This series serves several purposes. It not only includes books at the leading edge of research in each field but also emphasizes theoretical research, analytical techniques, and applications that merit the attention of workers in all disciplines. In this way the series should help acquaint researchers in each field with other perspectives and techniques and provide cornerstones for the development of new research areas within each discipline and across the boundaries.

Mathukumalli Vidyasagar's book *Control System Synthesis: A Factorization Approach* is a timely and important addition to the series. In recent years the development of reliable methods for designing multivariable feedback control systems has been the subject of intensive study. Concerns such as sensitivity minimization, robustness to model uncertainties, and fault-tolerance have spurred the increased level of research activity in this area, and out of these efforts have emerged several major innovations and frameworks for design. One of these is the algebraic, factorization approach pioneered by Vidyasagar; researchers and practitioners alike are extremely fortunate that he has chosen to produce this cohesive and well-written treatment so soon after the introduction and development of the methodology.

The approach described in this book can certainly be described as elegant. Once the reader has mastered the modest amount of algebraic formalism required, he or she will be startled by the ease with which the solutions to seemingly difficult and extremely important control synthesis problems can be characterized. Furthermore, not only do these characterizations directly suggest computational algorithms for system synthesis but the machinery itself suggests a variety of other control design problems that can be formulated and readily solved.

Since the methods described in this book are of so recent vintage, one is naturally led to ask several questions. Have the essential ideas behind this approach crystallized sufficiently to offer even the possibility of presenting a treatment that can serve for some time as a worthwhile primer on the fundamentals of the methodology, or is this book simply a research monograph that will spur further work in this area? Is the nature of the material such that the audience for this book consists of a small group of specialists, or could it be used as a graduate text or as a useful reference for practitioners in the field of advanced control system design? Thanks to his outstanding skills in exposition and the great deal of thought that he put into choosing and organizing the contents, Vidyasagar has provided us with a book that should be of value to a broad audience for many years. It is true that there are exciting and important directions for future research, and those interested in pursuing such directions will find this treatment to be a treasure house of ideas. On the other hand, by focusing on a thorough development of fundamentals and by using several important control problems to illustrate the use of the machinery, Vidyasagar has ensured the place of this book as a valued reference for years to come. Furthermore, while there is certainly much to be appreciated and pursued by individuals specifically interested in research in this area, Vidyasagar has worked hard to make the development of his approach appropriate for a graduate course on the subject and accessible to as broad an audience as possible. To be sure, some investment in time is required to become familiar with the basic mathematical framework, but I believe that those who do make the investment will find it to have been most worthwhile.

Alan S. Willsky

Preface

The objective of this book is to present a comprehensive treatment of some recent results in the area of linear multivariable control that can be obtained using the so-called "factorization" approach. It is intended as a second level graduate text in linear systems and as a research monograph. The prerequisites for reading this book are covered in three appendices, but a reader encountering these topics for the first time would undoubtedly have difficulty in mastering this background on the basis of these appendices alone. Moreover, the appendices concentrate on the *mathematical* background needed to understand the material covered here, but for *motivational* background a standard first course in graduate level linear system theory would be desirable.

The central idea that is used repeatedly in the book is that of "factoring" the transfer matrix of a (not necessarily stable) system as the "ratio" of two *stable* rational matrices. This idea was first used in a paper published in 1972 (see [88]), but the emphasis there was on analyzing the stability of a *given* system rather than on the *synthesis* of control systems as is the case here. It turns out that this seemingly simple stratagem leads to conceptually simple and computationally tractable solutions to many important and interesting problems; a detailed description can be found in Chapter 1.

The starting point of the factorization approach is to obtain a simple parametrization of *all* compensators that stabilize a given plant. One could then, in principle, choose the best compensator for various applications. This idea was presented in the 1976 paper by Youla, Jabr, and Bongiorno entitled "Modern Wiener-Hopf Design of Optimal Controllers, Part II: The Multivariable Case," which can be considered to have launched this entire area of research. The viewpoint adopted in this book, namely that of setting up all problems in a ring, was initially proposed in a 1980 paper by Desoer, Liu, Murray, and Saeks. This paper greatly streamlined the Youla et al. paper and reduced the problem to its essentials. Thus virtually all of the research reported here is less than five years old, which bears out the power of this approach to formulate and solve important problems.

In writing the book, some assumptions have been made about the potential readership. First, it is assumed that the reader is already well versed in the aims and problems of control system analysis and design. Thus, for example, the book starts off discussing the problem of stabilization without any attempt to justify the importance of this problem; it is assumed that the reader already knows that stable systems are better than unstable ones. Also, as the book is aimed at professional researchers as well as practitioners of control system synthesis, a theorem-proof format has been adopted to bring out clearly the requisite conditions under which a particular statement is true, but at the same time, the principal results of each section have been stated as close to the start of the section as is practicable. In this way, a reader who is interested in pursuing the topics presented here in greater depth is enabled to do so; one who is only interested in using the results can rapidly obtain an idea of what they are by scanning the beginnings of various sections and by skipping proofs. In this connection it is worth noting that Chapter 3 is a painless introduction to the factorization approach to scalar systems that could, in my opinion, be taught to undergraduates without difficulty.

At various times, I have taught the material covered here at Waterloo, Berkeley, M.I.T., and the Indian Institute of Science. Based on these experiences, I believe that the appendices plus the first five chapters can be covered in a thirteen-week period, with three hours of lectures per week. In a standard American semester consisting of eighteen weeks of lectures with three hours of lectures per week, it should be possible to cover the entire book, especially if one starts directly from Chapter 1 rather than the appendices. Most of the sections in the appendices and the first five chapters contain several problems, which contain various ancillary results. The reader is encouraged to attempt all of these problems, especially as the results contained in the problems are freely used in the subsequent sections.

It is now my pleasure to thank several persons who aided me in this project. My wife Shakunthala was always a great source of support and encouragement during the writing of this book, which took considerably more time than either of us thought it would. Little Aparna came into being at about the same time as the third draft, which gave me a major incentive to finish the book as quickly as possible, so that I might then be able to spend more time with her. Several colleagues gave me the benefit of their helpful comments on various parts of the manuscript. Of these, I would like to mention Ken Davidson, Charlie Desoer, John Doyle, Bruce Francis, Allen Tannenbaum, George Verghese, N. Viswanadham, and Alan Willsky. My students Chris Ma and Dean Minto went over large parts of the material with a great deal of care and exposed more than one serious mistake. I would like to acknowledge my indebtedness to all of these individuals. The final camera ready copy was produced by me using the troff facility and the pic preprocessor to generate the diagrams. In this connection, I would like to thank Brian Haggman, a differential topologist turned computer hack, for much valuable advice. Finally, at the outset of this project, I was fortunate enough to receive an E. W. R. Steacie Memorial Fellowship awarded by the Natural Sciences and Engineering Research Council of Canada, which freed me from all teaching duties for a period of two years and enabled me to concentrate on the research and writing that is reflected in these pages.

I would like to conclude this preface with a historical aside, which I hope the reader will find diverting. It is easy enough to discover, even by a cursory glance at the contents of this book, that one of the main tools used repeatedly is the formulation of the general solution of the matrix equation $XN + YD = I$, where all entities are matrices with elements in a principal ideal domain. Among recent writers, Tom Kailath [49] refers to this equation as the *Bezout identity*, while V. Kučera [56] refers to it as the *Diophantine equation.* In an effort to pin down just exactly what it should be called, I started searching the literature in the area of the history of mathematics for the details of the person(s) who first obtained the general solution to the equation $ax + by = 1$, where a and b are given integers and one seeks integer-valued solutions for x and y.

It appears that the equation "Diophantine equation" was commonly applied by European mathematicians of the seventeenth century and later to any equation where the unknowns were required to assume only integer values.

The phrase honors the Greek mathematician Diophantus, who lived (in Alexandria) during the latter part of the third century A. D. However, the general solution of the linear equation in integer variables mentioned above was never studied by him. In fact, Smith [86, p. 134] states that Diophantus never studied indeterminate equations, that is equations that have more than one solution. According to Colebrooke [17, p. xiii], the first occidental mathematician to study this equation and to derive its general solution was one Bachet de Meziriac in the year 1624. The first mathematician of antiquity to formulate and find *all* solutions to this equation was an ancient Hindu by the name of Aryabhatta, born in the year 476. A detailed and enjoyable exposition of this subject can be found in the recent comprehensive book by van der Waerden [100]. Thus, in order to respect priority, I submit that the equation in question should henceforth be referred to as *Aryabhatta's identity.*

Glossary of Symbols

A 36
A_s 36
$\mathbf{A}$ 341
$\mathbf{B}$ 361
C_+ 9
C_{+e} 9
C_- 9
$\mathbf{D}$ 36
$\mathbf{F}$ 65,371
$\mathbf{F}[s]$ 387
$H(p,c)$ 44
$H(P,C)$ 101
$\mathbf{H}_\infty$ 161
$\mathbf{H}_\infty(C_+)$ 165
$\mathbf{H}_p$ 160
$\mathbf{H}_p(C_+)$ 165
$\mathbf{H}_2$ 161
$\mathbf{H}_2(C_+)$ 165
$\mathbf{H}_2^+$ 168
$\mathbf{H}_2^-$ 168
$\mathbf{M}(\mathbf{F})$ 65
$\mathbf{M}(\mathbf{H}_\infty)$ 165
$\mathbf{M}(\mathbf{H}_2)$ 165
$\mathbf{M}(\mathbf{R})$ 65
$\mathbf{M}(\mathbb{R}(s))$ 100
$\mathbf{M}(\mathbf{S})$ 17
$\|\cdot\|_p$ 161
$\|\cdot\|_\infty$ 161
$\mathbf{R}$ 65,369
$\mathbb{R}$ 9
$\mathbf{R}[s]$ 387
$\mathbb{R}[s]$ 9
$\mathbb{R}(s)$ 9
$\mathbf{R}_\infty$ 16
$\mathbf{S}$ 9
$\mathbf{S}_\mathbf{D}$ 14
$S(p)$ 46
$S(P)$ 108
$S(m,n)$ 391
T_f 162
T_F 164
$T(P,C)$ 290
$\mathbf{U}$ 19
$\mathbf{U}(\mathbf{R})$ 65
$\mathbf{U}(\mathbf{S})$ 17
$W(P,C)$ 102
$\sim$ 396
$=_L$ 396
$=_R$ 396
$\mathbf{Z}$ 370

हरिः ॐ · · ·

सहनाववतु

सहनौभुनक्तु

सहवीर्यम् करवावहैः

तेजस्विनावधीतमस्तु

माविद्विषावहैः

ॐ शान्तिः शान्तिः शान्तिः

Dear Reader,

Hari Om · · ·

May He protect us both

May He nourish us both

May we work together with great energy

May our study be vigorous and effective

May we not have ill feelings

Let there be peace peace peace

(From *Taittiriyopanishad*, c. 700 B. C.)

Chapter 1. INTRODUCTION

The objective of this book is to present a comprehensive treatment of various recent results in the area of linear multivariable control that can be obtained using the so-called *factorization approach.* The central idea of this approach is to "factor" the transfer matrix of a (not necessarily stable) system as the "ratio" of stable matrices. This seemingly simple idea gives rise to an elegant methodology which leads in a simple and natural way to the resolution of several important control problems. Moreover, the approach given here is very general, encompassing continuous-time as well as discrete-time systems, lumped as well as distributed systems, and one-dimensional as well as multi-dimensional systems, all within a single framework.

The basic problem studied in this book can be stated as follows: Given a plant P, which is assumed to be linear and time-invariant (or, in the case of multi-dimensional systems, shift-invariant), together with a set of performance specifications, design a compensator C such that the plant-compensator pair meets the performance requirements. The performance requirements studied can be classed into two types: (i) The transfer matrix of the compensated system is "desirable" in some sense to be made precise shortly. (ii) The transfer matrix of the compensated system is optimal in some sense. To put it another way, the class of compensator design problems studied here can be classed into two categories: In the first category, one is given a plant, together with a specification of a class of desirable transfer functions (or matrices), and any compensator is deemed to be acceptable so long as it results in a compensated system whose transfer matrix lies within the desirable class. A typical example of this sort of problem is that of stabilizing an unstable system. In the second category, one is given a plant, a class of desirable transfer matrices, as well as a performance measure, and the objective is to choose, among all the compensators that result in a desirable closed-loop transfer matrix, one that optimizes the performance measure. Illustrations of this class of problems include those of designing a compensator for a system so as to achieve optimal disturbance rejection, optimal robustness, optimal tracking etc., all while maintaining closed-loop stability.

Within the first class of problems, the important specification is that of the class of transfer functions that are deemed to be "desirable." This class can vary from one application to another. For instance, in the case of lumped continuous-time systems, if one is interested in the bounded-input-bounded-output (BIBO) stability of the compensated system, then the set of desirable transfer functions consists of all proper rational functions in the Laplacian variable s whose poles all lie in the open left half-plane. However, in many applications, mere BIBO stability is not enough, and it is required that the poles of the transfer matrix of the compensated system lie in some more restrictive region of stability $\mathbf{D}$, as shown in Figure 1.1.

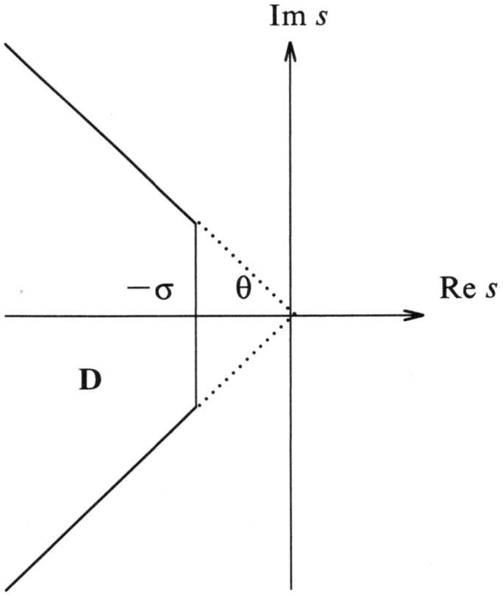

Figure 1.1. Generalized Region of Stability.

If this requirement is satisfied, then the step response of the compensated system exhibits a settling time of no more than $4/\sigma$ and a maximum overshoot corresponding to the angle θ. In the case of discrete-time systems, the stable region is the exterior of the closed unit disc, and the set of "stable" transfer functions are those rational functions that are analytic over the closed unit disc. If a more stringent form of stability is required, the "forbidden region" where the poles of desirable transfer functions may not lie can be enlarged.

In all of the above illustrations, there is one constant feature, namely: cascade and parallel connections of desirable systems are themselves desirable. In more mathematical terms, this can be restated as follows: Let $\mathbf{R}$ denote the set of desirable transfer functions in a particular application. Then sums and products of functions in the set $\mathbf{R}$ once again belong to $\mathbf{R}$. To put it another way, the set $\mathbf{R}$ is a *ring* (see Appendix A for a precise definition). This now leads to a slightly more formal statement of the first category of compensator design problems, which is referred to hereafter as the *stabilization problem:* Given a plant P, together with a ring $\mathbf{R}$ of stable transfer functions,[1] find a compensator C such that the plant-compensator pair (P,C) is stable. There are several possible compensation schemes, but it is by now well-known that the feedback type of compensation is best from the viewpoints of sensitivity reduction, robustness against modeling uncertainty, etc. There are two types of feedback systems studied in this book, namely the "one-parameter" scheme shown in Figure 1.2 and the "two-parameter" scheme shown in Figure 1.3. (The rationale behind the names is explained in Chapter 5.)

In the above problem statement, we have specified that the set $\mathbf{R}$ of desirable transfer functions is a ring, but have not specified the class of transfer functions within which the plant transfer matrix P may lie. To put it in other words, we have specified that the universe of "stable" transfer functions must be a ring, but have not specified the universe of "unstable" (meaning not necessarily stable) systems. Actually, this latter class is specified automatically as a by-product of choosing the compensation scheme.

---

[1] It is much more evocative to use the term "desirable transfer function" rather than "stable" transfer function, but the use of the latter term is by now entrenched.

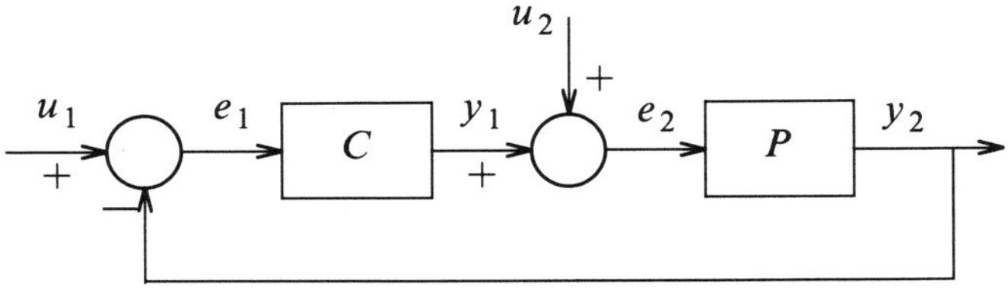

Figure 1.2. One-Parameter Compensation Scheme.

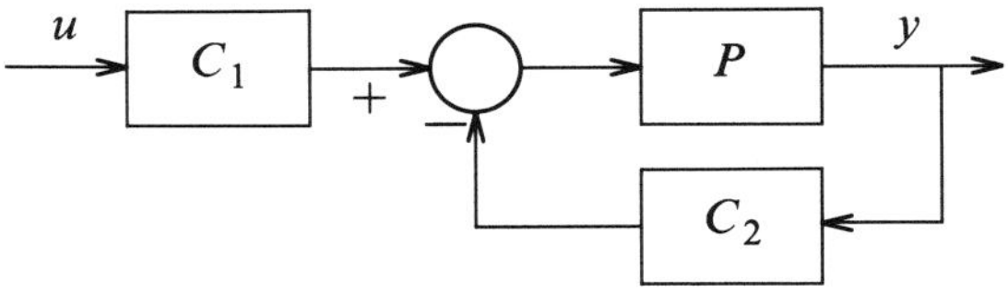

Figure 1.3. Two-Parameter Compensation Scheme.

Consider the feedback configuration shown in Figure 1.2. The transfer matrix from the two external inputs u_1 and u_2 to the outputs y_1 and y_2 of the compensator and plant, respectively, is

1 $$W(P,C) = \begin{bmatrix} C(I+PC)^{-1} & PC(I+PC)^{-1} \\ CP(I+CP)^{-1} & P(I+CP)^{-1} \end{bmatrix}.$$

This formula can be written in a more compact form. Define

2 $$G = \begin{bmatrix} C & 0 \\ 0 & P \end{bmatrix}, F = \begin{bmatrix} 0 & I \\ -I & 0 \end{bmatrix}.$$

Then

3 $$W(P,C) = G(I+FG)^{-1}.$$

Now, as discussed above, the system of Figure 1.2 is "stable" (we drop the quotation marks hereafter) if every element of the matrix $W(P,C)$ belongs to the specified ring $\mathbf{R}$ of stable transfer functions. Let W denote the matrix $W(P,C)$, and solve (3) for G. This gives

4 $$G = W(I-FW)^{-1}.$$
$$= W(I-FW)^{adj} / \mid I-FW \mid .$$

where M^{adj} and $\mid M \mid$ denote respectively the adjoint and the determinant of a square matrix M. The last expression in (4) shows that every element of the matrix G can be expressed as a ratio of two functions belonging to $\mathbf{R}$. In particular, every element of the matrix P can be expressed as a ratio of two functions in $\mathbf{R}$. Thus the conclusion is that, once a set $\mathbf{R}$ of stable transfer functions has been specified, the class of all (possibly unstable) transfer functions that can be encompassed within the theory of stabilization consists of those that can be expressed as a *ratio* of stable transfer functions.

This last observation leads to a final and precise statement of the stabilization problem, as follows: Given a ring $\mathbf{R}$ of stable transfer functions, together with a plant transfer matrix P where each element of P is a ratio of functions in $\mathbf{R}$, find *all* compensators C that result in a closed-loop transfer matrix whose elements all belong to $\mathbf{R}$.

The stabilization problem as stated above differs in a fundamental manner from those treated in earlier design theories such as LQG (linear-quadratic-Gaussian), INA (inverse Nyquist array), CL (characteristic loci) etc. The latter techniques are all addressed to producing *several*, but by no means *all*, compensators that stabilize a given plant. In contrast, the present problem has as its objective the determination of *all* stabilizing compensators corresponding to a particular plant. The reasons for formulating the problem in this manner become clear shortly.

We are now in a position to describe concisely the contents of the book. Chapter 2 is devoted to a study of the set $\mathbf{S}$, consisting of all proper rational functions whose poles are all in the open left half-plane. This is precisely the set of transfer functions of BIBO-stable lumped continuous-time systems. By examining various features of this set, such as its algebraic properties and a suitable definition of a norm, we are then able to give in Chapter 3 a quick introduction to the factorization approach for the special case of single-input-single-output systems. It is by now widely recognized that the design of multivariable control systems poses challenges that are fundamentally different in nature from those encountered in the design of scalar systems. The techniques required to extend the simple methods of Chapter 3 to multivariable systems are presented in Chapter 4, which deals with various properties of matrices whose elements all belong to a ring.

With all of this background material in hand, we then proceed to the main body of the book, which commences with Chapter 5. This chapter is addressed to the stabilization problem stated above. The first principal result of this chapter is given in Section 5.2, and consists of a simple parametrization of all compensators that stabilize a given plant, together with a corresponding parametrization of all stable transfer matrices that can be obtained as a result of stabilizing this plant. The utility of this result lies in the fact that both sets can be parametrized in terms of a single "free" parameter which ranges over all matrices (of appropriate dimensions) whose elements are stable functions, and that the stable closed-loop transfer matrix corresponding a particular compensator is an *affine* function of this parameter. This parametrization can be used to solve a variety of design problems where one is required to find a stabilizing compensator that also satisfies some additional constraints. Two typical problems are mentioned here, namely those of reliable stabilization, and regulation. *Reliable stabilization* refers to a situation where it is desired to maintain closed-loop stability in the face of plant/controller failures; this is referred to elsewhere as the problem of designing control systems with "integrity." For example, suppose one is provided with a transfer matrix P_0 that describes the plant under study during normal operating conditions, together with a set $P_1, \cdots, P_l$ of transfer matrices that describe the same plant in various failed modes, e.g. the loss of sensors and/or actuators, changes in operating point, etc. The problem of finding a compensator that stabilizes this plant reliably against the specified list of contingencies is mathematically one of finding a compensator that stabilizes *each* of the plants $P_0, \cdots, P_l$. This is called *simultaneous stabilization*, and is solved in Section 5.4. *Regulation* refers to a situation where a compensated plant is not only required to be stable, but also to track a given reference signal (such as a step, sinusoid, etc.) or to reject a disturbance signal occuring at the output of the plant or compensator. This problem need not always have a solution, and conditions for it to be solvable are given in Section 5.7. In both of the above applications, as well as others that are studied in Chapter 5, the parametrization given in Section 5.2 plays a central role, especially in concluding the nonexistence of a solution. In fact, problems such as simultaneous stabilization are almost impossible to analyze without a knowledge of *all* stabilizing compensators corresponding to a given plant, while the regulation problem is reduced to the analysis of a linear equation over a ring with the aid of the parametrization of all stable transfer matrices obtainable from a given plant.

Chapter 6 is addressed to problems of filtering and sensitivity minimization. The theme here is that one is interested in selecting, among all compensators that stabilize a given plant, one that is optimal in some sense. The criterion of optimality used here is the ability of the compensated system to track a (stable) reference signal, or to reject a (stable) disturbance signal. Depending on the norm used to measure the tracking error, one is led to two distinct classes of mathematical problems, these being filtering and sensitivity minimization. The filtering problem is closely related to the classical Wiener-Hopf problem. The sensitivity minimization problem is very recent in origin, and is also referred to as $\mathbf{H}_\infty$-norm minimization.

Chapter 7 is devoted to a study of the robustness of stabilization. This question arises because there are always errors and/or simplifications in formulating the model of the plant to be stabilized, and in implementing the compensator. Thus it is important to have an idea of the nature and extent of the uncertainties in the plant and/or compensator that can be permitted without destroying the stability of the compensated system. This issue is fully resolved in this chapter. In addition, the problems of optimal tracking and disturbance rejection are revisited, this time without the assumption that the signals being tracked or rejected are stable.

In the last chapter of the book, some of the contents of the preceding chapters are extended to more general situations. Once one leaves the realm of lumped systems (be they continuous-time or discrete-time), the mathematical structure of the underlying set of stable functions becomes much more complicated. As a consequence, some of the results of the earlier chapters do not carry over smoothly to the more general case. Several do, however, and this chapter contains an exposition of these.

In conclusion, this book presents some significant contributions to the *theory* of linear multivariable control, which make it possible to resolve several important synthesis problems in a logically simple manner. The next step is to develop computationally efficient and numerically robust algorithms to carry out the various procedures advanced here. Much work has already been done in this direction, and once this task is completed, the factorization approach will provide an attractive alternative to existing design methodologies.

Chapter 2. PROPER STABLE RATIONAL FUNCTIONS

In this chapter, we study the ring of proper stable rational functions, which plays a central role in the synthesis theory developed in this book. It is shown that this ring is a proper Euclidean domain, and the Euclidean division process is characterized. A topology is defined on this ring, and it is shown that the set of units is open. Certain generalizations of this ring are also discussed.

2.1 THE RING S AS A EUCLIDEAN DOMAIN

Let $\mathbb{R}[s]$ denote the set of polynomials in the indeterminate s with coefficients in the field $\mathbb{R}$ of real numbers. Then $\mathbb{R}[s]$ is a Euclidean domain if the degree of a polynomial is defined in the usual way (see Section A.4). The field of fractions associated with $\mathbb{R}[s]$ is denoted by $\mathbb{R}(s)$, and consists of rational functions in s with real coefficients. One can think of $\mathbb{R}(s)$ as the set of all possible transfer functions of scalar, lumped, linear time-invariant systems.

Now let $\mathbf{S}$ denote the subset of $\mathbb{R}(s)$ consisting of all rational functions that are bounded at infinity, and whose poles all have negative real parts; i.e. $\mathbf{S}$ consists of *all proper stable rational functions*. A function $p(\cdot)$ belongs to $\mathbf{S}$ if and only if it is the transfer function of a BIBO (bounded-input-bounded-output)-stable system. As is customary, let C_+ denote the closed right half-plane $\{s\colon \text{Re } s \geq 0\}$, and let C_{+e} denote the extended right half-plane, i.e. C_+ together with the point at infinity. Then a rational function belongs to $\mathbf{S}$ if and only if it has no poles in C_{+e}.

It is left to the reader to verify that, under the usual definitions of addition and multiplication in the field $\mathbb{R}(s)$, the set $\mathbf{S}$ is a commutative ring with identity, and is a domain (Problem 1). Moreover, the field of fractions associated with $\mathbf{S}$ is precisely $\mathbb{R}(s)$. It is clear that the ratio of any two elements $a, b \in \mathbf{S}$ with $b \neq 0$ belongs to $\mathbb{R}(s)$. To prove the converse, suppose $h \in \mathbb{R}(s)$, and that $h = \alpha/\beta$ where α, β are polynomials. Let n equal the larger of the degrees of α and β, and define

1 $$f(s) = \frac{\alpha(s)}{(s+1)^n},\ g(s) = \frac{\beta(s)}{(s+1)^n}.$$

Then $h(s) = f(s)/g(s)$ is a ratio of two elements of $\mathbf{S}$.

It is easy to show that a function in $\mathbf{S}$ is a unit of $\mathbf{S}$ (i.e. has an inverse in $\mathbf{S}$; see Section A.1) if and only if it has no zeros in the extended right half-plane C_{+e} (Problem 2). The units of $\mathbf{S}$ are sometimes referred to as *miniphase functions.* If $f, g \in \mathbf{S}$, then f divides g in $\mathbf{S}$ if and only if every zero of f in C_{+e} is also a zero of g with at least the same multiplicity (Problem 3). As an illustration of this, let

2 $$f_1(s) = \frac{s+1}{(s+2)^2},\ f_2(s) = \frac{s-1}{s+1},\ f_3(s) = \frac{s-1}{(s+1)^2},$$

3 $$g_1(s) = \frac{s-1}{s+4},\ g_2(s) = \frac{1}{3},\ g_3(s) = \frac{s-1}{(s+2)^3}.$$

Then f_1 divides g_2, g_3 but not g_1; f_2 divides g_1, g_3 but not g_2; and f_3 divides g_3 but not g_1, g_2.

We now define a degree function on the ring $\mathbf{S}$ such that it becomes a Euclidean domain.

4 *Theorem* Define a function $\delta : \mathbf{S} - 0 \rightarrow \mathbf{Z}_+$ as follows: If $f \in \mathbf{S} - 0$, then

5 $$\delta(f) = \text{no. of zeros of } f \text{ in } C_{+e}$$
$$= \text{no. of zeros of } f \text{ in } C_+ + \text{relative degree of } f.$$

Then $\mathbf{S}$ is a proper Euclidean domain with degree function δ.

The proof of Theorem (4) requires a preliminary result.

6 *Lemma* Suppose $\alpha,\beta,\omega \in \mathbb{R}[s]$, with α and β coprime. Then there exist $\phi,\psi \in \mathbb{R}[s]$, with the degree of ϕ less than the degree of β, such that

7 $$\alpha\phi + \beta\psi = \omega.$$

Proof Since α and β are coprime, there exist $\phi_1,\psi_1 \in \mathbb{R}[s]$ such that

8 $$\alpha\phi_1 + \beta\psi_1 = 1.$$

Multiplying both sides of (8) by ω and denoting $\phi_1\omega$ by ϕ_2, $\psi_1\omega$ by ψ_2 gives

9 $$\alpha\phi_2 + \beta\psi_2 = \omega.$$

Now, for *any* $\theta \in \mathbb{R}[s]$, it follows from (9) that

10 $$\alpha(\phi_2 - \beta\theta) + \beta(\psi_2 + \alpha\theta) = \omega.$$

Since $\mathbb{R}[s]$ is a Euclidean domain, by proper choice of θ we can make the degree of $\phi_2 - \beta\theta$ lower than that of β. Choosing such a θ and defining $\phi = \phi_2 - \beta\theta$, $\psi = \psi_2 + \alpha\theta$ gives (7). □

Proof of Theorem (4) Let $f,g \in \mathbf{S}$ with $g \neq 0$. It is shown that there exists a q such that $\delta(f-gq) < \delta(g)$, where $\delta(0)$ is taken as $-\infty$. If $\delta(g) = 0$, then g is a unit. In this case, take $q = fg^{-1}$; this gives $f-gq = 0$, $\delta(f-gq) = -\infty < \delta(g)$.

Now suppose $\delta(g) > 0$, and suppose $g = \alpha_g/\beta_g$, where $\alpha_g,\beta_g \in \mathbb{R}[s]$ are coprime. Factor α_g as a product $\alpha_-\gamma$ where all zeros of α_- are in C_- and all zeros of γ are in C_+. Express g in the form

11 $$g(s) = e(s)\frac{\gamma(s)}{(s+1)^n}, \text{ where } e(s) = \frac{\alpha_-(s)(s+1)^n}{\beta_g(s)},$$

with $n = \delta(g)$, and observe that e is a unit of **S**. Let $f = \alpha_f/\beta_f$, where $\alpha_f,\beta_f \in \mathbb{R}[s]$ are coprime. Since all zeros of β_f are in C_- and all zeros of γ are in C_+, it follows that β_f,γ are coprime in $\mathbb{R}[s]$. Hence there exist $\phi,\psi \in \mathbb{R}[s]$ such that the degree of ϕ is less than that of β_f, and

12 $$\gamma(s)\phi(s) + \beta_f(s)\psi(s) = \alpha_f(s)(s+1)^{n-1}.$$

Dividing both sides of (12) by $\beta_f(s)(s+1)^{n-1}$ gives

13 $$\frac{\gamma(s)}{(s+1)^n}\cdot\frac{\phi(s)(s+1)}{\beta_f(s)} + \frac{\psi(s)}{(s+1)^{n-1}} = \frac{\alpha_f(s)}{\beta_f(s)}.$$

So if we define

14 $$q(s) = \frac{\phi(s)(s+1)}{\beta_f(s)}\cdot\frac{1}{e(s)},$$

15 $$r(s) = \frac{\psi(s)}{(s+1)^{n-1}},$$

then it follows from (13) that $f = gq + r$. Moreover, since the degree of ϕ is *less* than that of β_f, q is proper; q is also stable, since all zeros of β_f are in C_-. Hence $q \in \mathbf{S}$. Since $r = f - gq$, r also belongs to $\mathbf{S}$. In particular, r is proper, which shows that the degree of ψ is less than or equal to $n-1$. Hence, from (5), $\delta(r) < \delta(g)$. Since the above procedure can be repeated for every $f, g \in \mathbf{S}$ with $g \neq 0$, we have shown that $\mathbf{S}$ is a Euclidean domain with degree function δ.

It is left to the reader to show that $\mathbf{S}$ is not a field and that $\delta(fg) = \delta(f) + \delta(g)\ \forall f, g \in \mathbf{S}$. Hence $\mathbf{S}$ is a *proper* Euclidean domain (see Problem 4). □

Let

16 $$f(s) = \frac{s+4}{s+1},\quad g(s) = -\frac{2s+3}{s+1}.$$

Then both f and g are units of $\mathbf{S}$ and accordingly $\delta(f) = \delta(g) = 0$. However,

17 $$f(s) + g(s) = \frac{-s+1}{s+1}$$

has a degree of 1. Hence in general it is *not* true that $\delta(f+g) \leq \max\{\delta(f), \delta(g)\}$. As a result, the Euclidean division process does not result in a unique remainder as is the case with polynomials; i.e., given $f,g \in \mathbf{S}$ with $g \neq 0$, there may exist *several* $q \in \mathbf{S}$ such that $\delta(f-gq) < \delta(g)$. For example, let

18 $$f(s) = \frac{(s-1)(s^2+2s+2)}{(s+1)^3}, \quad g(s) = \frac{s+5}{(s+1)^3}.$$

Then $\delta(g) = 2$ since g has no zeros in C_+ but a double zero at infinity. Now suppose c is a nonnegative constant, and consider $\delta(f+cg)$. It is easy to verify using root-locus arguments that

19 $$\delta(f+cg) = \begin{cases} 1 & \text{if } 0 \leq c \leq 0.4 \\ 0 & \text{if } 0.4 \leq c < 0.5 \\ 2 & \text{if } c \geq 0.5 \end{cases}$$

Thus there exist *infinitely many* c such that $\delta(f+cg) < \delta(g)$; moreover the degree of the remainder $\delta(f+cg)$ can be made to assume *several* distinct values less than $\delta(g)$ by suitable choice of c. These phenomena have no analogs in the case of the polynomial ring $\mathbb{R}[s]$. In view of this, an important question is the following: Given $f,g \in \mathbf{S}$ with $g \neq 0$, what is the *smallest* value that $\delta(f-gq)$ can achieve as q varies over $\mathbf{S}$? This question is answered in Section 2.3 (see Theorem (2.3.5)).

The primes in the ring $\mathbf{S}$ are functions of the form $1/(s+a)$ where $a>0$; $(s-a)/(s+b)$ where $a \geq 0, b>0$; $[(s-a)^2+b^2]/(s+c)^2$ where $a \geq 0$, $b,c>0$; and their associates. Since two functions f and g are coprime if and only if they have no common prime divisors (Fact (A.3.15)), it follows that f and g are coprime if and only if they have no common zeros in the extended RHP C_{+e}, or equivalently, if they have no common zeros in C_+ and at least one of them has relative degree zero. This leads to the following observation.

20 *Fact* Suppose $f \in \mathbb{R}(s)$ and express f as α/β where $\alpha,\beta \in \mathbb{R}[s]$ have no common zeros in C_+. Let n equal the larger of the degrees of α and β, and define

21 $$a(s) = \frac{\alpha(s)}{(s+1)^n}, \; b(s) = \frac{\beta(s)}{(s+1)^n}.$$

Then $a,b \in \mathbf{S}$ are coprime, and $f = a/b$ is a reduced form for f viewed as a fraction over $\mathbf{S}$. A point $s \in C_{+e}$ is a zero (resp. pole) of f if and only if it is a zero of a (resp. a zero of b). If so, the order of s as a zero (resp. pole) of f equals its multiplicity as a zero of a (resp. zero of b).

Note that the polynomials α,β defined above *can* have common zeros in the *left* half-plane without violating the above conclusions. As an illustration, suppose

22 $$f(s) = \frac{(s-1)(s+2)}{(s-2)(s-3)}$$

and let

23 $$\alpha(s) = (s-1)(s+2)(s+4), \; \beta(s) = (s-2)(s-3)(s+4).$$

Then

24 $$a(s) = \frac{(s-1)(s+2)(s+4)}{(s+1)^3}, b(s) = \frac{(s-2)(s-3)(s+4)}{(s+1)^3}$$

are coprime in $\mathbf{S}$. Note that $s = -4$ is a zero of a even though it is not a zero of f. But this does not contradict Fact (20), as the latter only applies to points $s \in C_{+e}$.

Quite often, the objective of control system design is not merely to stabilize a given plant but to place the closed-loop poles in some pre-specified region of the left half-plane. In such applications, it is desirable to replace the open left half-plane C_- by a more general *domain of stability* $\mathbf{D}$ which is open and has the property that $\bar{s} \in \mathbf{D}$ whenever $s \in \mathbf{D}$ (here the bar denotes complex conjugation). In this case, one replaces the set $\mathbf{S}$ by the set $\mathbf{S_D}$, which consists of all proper rational functions whose poles all lie in $\mathbf{D}$. Thus $\mathbf{S}$ is a special case of $\mathbf{S_D}$, with $\mathbf{D} = C_-$. It is left to the reader to verify that $\mathbf{S_D}$ is a commutative domain with identity; moreover, if we define a function $\delta: \mathbf{S_D} - 0 \rightarrow \mathbf{Z}_+$ by

25 $\delta(f)$ = Relative degree of f + no. of zeros of f outside **D**,

Then $\mathbf{S_D}$ is a proper Euclidean domain provided the region **D** contains at least one real number (see Problem 5).

For instance, if it is desired to design continuous-time control systems with prescribed maximum settling time and miminum damping, the domain of stability **D** should be taken as in Figure 2.1.

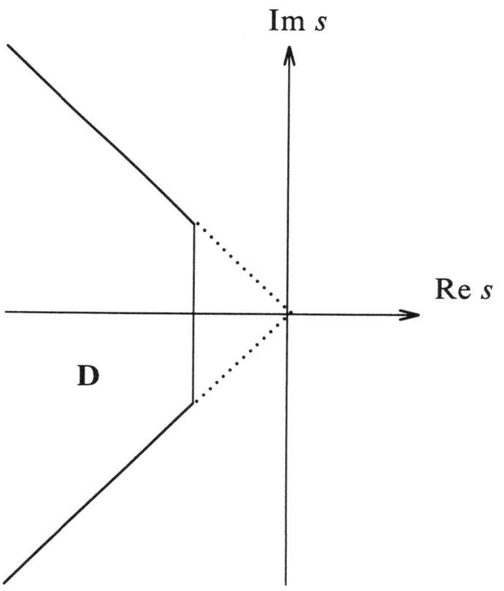

Figure 2.1. Generalized Region of Stability.

In the study of discrete-time control systems, the usual notion of stability is that the unit pulse response of the system is absolutely summable. Thus, if a system has the unit pulse response sequence $\{h_i\}_{i=0}^{\infty}$, the system is BIBO-stable if and only if [D1, Chapter 6]

26 $$\sum_{i=0}^{\infty} |h_i| < \infty.$$

Define the transfer function of the system by

27 $$f(z) = \sum_{i=0}^{\infty} h_i z^i.$$

Then a system is BIBO-stable if and only if the power series defining its transfer function converges absolutely for all z with $|z| \leq 1$. If the system is lumped, then the transfer function $f(z)$ is rational. In this case the system is BIBO-stable if and only if $f(\cdot)$ has no poles on or inside the unit circle in the complex plane. The set of such rational functions is denoted by $\mathbf{R}_\infty$ (the rationale for this notation can be found in Chapter 6). If we define a function $\delta : \mathbf{R}_\infty - 0 \rightarrow \mathbf{Z}_+$ by

28 $$\delta(f) = \text{no. of zeros of } f \text{ inside the closed unit circle},$$

then $\mathbf{R}_\infty$ becomes a proper Euclidean domain (Problem 7).

The reader is cautioned that the definition (27) of the z-transform of a sequence differs from the conventional one, which is

29 $$g(z) = \sum_{i=0}^{\infty} h_i z^{-i}.$$

Clearly $g(z) = f(z^{-1})$. The current definition has some advantages, one of which is that every polynomial in $\mathbb{R}[z]$ belongs to $\mathbf{R}_\infty$; in fact, elements of $\mathbb{R}[z]$ correspond to so-called finite impulse response (FIR) filters.

There is a close connection between the sets $\mathbf{S}$ and $\mathbf{R}_\infty$, which represent the sets of stable transfer functions in the case of continuous-time and discrete-time systems, respectively. Specifically, if we define the bilinear mapping

30 $$z = \frac{s-1}{s+1}$$

that takes the extended right half-plane C_{+e} into the closed unit disc $\mathbf{D}$, then there is a one-to-one correspondence between functions in $\mathbf{S}$ and in $\mathbf{R}_\infty$ via the association

31 $$f(s) \rightarrow g(z) = f((1+z)/(1-z)).$$

The inverse correspondence is of course given by

32 $$g(z) \rightarrow f(s) = g((s-1)/(s+1)).$$

Problems

1. Show that **S** is a commutative domain with identity.

2. Show that a function in **S** has an inverse in **S** if and only if it has no zeros in C_{+e}.

3. Suppose $f, g \in \mathbf{S}$. Show that f divides g in **S** (i.e. $g/f \in \mathbf{S}$) if and only if every zero of f in C_{+e} is also a zero of g of at least the same multiplicity.

4. Show that **S** is a proper Euclidean domain by showing that $\delta(fg) = \delta(f) + \delta(g)\ \forall f, g \in \mathbf{S}-0$.

5. Show that if the region **D** contains at least one real number, then $\mathbf{S_D}$ is a proper Euclidean domain with degree function δ defined in (25).

6. Let $\mathbb{R}^{-}(s)$ consist of those functions in $\mathbb{R}(s)$ whose poles are all in the open left half-plane (but which need not be proper). Show that $\mathbb{R}^{-}(s)$ is a proper Euclidean domain if we define $\delta(f)$ = no. of zeros of f in C_{+}.

7. Show that $\mathbf{R}_{\infty}$ is a proper Euclidean domain with the degree function δ defined in (28).

2.2 TOPOLOGY ON S AND M(S)

In this section, we define a topology on the ring **S** of proper stable rational functions, and show that the set **U** of units of **S** is open in the proposed topology. This topology is extended to the set **M(S)** of matrices with elements in **S**, and it is shown that the set **U(S)** of unimodular matrices is open in **M(S)**. The generalizations to the ring $\mathbf{S_D}$ are also indicated.

1 *Definition* The norm function $\|\cdot\|:\mathbf{S}\rightarrow\mathbb{R}$ is defined by

2 $$\|f\| = \sup_{s\in C_+} |f(s)|.$$

Since every function in $\mathbf{S}$ is analytic over the open right half-plane, it follows from the maximum modulus theorem [77, p.229] that

3 $$\|f\| = \sup_{\omega\in\mathbb{R}} |f(j\omega)|.$$

The quantity in (3) can be interpreted simply in terms of the Nyquist plot of $f(\cdot)$: The norm of f is the radius of the smallest circle (centered at the origin) that contains the Nyquist plot of f (see Figure 2.2).

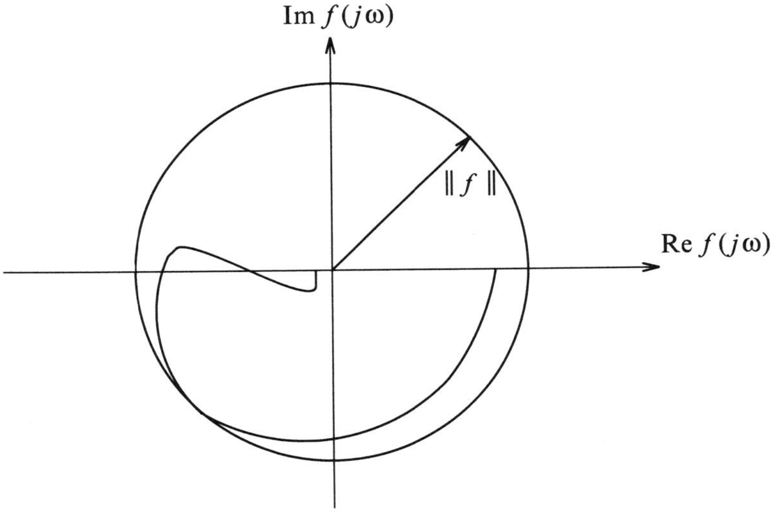

Figure 2.2 Graphical Interpretation of the Norm of a Function.

It can be verified that, with this norm, $\mathbf{S}$ becomes a normed algebra over the field of real numbers $\mathbb{R}$. Hence there is a natural topology on $\mathbf{S}$ induced by this norm. For every $f\in\mathbf{S}$ and $\varepsilon>0$, define the ball $\mathbf{B}(f,\varepsilon)$ by

4 $$\mathbf{B}(f,\varepsilon) = \{g \in \mathbf{S}: \|f-g\| < \varepsilon\}.$$

Then the collection of balls $\mathbf{B}(f,\varepsilon)$ forms a base for a topology on $\mathbf{S}$. In this topology, a set $\mathbf{A}$ is open if (and only if), corresponding to every $f \in \mathbf{A}$, there is a ball $\mathbf{B}(f,\varepsilon) \subseteq \mathbf{A}$. A set $\mathbf{B}$ is closed if and only if its complement is open, or equivalently, the limit of any convergent sequence in $\mathbf{B}$ again belongs to $\mathbf{B}$. Also, a sequence $\{f_i(\cdot)\}$ in $\mathbf{S}$ converges to $f \in \mathbf{S}$ if and only if $f_i(s) \rightarrow f(s)$ uniformly for all $s \in C_+$, or equivalently, if $f_i(j\omega) \rightarrow f(j\omega)$ uniformly for all $\omega \in \mathbb{R}$; this follows from (2) and (3).

Recall from Section 2.1 that a function $f \in \mathbf{S}$ is a unit of $\mathbf{S}$ (i.e. $1/f \in \mathbf{S}$) if and only if f has no zeros in the extended right half-plane C_{+e}. An equivalent statement is the following: $f \in \mathbf{S}$ has an inverse in $\mathbf{S}$ if and only if

5 $$\inf_{s \in C_+} |f(s)| =: c(f) > 0.$$

Further, if f is a unit, then it is immediate that

6 $$\sup_{s \in C_+} |1/f(s)| = [\inf_{s \in C_+} |f(s)|]^{-1},$$

so that $\|f^{-1}\| = 1/c(f)$. In particular, $1+f$ is a unit whenever $\|f\| < 1$, since

7 $$c(1+f) = \inf_{s \in C_+} |1+f(s)|$$

$$\geq 1 - \sup_{s \in C_+} |f(s)| = 1 - \|f\| > 0.$$

Moreover, $\|(1+f)^{-1}\| = 1/c(1+f) \leq 1/(1 - \|f\|)$. The next lemma presents a generalization of this fact and also establishes one of the most useful properties of the topology on $\mathbf{S}$.

8 *Lemma* Let $\mathbf{U}$ denote the set of units of $\mathbf{S}$, and suppose $f \in \mathbf{U}$. Then $g \in \mathbf{U}$ whenever $\|g-f\| < 1/\|f^{-1}\|$, in which case

9 $$\| g^{-1} - f^{-1} \| \leq \| f^{-1} \| \frac{\| f^{-1} \| \cdot \| g-f \|}{1- \| f^{-1} \| \cdot \| g-f \|}.$$

As a consequence, **U** is an open subset of **S** and the map $f \rightarrow f^{-1}$ mapping **U** into itself is continuous. Finally, the above bound is the best possible in the sense that if f is any unit, then there exists a nonunit g such that $\| g-f \| = 1/ \| f^{-1} \|$.

Proof Suppose $f \in \mathbf{U}$ and that $\| g-f \| < 1/ \| f^{-1} \|$. Then $\| gf^{-1} - 1 \| \leq \| g-f \| \cdot \| f^{-1} \| < 1$. It follows from the preceding discussion that $1+[gf^{-1}-1] = gf^{-1}$ is a unit, whence g is also a unit. To establish the bound (9), observe that

10 $$\| g^{-1} - f^{-1} \| \leq \| g^{-1} \| \cdot \| f-g \| \cdot \| f^{-1} \|$$
$$\leq [\| f^{-1} \| + \| g^{-1} - f^{-1} \|] \cdot \| f-g \| \cdot \| f^{-1} \|.$$

11 $$[1 - \| f-g \| \cdot \| f^{-1} \|] \, \| g^{-1} - f^{-1} \|$$
$$\leq \| f^{-1} \|^2 \cdot \| f-g \|.$$

Now (9) follows readily from (11).

The proof that the bound is the best possible is somewhat involved. Since f is a unit, f^{-1} is analytic over C_+, which implies that f attains its *minimum* modulus over C_+ on the $j\omega$-axis or at infinity. Thus, either $|f(\infty)| = c(f)$ or else there exists a finite ω_0 such that $|f(j\omega_0)| = c(f)$.

Case (i) $|f(\infty)| = c(f)$. Since $f(\infty)$ is real, it equals $\pm c(f)$. Let h equal the *constant* function *sign* $c(f) \cdot c(f)$. Then $\| h \| = c(f)$ but $g = f - h$ is not a unit since $g(\infty) = (f-h)(\infty) = 0$.

Case (ii) $|f(0)| = c(f)$. Once again $f(0)$ is real and must equal $\pm c(f)$. This is similar to the above case.

Case (iii) $|f(j\omega_0)| = c(f)$ for some finite nonzero ω_0. The additional complication in this case is that $f(j\omega_0)$ could be nonreal. If $f(j\omega_0)$ is real simply choose h as in the above two cases. Otherwise, assume without loss of generality that $\omega_0 > 0$ (since $|f(j\omega_0)| = |f(-j\omega_0)|$), and express $f(j\omega_0)$ in the form $\pm c(f)\exp(j\theta_0)$ where $\theta_0 \in (-\pi, 0]$. Let h be the all-pass function

12 $$h(s) = \pm c(f) \frac{s-\alpha}{s+\alpha}$$

where $\alpha \geq 0$ is adjusted so that $h(j\omega_0) = f(j\omega_0)$. Then $\| h \| = c(f)$, but $g = f - h$ is not a unit. □

Consider the set $\mathbf{M}(\mathbf{S})$ of matrices with elements in the ring $\mathbf{S}$. The norm of a matrix-valued function $A \in \mathbf{M}(\mathbf{S})$ is defined by

13 $$\| A \| = \sup_{s \in C_+} \bar{\sigma}(A(j\omega)),$$

where $\bar{\sigma}(\cdot)$ denotes the largest singular value of a matrix.[1] Several comments are in order concerning this norm.

(i) Recall that the *Euclidean norm* of a vector x in C^n is defined by

14 $$\| x \|_2 = (x^* x)^{1/2}.$$

where * denotes the conjugate transpose. If $M \in C^{m \times n}$, its *Euclidean induced norm* is defined by

15 $$\| M \|_2 = \sup_{x \in C^n - 0} \frac{\| Mx \|_2}{\| x \|_2}$$

and equals $\bar{\sigma}(M)$. Thus the norm of a matrix $A \in \mathbf{M}(\mathbf{S})$ can also be defined by

16 $$\| A \| = \sup_{s \in C_+} \| A(s) \|_2.$$

If M, N are complex matrices of compatible dimensions, then $\| MN \|_2 \leq \| M \|_2 \cdot \| N \|_2$. Based on this and (16), it is immediate if $A, B \in \mathbf{M}(\mathbf{S})$ and the product AB is defined, then

---

[1] Recall that the largest singular value of a matrix M is defined as the square-root of the largest eigenvalue of the symmetric matrix M^*M.

17 $$\| AB \| \leq \| A \| \cdot \| B \| .$$

(ii) In general, even if $A(\cdot)$ is analytic over C_+, the function $s \rightarrow \bar{\sigma}(A(s))$ is not. Thus the maximum modulus principle does not apply to $\bar{\sigma}(A(\cdot))$. Nevertheless, it is true that

18 $$\sup_{\omega \in \mathbb{R}} \bar{\sigma}(A(j\omega)) = \sup_{s \in C_+} \bar{\sigma}(A(s)) = \| A \| .$$

Thus the norm of $A \in \mathbf{M(S)}$ can be computed based on the behavior of $A(\cdot)$ on the $j\omega$-axis alone.

The topology on $\mathbf{M(S)}$ is the one induced by the norm (13). It can be shown to be the same as the product topology on $\mathbf{M(S)}$ obtained from the one on $\mathbf{S}$. In simpler terms, this means the following: Suppose $\{A^l\}$ is a sequence of matrices, and A is another matrix, all of the same order, in $\mathbf{M(S)}$. Then $\| A^l - A \| \rightarrow 0$ as $l \rightarrow \infty$ if and only if $\| a_{ij}^l - a_{ij} \| \rightarrow 0$ as $l \rightarrow \infty$ for each i,j. In other words, the sequence of *matrices* $\{A^l\}$ converges to the *matrix* A in the sense of the norm (13) if and only if each component sequence of *scalars* a_{ij}^l converges to the *scalar* a_{ij} in the sense of the norm (2). Thus convergence in $\mathbf{M(S)}$ is equivalent to component-wise uniform convergence over C_+ (or over the $j\omega$-axis).[2]

The next result generalizes Lemma (8) to the matrix case.

19 *Lemma* Suppose F is a unit of $\mathbf{M(S)}$ and that $\| G - F \| < \| F^{-1} \|$. Then G is also a unit, and

20 $$\| G^{-1} - F^{-1} \| \leq \| F^{-1} \| \frac{\| F^{-1} \| \cdot \| G - F \|}{1 - \| F^{-1} \| \cdot \| G - F \|} .$$

Moreover, this bound is the best possible in the sense that if F is any unit of $\mathbf{M(S)}$, then there exists a nonunit matrix G such that $\| G - F \| = \| F^{-1} \|$.

---

[2] The reason for drawing attention to this point at great length is that this is a property special to *stable* systems, and is not true for sequences of *unstable* transfer matrices; see Section 7.2.

Proof Recall that G is a unit of $\mathbf{M}(\mathbf{S})$ if and only if $|G|$ is a unit of $\mathbf{S}$. We begin by showing that if $R \in \mathbf{M}(\mathbf{S})$ and $\| R \| < 1$, then $I+R$ is a unit of $\mathbf{M}(\mathbf{S})$. For each $s \in C_{+e}$, the spectrum of the matrix $I+R(s)$ consists of $\{1+\lambda_i(s), i=1, \cdots ,n\}$, where $\lambda_1(s), \cdots ,\lambda_n(s)$ are the eigenvalues of $R(s)$. Since $\| R \| < 1$, it follows that the norm of the *matrix* $R(s)$ is less than one for all $s \in C_{+e}$. Hence $|\lambda_i(s)| \leq \| R(s) \| < 1 \ \forall i, \forall s \in C_{+e}$. As a result $1+\lambda_i(s) \neq 0 \ \forall i, \forall s \in C_{+e}$, whence $|I+R(s)| = \prod(1+\lambda_i(s)) \neq 0 \ \forall s \in C_{+e}$. This shows that $|I+R|$ is a unit of $\mathbf{S}$, so that $I+R$ is a unit of the matrix ring $\mathbf{M}(\mathbf{S})$.

Now if F is a unit and $\| G-F \| < 1/\| F^{-1} \|$, then $\| GF^{-1} - I \| < \| G-F \| \cdot \| F^{-1} \| < 1$, so that $GF^{-1} = I+(GF^{-1}-I)$ is a unit of $\mathbf{M}(\mathbf{S})$, by the preceding argument. Thus G is also a unit.

The proof of the inequality (20) is entirely analogous to that of (9) and is left to the reader.

To prove that the bound is the best possible, we construct a matrix $H \in \mathbf{M}(\mathbf{S})$ such that $\| H \| = \| F^{-1} \|$ but $G = F-H$ is not a unit. The argument exactly follows that in the scalar case after taking into account some additional complications arising due to the fact of dealing with matrices. Now (16) shows that either $\| F^{-1}(\infty) \|_2 = \| F^{-1} \|$, or else $\| F^{-1}(j\omega) \|_2 = \| F^{-1} \|$ for some finite ω.

Case (i) $\| F^{-1}(\infty) \|_2 = \| F^{-1} \|$. Since $F^{-1}(\infty)$ is a *real* matrix, there exists a *real* vector v such that $\| v \|_2 = 1$, and $\| F^{-1}(\infty)v \|_2 = \| F^{-1}(\infty) \|_2$. For brevity let c denote the constant $\| F^{-1}(\infty) \|_2 = \| F^{-1} \|$, and let u denote the vector $F^{-1}(\infty)v$. Now let H equal the constant matrix vu'/c^2. Then routine calculations show that $\| H \| = 1/c$. However, $[F(\infty)-H]u = 0$, which shows that $|(F-H)(\infty)| = 0$. Thus $G = F-H$ is not a unit.

Case (ii) $\| F^{-1}(0) \|_2 = \| F^{-1} \|$. Since $F^{-1}(0)$ is a real matrix, the above proof applies.

Case (iii) $\| F^{-1}(j\omega) \|_2 = \| F^{-1} \|$ for some finite nonzero ω, which can be assumed to be positive without loss of generality. As in the scalar case, the possible complication is that $F(j\omega)$ may not be a real matrix. Select a complex vector v such that $\| v \|_2 = 1$ and $\| F^{-1}(j\omega)v \|_2 = \| F^{-1} \|$. As before let c denote $\| F^{-1} \|$ and let u denote $F^{-1}(j\omega)v$. Express each component of u,v in the form

21 $$u_i = \bar{u}_i \exp(j\theta_i),\ v_i = \bar{v}_i \exp(j\phi_i),\ i=1, \cdots, n.$$

where $\bar{u}_i, \bar{v}_i$ are real and $\theta_i, \phi_i \in (-\pi, 0]\ \forall i$. Define the vector-valued functions $a(s)$, $b(s)$ by

22 $$a_i(s) = \bar{v}_i \frac{s-\alpha_i}{s+\alpha_i},\ b_i(s) = \bar{u}_i \frac{s-\beta_i}{s+\beta_i},$$

where the nonnegative constants α_i, β_i are chosen such that $a(j\omega) = v$, $b(j\omega) = u$. Finally, define $H(s) = c^{-2}a(s)b'(s)$. Then $\| H \| = c$, but $(F-H)(j\omega)$ is singular, so that $G = F-H$ is not a unit. □

All of the preceding development can be carried out in the more general ring $\mathbf{S_D}$, consisting of all proper rational functions whose poles all lie in the domain of stability $\mathbf{D}$, subject to a few technical conditions. Specifically, the norm of a function f in $\mathbf{S_D}$ is defined as

23 $$\| f \| = \sup_{s \in \mathbf{D}^c} | f(s) |,$$

where the superscript "c" denotes the complement of a set. In order for this norm to be well-defined, it is necessary for the set $\mathbf{D}$ to be open, or equivalently, for its complement to be closed.

Finally, consider the case of discrete-time systems, where one is interested in the set $\mathbf{R}_\infty$, consisting of all rational functions that are analytic on the closed unit disc. In this case one can define the norm of a function f in $\mathbf{R}_\infty$ as

24 $$\|f\| = \sup_{|z| \leq 1} |f(z)|.$$

All of the previously derived bounds etc. carry over to either of these more general situations.

In conclusion, note that the correspondence between the sets **S** and $\mathbf{R}_\infty$ defined in (2.1.30-31) is actually norm-preserving if the norms on these two algebras are defined as in (2) and (24) respectively.

2.3 EUCLIDEAN DIVISION IN S

As shown in Section 2.1, the ring **S** is a proper Euclidean domain if the degree of a function f is defined as the number of zeros of f in the extended right half-plane C_{+e}. Thus, given $f, g \in \mathbf{S}$ with $g \neq 0$, there exists a $q \in \mathbf{S}$ such that $\delta(f-gq) < \delta(g)$. However, such a q need not be unique. The main objective of this section is to answer the question: What is the smallest value that $\delta(f-gq)$ can achieve as q is varied? As a prelude to answering this question, we derive a very useful interpolation result. It turns out that this result can be proved in two ways: one proof is elementary but very tedious, and this is the one given in this section. A much more elegant proof requiring more advanced concepts is given in the next section.

Recall that a function f in **S** is a unit if and only if f has no zeros in C_{+e}, or equivalently $\delta(f) = 0$. Now suppose $S = \{s_1, \cdots, s_n\}$ is a set of points in C_{+e}, $M = \{m_1, \cdots, m_n\}$ is a corresponding set of postive integers, and $R = \{r_{ij}, j = 0, \cdots, m_i - 1, i = 1, \cdots, n\}$ is a corresponding set of complex numbers. We are interested in knowing whether or not there exists a unit $u(\cdot)$ in **S** that satisfies the interpolation constraints

1 $$\frac{d^j}{ds^j} u(s_i) = r_{ij}, j = 0, \cdots, m_i - 1; i = 1, \cdots, n,$$

where the zeroth order derivative of u is taken as u itself. In other words, we would like to know whether or not there exists a unit u such that its functional and derivative values at specified points s_i in C_{+e} are equal to the specified values r_{ij}.

Before tackling the problem, a few simplifying observations are made. First, since u is a rational function with *real* coefficients, as are all its derivatives,

$$2 \qquad \frac{d^j}{ds^j}\, u(s_0) = \overline{\{\frac{d^j}{ds^j}\, u(\bar{s}_0)\}} \;\; \forall s_0 \in C_{+e},$$

where the bar denotes complex conjugation. Thus, in order for a unit u to exist that satisfies (1), two necessary conditions must be satisfied: (i) If s_i is real for some i, then the corresponding r_{ij} must be real for all j. (ii) If $s_i = \bar{s}_k$ for some i,k, then $r_{ij} = \bar{r}_{kj}$ for all j. Thus we may as well assume at the outset that $\text{Im } s_i \geq 0$ for all i, and that r_{ij} is real whenever s_i is real. There is one other obvious necessary condition: If u is a unit, it can have no zeros in C_{+e}. Hence the *function value* $r_{i0} \neq 0$ for all i. However, the constants r_{ij}, $j \geq 1$, which are the prescribed values of the derivatives of u, may be zero for some i and j.

Theorem (3) below gives an extremely simple necessary and sufficient condition for the existence of a unit u satisfying (1). However, the proof is anything but simple.

3 *Theorem* Let $\sigma_1, \cdots, \sigma_l$ be distinct nonnegative extended real numbers,[3] and let $s_{l+1}, \cdots, s_n$ be distinct complex numbers with positive imaginary part. Let $S = \{\sigma_1, \cdots, \sigma_l, s_{l+1}, \cdots, s_n\}$, let $M = \{m_1, \cdots, m_n\}$ be a corresponding set of positive integers, and let $R = \{r_{ij}, j = 0, \cdots, m_i - 1;\; i = 1, \cdots, n\}$ be a set of complex numbers with r_{ij} real whenever $j = 0, \cdots, m_i - 1$, $i = 1, \cdots, l$, and $r_{i0} \neq 0$ for all i. Under these conditions, there exists a unit u in $\mathbf{S}$ satisfying (1) if and only if the numbers $r_{10}, \cdots, r_{l0}$ are all of the same sign.

The proof of Theorem (3) is given at the end of the section, as it is long and tedious. The next step is to study Euclidean division in the ring $\mathbf{S}$.

[3] This implies that at most one of the σ_i can be infinite.

Suppose $f,g \in \mathbf{S}$ are given, with $g \neq 0$. With the aid of Theorem (3) it is possible to compute the minimum value attained by $\delta(f-gq)$ as q varies over $\mathbf{S}$. For this purpose, we may as well assume at the outset that f and g are coprime. To see why this is so, suppose f and g are not coprime, and let $f = f_1 d$, $g = g_1 d$, where d is a greatest common divisor of f and g. Then, since $\mathbf{S}$ is a proper Euclidean domain,

4 $$\delta(f-gq) = \delta(d) + \delta(f_1 - qg_1).$$

Thus the problem of minimizing $\delta(f-gq)$ is equivalent to minimizing $\delta(f_1 - qg_1)$, where f_1 and g_1 are coprime.

5 *Theorem* Suppose $f,g \in \mathbf{S}$ are coprime, with $g \neq 0$, and let $\sigma_1, \cdots, \sigma_l$ be the real nonnegative zeros of $g(\cdot)$ in C_{+e} (if any), arranged in ascending order, with $\sigma_l = \infty$ if appropriate. Let ν equal the number of sign changes in the sequence $\{f(\sigma_1), \cdots, f(\sigma_l)\}$.[4] Then

6 $$\min_q \delta(f-gq) = \nu.$$

Proof The proof is divided into two parts: In the first part, it is shown that $\delta(f-gq) \geq \nu$ for all q. In the second part, it is shown that this lower bound is exact.

To show that $\delta(f-gq) \geq \nu$ for all q, note that

7 $$(f-gq)(\sigma_i) = f(\sigma_i) \text{ for } i = 1, \cdots, l.$$

Therefore the sequence $\{(f-gq)(\sigma_1), \cdots, (f-gq)(\sigma_l)\}$ contains ν sign changes, whatever be q. Hence the function $f-gq$ has at least ν real zeros in C_{+e}. This shows that $\delta(f-gq) \geq \nu$ for all q.

Next it is shown that there exists a q such that $\delta(f-gq) = \nu$. Towards this end, select $a,b \in \mathbf{S}$ such that

[4] Note that $f(\sigma_i) \neq 0$ for all i, since f and g are coprime; hence the sign of $f(\sigma_i)$ is well-defined.

8 $af + bg = 1.$

and note that a and g are also coprime as a consequence. Now (8) shows that $(af)(\sigma_i) = 1$ for all i, which shows that $a(\sigma_i)$ and $f(\sigma_i)$ have the same sign for all i. Now select a function $h \in \mathbf{S}$ such that $\delta(h) = \nu$ and such that the sequence $\{(ah)(\sigma_1), \cdots, (ah)(\sigma_l)\}$ has no sign changes.[5] Let S denote the set of C_{+e}-zeros of the function g. (This set of course includes the points $\sigma_1, \cdots, \sigma_l$). By construction, the function ah has the same sign at all real C_{+e}-zeros of the function g. Hence, by Theorem (3), there exists a unit, say u, that interpolates the values of ah and its derivatives at the C_{+e}-zeros of g, or equivalently, such that $u-ah$ is divisible by g in $\mathbf{S}$. Suppose $u-ah = cg$, and rewrite this in the form

10 $ah + cg = u.$

Dividing both sides of (10) by u gives

11 $ar + pg = 1,$

where $r = h/u, p = c/u$. Observe also that

12
$$\begin{aligned}\delta(r) &= \delta(h/u)\\ &= \delta(h) + \delta(1/u)\\ &= \delta(h) \text{ since } 1/u \text{ is a unit}\\ &= \nu \text{ by construction.}\end{aligned}$$

The proof is concluded by showing that $r = f-gq$ for some $q \in \mathbf{S}$, or equivalently, that g divides $f-r$. Subtracting (11) from (8) gives

13 $a(f-r) = g(p-b).$

Thus g divides the product $a(f-r)$. Since a and g are coprime, it must be that g divides $f-r$ (see Lemma (A.3.21)). □

---

[5] This can be accomplished by choosing the function h to have a simple real zero lying between σ_i and σ_{i+1} whenever $a(\sigma_i)$ and $a(\sigma_{i+1})$ have different signs.

This section is concluded by giving a proof of Theorem (3).

Proof of Theorem (3) "only if" Suppose $r_{i0} > 0$ and $r_{k0} < 0$ for some $i,k \in \{1, \cdots, l\}$. Then *any* function satisfying (1) must have a real zero between σ_i and σ_k and therefore cannot be a unit.

"if" This part of the proof is constructive.[6] We show that, if $r_{10}, \cdots, r_{l0}$ are all of the same sign, then it is possible to construct a unit u satisfying (1) after $m = m_1 + \cdots + m_n$ repetitions of the following algorithm. Because of the extremely involved nature of the proof, it is divided into four parts.

First, suppose we have found a unit u such that

14 $$u^{(j)}(\sigma_i) = r_{ij},\ j = 0, \cdots, m_i - 1;\ i = 1, \cdots, k-1,\ k \le l.$$

We show how to construct a unit v such that v satisfies (14) (with u replaced by v) and in addition, satisfies $v(\sigma_k) = r_{k0}$. In other words, v interpolates all the same values as u, plus one extra function value at σ_k, which happens to be a real number. If $k = 1$ so that the condition (14) is vacuous, simply choose $v = r_{k0}$. Otherwise, proceed as follows: Define

15 $$f(s) = \prod_{i=1}^{k-1} \left[\frac{s - \sigma_i}{s+1} \right]^{m_i}.$$

Then $f \in \mathbf{S}$; moreover, $1 + bf$ is a unit whenever $|b| < 1/\|f\|$. For such a b, the function $(1+bf)^a$ is also a unit for all positive integers a. We shall show how to select b and a such that $|b| < 1/\|f\|$ and such that

16 $$v = (1+bf)^a u$$

satisfies $v(\sigma_k) = r_{k0}$ as well as (14).

Using the binomial expansion,

[6] The reader is reminded that Section 2.4 contains a simple but advanced proof of Theorem (3).

17 $$v = u + u \sum_{j=1}^{a} q_{aj} (bf)^j,$$

where $q_{aj} = a!/(a-j)!j!$ is the binomial coefficient. It is immediate from (15) that f and its first $m_i - 1$ derivatives vanish at σ_i, for all $i \leq k-1$. The same is true of all higher powers of f and thus of uf^j for all j. Hence v continues to interpolate the same function and derivative values as u; that is,

18 $$v^{(j)}(\sigma_i) = r_{ij},\ j = 0, \cdots, m_i - 1;\ i = 1, \cdots, k-1.$$

Next,

19 $$v(\sigma_k) = [1 + bf(\sigma_k)]^a\ u(\sigma_k).$$

It is routine to compute from (19) that, to achieve $v(\sigma_k) = r_{k0}$, one must have

20 $$b = \frac{1}{f(\sigma_k)} \left\{ [r_{k0}/u(\sigma_k)]^{1/a} - 1 \right\}.$$

Note that $f(\sigma_k) \neq 0$ from (15). Also, since u is a unit and satisfies (14), both $u(\sigma_k)$ and r_{k0} have the same sign, so that the fractional power in (20) is well-defined. Now, the right side of (20) approaches zero as $a \to \infty$. Hence, for large enough a, b satisfies $|b| < 1/\|f\|$. For any such b and a, the function $(1+bf)^a$ is a unit, so that v given by (16) is a unit and also satisfies $v(\sigma_k) = r_{k0}$.

The remaining steps in the proof follow the same general lines. In the second step, suppose we are still working on the interpolation conditions involving the real values σ_i, and suppose we have found a unit u that satisfies

21 $$u^{(j)}(\sigma_i) = r_{ij},\ j = 0, \cdots, m_i - 1;\ i = 1, \cdots, k-1,\ k \leq l,$$

22 $$u^{(j)}(\sigma_k) = r_{kj},\ j = 0, \cdots, t-1,\ t \leq m_k.$$

In other words, u fully meets the interpolation conditions at the first $k-1$ points and partially meets them at the k-th point. We shall show how to construct a unit that meets the next interpolation constraint, namely the next derivative constraint at σ_k.

Define

23 $$f(s) = \left\{ \frac{s-\sigma_k}{s+1} \right\}^t \prod_{i=1}^{k-1} \left\{ \frac{s-\sigma_i}{s+1} \right\}^{m_i}.$$

Then $f \in \mathbf{S}$. Moreover, all derivatives of f up to order $m_i - 1$ vanish at σ_i, $1 \le i \le k-1$, and all derivatives of f up to order $t-1$ vanish at σ_k. Let

24 $$v = (1 + bf)^a u.$$

As before, it follows from (17) that v satisfies (21)-(22). Now expand (17) further as

25 $$v = u + aubf + u \sum_{j=2}^{a} q_{aj} \, (bf)^j.$$

Then

26 $$v^{(t)}(\sigma_k) = u^{(t)}(\sigma_k) + (aubf)^{(t)}(\sigma_k),$$

because higher powers of bf have a zero of multiplicity greater than t at σ_k. Now, by Leibnitz' rule,

27 $$(aubf)^{(t)}(\sigma_k) = ab \sum_{j=0}^{t} q_{aj} \, u^{(t-j)}(\sigma_k) \, f^{(j)}(\sigma_k)$$

$$= ab \; u(\sigma_k) \, f^{(t)}(\sigma_k).$$

Hence, if[7]

28 $$b = \frac{r_{kt} - u^{(t)}(\sigma_k)}{af^{(t)}(\sigma_k) \, u(\sigma_k)},$$

then $v^{(t)}(\sigma_k) = r_{kt}$. Finally, if a is sufficiently large, then $|\, b \,| < \| f \|^{-1}$, so that $1 + bf$, $(1+bf)^a$, and v are all units.

---

[7] Note that $f^{(t)}(\sigma_k) \ne 0$ from (23) and $u(\sigma_k) \ne 0$ because u is a unit.

Third, suppose we have found a unit u such that

29 $$u^{(j)}(\sigma_i) = r_{ij}, j = 0.\cdots,m_{i-1;}\ i = 1,\cdots,l,$$

$$u^{(j)}(s_i) = r_{ij}, j = 0,\cdots,m_i-1; i = l+1,\cdots,k-1.$$

where $k \leq n$. We now show how to find a unit v such that v satisfies (28) and also $v(s_k) = r_{k0}$.

Define

30 $$f(s) = \frac{s-\beta}{s+\beta} \prod_{i=1}^{l} \left[\frac{s-\sigma_i}{s+1}\right]^{m_i} \prod_{i=l+1}^{k-1} \left[\frac{(s-s_i)(s-\bar{s}_i)}{(s+1)^2}\right]^{m_i}.$$

31 $$v = (1+bf)^a u,$$

where the constants $\beta > 0$, b (real) and a (integer) are yet to be chosen. As before v satisfies (29), and it only remains to select these constants such that v is a unit and $v(s_k) = r_{k0}$.

Since $\text{Im } s_i > 0$, the quantity

32 $$q := \inf_{\beta \geq 0} \left| \frac{s_k - \beta}{s_k + \beta} \right|$$

is well-defined and positive. Let $f_1(s) = (s-\beta)/(s+\beta)$ and define $f_2(s) = f(s)/f_1(s)$. Thus $f_2(s)$ is the product of all terms on the right side of (30) except $(s-\beta)/(s+\beta)$. Now choose the integer a sufficiently large that

33 $$\left| \left\{ \frac{r_{k0}}{u(s_k)} \right\}^{1/a} - 1 \right| \, \|f_2\| < q \, |f_2(s_k)|,$$

where $(\cdot)^{1/a}$ denotes the a-th principal root. Note that such an a can always be found because the left side of (33) approaches zero as a approaches infinity. Define

34 $$b = \frac{[r_{k0}/u(s_k)]^{1/a} - 1}{f_1(s_k) f_2(s_k)},$$

where by varying the constant β the argument of $f_1(s_k)$ can be adjusted so as to make b a real number. Since f_1 is an "all-pass" function, $|f_1(j\omega)| = 1$ for all ω, and as a result $\| f \| = \| f_2 \|$. The inequalities (32) and (33) thus imply that $\| bf \| < 1$. Therefore, $1+bf$, $(1+bf)^a$, and v are all units, and moreover $v(s_k) = r_{k0}$ from (34).

Fourth and last, suppose we have found a unit u such that

35 $$u^{(j)}(s_i) = r_{ij},\ j = 0, \cdots, m_i - 1;\ i = 1, \cdots, k-1,$$

36 $$u^{(j)}(s_k) = r_{kj},\ j = 0, \cdots, t-1.$$

where $l < k \le n$, $t < m_k$ and we have used s_i as a shorthand for σ_i when $i = 1, \cdots, l$. We will find a unit v that satisfies (35)-(36) and in addition satisfies $v^{(t)}(s_k) = r_{kt}$.

Define

37 $$f(s) = \frac{s-\beta}{s+\beta} \prod_{i=1}^{l} \left[\frac{s-\sigma_i}{s+1} \right]^{m_i} \prod_{i=l+1}^{k-1} \left[\frac{(s-s_i)(s-\bar{s}_i)}{(s+1)^2} \right]^{m_i}$$
$$\cdot \left[\frac{(s-s_k)(s-\bar{s}_k)}{(s+1)^2} \right]^t.$$

38 $$v = (1+bf)^a u.$$

Then, as before, v satisfies (35)-(36). Using the same reasoning as in (25)-(27), it follows that

39 $$v^{(t)}(s_k) = u^{(t)}(s_k) + ab\, u(s_k) f^{(t)}(s_k).$$

As before, let $f(s) = f_1(s) f_2(s)$, where $f_1(s) = (s-\beta)/(s+\beta)$. By Leibnitz' rule,

40 $$f^{(t)}(s_k) = \sum_{i=0}^{t} q_{ti}\, f_1^{(t-i)}(s_k)\, f_2^{(i)}(s_k)$$

$$= f_1(s_k)\, f_2^{(t)}(s_k),$$

since $f_2^{(i)}(s_k) = 0$ for $i < t$. Substitution of (40) into (39) gives

41 $$v^{(t)}(s_k) = u^{(t)}(s_k) + ab\; u(s_k)\, f_1(s_k)\, f_2^{(t)}(s_k).$$

Accordingly, $v^{(t)}(s_k) = r_{kt}$ if we choose

42 $$b = \frac{r_{kt} - u^{(t)}(s_k)}{a\; u(s_k)\, f_1(s_k)}.$$

By selecting a sufficiently large and by adjusting β, we can ensure that b is real and has magnitude less than $\| f \|$. For such a choice of b, v is given by (38). □

2.4 INTERPOLATION IN THE DISC ALGEBRA

The main objective of this section is to give a very simple proof of Theorem (2.3.3). Though the proof is very simple in detail, it uses a few advanced concepts, specifically that of a logarithm of an element of a Banach algebra.

Recall that a pair $(\mathbf{B}, \|\cdot\|)$ is a *Banach algebra* if

(i) $(\mathbf{B}, \|\cdot\|)$ is a Banach space,

(ii) $\mathbf{B}$ is an associative algebra over the real or complex field, and

(iii) $\| ab \| \leq \| a \| \cdot \| b \|$ for all $a, b \in \mathbf{B}$.

$\mathbf{B}$ is *commutative* if $ab = ba$ for all $a, b \in \mathbf{B}$, and *has an identity* if there is an element $1 \in \mathbf{B}$ such that $1 \cdot a = a \cdot 1 = a$ for all $a \in \mathbf{B}$. An element $a \in \mathbf{B}$ is a *unit* of $\mathbf{B}$ if there exists a $b \in \mathbf{B}$ such that $ab = ba = 1$ (assuming of course that $\mathbf{B}$ has an identity).

Now suppose that $\mathbf{B}$ is a commutative Banach algebra with identity, and let $\mathbf{U}$ denote the set of its units. This set is nonempty since $1 \in \mathbf{U}$. For each $f \in \mathbf{B}$, the element

1 $$exp(f) = \sum_{i=0}^{\infty} f^i/i!$$

is well-defined. An element $f \in \mathbf{B}$ is said to *have a logarithm* if there exists a $g \in \mathbf{B}$ such that $f = exp(g)$. If f has a logarithm g, then $f \cdot exp(-g) = 1$, so that f must necessarily be a unit of $\mathbf{B}$. Thus only units can have logarithms. This naturally leads one to ask whether *every* unit has a logarithm.

2 *Lemma* A unit $f \in \mathbf{B}$ has a logarithm if and only if f is homotopic to the identity in $\mathbf{U}$, i.e. if and only if there exists a continuous function $h:[0,1] \rightarrow \mathbf{U}$ such that $h(0) = 1$, $h(1) = f$.

Proof "only if" Suppose f has a logarithm, say g. Define $h(\lambda) = \exp(\lambda g)$. Then $h(\cdot)$ is a continuous map from $[0,1]$ into $\mathbf{U}$ such that $h(0) = 1$, $h(1) = f$.

"if" This part of the proof proceeds in three steps.

Step 1 If $x \in \mathbf{B}$ and $\| x \| < 1$, then $1-x$ has a logarithm. In fact the standard Taylor series expansion

3 $$\sum_{i=1}^{\infty} x^i/i =: z$$

converges whenever $\| x \| < 1$, and it is routine to verify that $exp(z) = 1-x$.

Step 2 If x has a logarithm and $\| y-x \| < 1/\| x \|^{-1}$, then y has a logarithm. To show this, note that

4 $$yx^{-1} = [x-(x-y)]x^{-1} = 1-(x-y)x^{-1}$$

has a logarithm since $\| (x-y)x^{-1} \| < 1$. If $x = exp(a)$ and $yx^{-1} = exp(b)$, then since $\mathbf{B}$ is commutative, it follows that $y = (yx^{-1})x = exp(a+b)$.

Step 3 Suppose $h:[0,1] \rightarrow \mathbf{U}$ is continuous and satisfies $h(0) = 1$, $h(1) = f$. Since the function $u \rightarrow u^{-1}$ is continuous, it follows that the function $\lambda \rightarrow [h(\lambda)]^{-1}$ mapping $\mathbf{U}$ into itself is also continuous. Moreover, since $[0,1]$ is compact, this function is also bounded on $[0,1]$.

Accordingly, suppose r is a finite real number such that $\| [h(\lambda)]^{-1} \| < r$ for all $\lambda \in [0,1]$. Using once again the compactness of [0,1], we see that the function h is *uniformly* continuous on [0,1]. Thus there exists a $\delta > 0$ such that $\| h(\lambda) - h(\mu) \| < 1/r$ whenever $\lambda, \mu \in [0,1]$ and $|\lambda - \mu| < \delta$. Let m be an integer larger than $1/\delta$, and let $x_i = h(i\lambda/m)$. Then $x_0 = 1$, and $x_m = f$. Moreover, $\| x_i^{-1} \| \leq r$ and $\| x_{i+1} - x_i \| < 1/r$. From Step 2, we see that since x_0 has a logarithm, so does x_1; by induction, x_i has a logarithm for all i, and in particular f has a logarithm. □

Lemma (2) is false if **B** is noncommutative. In particular, Step 2 of the "if" part of the proof breaks down in the noncommutative case since the product of two units, each of which has a logarithm, need not itself have a logarithm (since $exp(a+b) \neq exp(a)exp(b)$ in general). If a unit $f \in \mathbf{B}$ is homotopic to the identity, then all one can say is that f is a finite product of units, each of which has a logarithm. In other words, f belongs to the group generated by the range of the "exp" function if and only if f is homotopic to the identity in **U**. This statement can be proved by slightly modifying the proof of Lemma (2).

Now we introduce the *disc algebra* A, which plays an important role in the control theory of linear distributed systems. The set A consists of all continuous functions mapping the closed unit disc **D** into the complex numbers which have the additional property that they are analytic on the interior of **D** (i.e. the open unit disc). If addition and multiplication of two functions in A are defined pointwise, then A becomes a commutative Banach algebra with identity over the complex field. Now let A_s denote the subset of A consisting of all symmetric functions; i.e.

5 $$A_s = \{ f \in A : \bar{f}(\bar{z}) = f(z) \ \forall z \in \mathbf{D} \}$$

where the bar denotes complex conjugation. Then A_s is a commutative Banach algebra with identity over the *real* field. The connection between A_s and the set **S** of proper stable rational functions is recalled from Section 2.1: Given $f \in \mathbb{R}(s)$, define

6 $g(z) = f((s-1)/(s+1))$.

Since the bilinear transformation $z = (s-1)/(s+1)$ maps the extended right half-plane C_{+e} onto the unit disc $\mathbf{D}$, we see that $g(\cdot)$ is a rational function belonging to A_s if and only if $f \in \mathbf{S}$. Thus if we define

7 $\mathbf{R}_\infty = \{f \in A_s : f \text{ is rational}\}$,

then $\mathbf{R}_\infty$ is isometrically isomorphic to $\mathbf{S}$. However, while $\mathbf{R}_\infty$ is a normed algebra, it is not complete; its completion is A_s. Thus every element in A_s can be approximated arbitrarily closely by a polynomial or a rational function in A_s.

In the discrete-time case, a lumped system is BIBO-stable if and only if its transfer function belongs to $\mathbf{R}_\infty$. Finally, note that every polynomial is in A_s.

It is easy to see that $f \in A_s$ is a unit of A_s if and only if $f(z) \neq 0$ for all $z \in \mathbf{D}$. The next lemma specifies which units in A_s have logarithms.

8 *Lemma* A unit f of A_s has a logarithm if and only if $f(z) > 0$ for all $z \in [-1,1]$.

Proof "only if" Suppose $f = exp(g)$ where $g \in A_s$. Since $\bar{g}(\bar{z}) = g(z) \; \forall z \in \mathbf{D}$, it follows that $g(z)$ is real whenever $z \in [-1,1]$. Hence $f(z) = exp(g(z)) > 0$ whenever $z \in [-1,1]$.

"if" Suppose $f(z) > 0$ for all $z \in [-1,1]$. It is shown that f is homotopic to the identity in the set $\mathbf{U}$ of units of A_s; it will then follow from Lemma (2) that f has a logarithm. Define

9 $h_1(z,\lambda) = f(\lambda z)$ for $\lambda \in [0,1]$.

Then $h_1(\cdot,1) = f(\cdot)$ and $h_1(\cdot,0)$ is the *constant* function $f(0)$. Clearly $h_1(\cdot,\lambda)$ is a continuous function of λ. Moreover, since f is a unit of A_s, we have $f(z) \neq 0$ whenever $|z| \leq 1$. Consequently $f(z) \neq 0$ whenever $|z| \leq \lambda \in [0,1]$, i.e. $h_1(\cdot,\lambda)$ is a unit for all $\lambda \in [0,1]$. Therefore f is homotopic to the constant function $f(0)$ in $\mathbf{U}$. Next, since $f(0) > 0$, the function

10 $$h_2(z,\lambda) = \lambda f(0) + (1-\lambda)$$

gives a homotopy between the constant function $f(0)$ and the identity of A_s. By transitivity, $f(\cdot)$ is homotopic to the identity and hence has a logarithm. □

Now we return to the problem of interpolation by a unit, first studied in Section 2.3. Suppose $\{s_1, \cdots, s_n\}$ is a set of points in C_{+e}, $\{m_1, \cdots, m_n\}$ is a corresponding set of positive integers, and $\{r_{ij}, j = 0, \cdots, m_i - 1, i = 1, \cdots, n\}$ is a corresponding set of complex numbers. The objective is to determine (if possible) a unit u of **S** such that

11 $$u^{(j)}(s_i) = r_{ij}, j = 0, \cdots, m_i - 1, i = 1, \cdots, n.$$

Using the bilinear transformation $z = (s-1)/(s+1)$, one can transform the above problem into an equivalent one of finding a rational unit f of A_s such that

12 $$f^{(j)}(z_i) = q_{ij}, j = 0, \cdots, m_i - 1, i = 1, \cdots, n,$$

where $z_i = (s_i - 1)/(s_i + 1)$, $q_{i0} = r_{i0}$, and q_{ij} for $j \geq 1$ is a more complicated function of s_i and of j (see Example (20)). The important point is that z_i is real if and only if s_i is real. One can always renumber the z_i's in such a way that $z_1, \cdots, z_l$ are real and $z_{l+1}, \cdots, z_n$ are nonreal.

Theorem (13) below shows that there exists a unit f of A_s satisfying (12) if and only if $q_{10}, \cdots, q_{l0}$ are all of the same sign, and gives a procedure for constructing such a unit when it exists. The unit so constructed is not rational in general, but an indication is also given of a way to construct a rational unit satisfying (12). Thus Theorem (13) contains the same result as Theorem (2.3.3), but the proof is considerably simpler.

13 *Theorem* Given elements $z_1, \cdots, z_n$ of **D**, positive integers $m_1, \cdots, m_n$, and complex numbers $q_{ij}, j = 0, \cdots, m_i - 1; i = 1, \cdots, n$, suppose $z_1, \cdots, z_l$ are real and that $z_{l+1}, \cdots, z_n$ are nonreal. Suppose also that q_{ij} is real for all j whenever z_i is real. Under these conditions, there exists a rational unit f of A_s satisfying (12) if and only if $q_{10}, \cdots, q_{l0}$ are all

of the same sign.

Proof "only if" If f is a unit of A_s, then $f(\sigma)$ does not change sign as σ varies over [-1,1]. Hence if a unit f satisfies (12), then the constants $q_{10}, \cdots, q_{l0}$ must all have the same sign.

"if" Suppose $q_{10}, \cdots, q_{l0}$ all have the same sign. We may assume without loss of generality that all these constants are positive; if not, the original problem can be replaced by the equivalent problem of finding a unit h that satisfies the interpolation conditions $h^{(j)}(z_i) = -q_{ij} \ \forall i,j$.

It is first shown that there exists a (not necessarily rational) unit $h \in A_s$ satisfying (12) (with f replaced by h, of course). If we can construct a function $g \in A_s$ satisfying

14 $$\frac{d^j}{dz^j} exp(g(z)) \,|_{z=z_i} = q_{ij}, \ \forall i,j,$$

then $h = exp(g)$ satisfies the conditions (12). Let us translate (14) into conditions on the values of g and its derivatives. First,

15 $$g(z_i) = \log q_{i0}, \ i = 1, \cdots, n.$$

The noteworthy point is that when $1 \leq i \leq l$ so that z_i is real, q_{i0} is real and positive so that log q_{i0} is (more precisely, can be chosen to be) real. When z_i is nonreal, log q_{i0} may be nonreal, but this does not matter. For higher derivatives of g, (14) leads to

16 $$g'(z_i) = q_{i1}/q_{i0},$$

17 $$g''(z_i) = \{q_{i2} - [g'(z_i)]^2\}/q_{i0},$$

and so on. Since $q_{i0} \neq 0$ for all i, the expressions in (16) and (17) are all well-defined. Moreover, the quantities $g^{(j)}(z_i)$ are all real whenever z_i is real. Thus the original interpolation problem has been reduced to one of constructing a function $g \in A_s$ (*not* required to be a unit) satisfying certain interpolation conditions. It is trivial to construct such a g; in fact, g can always be chosen to be a polynomial using the technique of Lagrange interpolation (see Example (20)). Setting $h = exp(g)$ gives a unit satisfying (12).

In general, h need not be rational. To construct a *rational* unit f satisfying (12), proceed as follows: Let d be any polynomial in A_s such that $d^{(j)}(z_i) = q_{ij}$. Let

18 $$e(z) = \prod_{i=1}^{n} (z - z_i)^{m_i}.$$

The fact that $h^{(j)}(z_i) = d^{(j)}(z_i) = q_{ij}$ for all i,j implies that $h-d$ and its first $m_i - 1$ derivatives vanish at z_i, for all i. Hence e divides $h-d$ in A_s. Let $c = (h-d)/e$. Since rational functions (and in fact polynomials) are dense in A_s, there exists a rational function $c_1 \in A_s$ such that $\| c - c_1 \| < 1/[\| h^{-1} \| \cdot \| e \|]$. Now define $f = d + c_1 e$. Then f is rational (polynomial if c_1 is polynomial); moreover, $\| h - f \| = \| (c - c_1)e \| < 1/ \| h^{-1} \|$. Hence f is a unit. Finally, since $d - f = c_1 e$ is a multiple of e and since e has a zero of order m_i at z_i, it follows that

19 $$f^{(j)}(z_i) = d^{(j)}(z_i) = q_{ij}, \forall i,j.$$

Thus f is rational and satisfies (12). □

The above proof shows that if $q_{10}, \cdots, q_{l0}$ are all of the same sign, then (12) can be satisfied by a *polynomial* $f(\cdot)$ with all of its zeros outside the unit disc **D** (and all of its poles at infinity). If we define $u(s) = f((s-1)/(s+1))$, then it follows that the s-domain conditions (11) can be satisfied by a unit $u(\cdot) \in \mathbf{S}$ that has *all* of its poles at $s = -1$ (or any other prespecified point on the negative real axis). Of course, the procedure given in Section 2.3 for constructing u also leads to the same conclusion.

20 *Example* Suppose it is required to construct a unit $u \in \mathbf{S}$ satisfying

21 $$u(1) = 1, u'(1) = 0.5, u(j) = 1 + j2, u(\infty) = 2.$$

Let $f(z) = u((1+z)/(1-z))$. Then f must satisfy

22 $f(0) = 1, f'(0) = -1, f(j) = 1+j2, f(1) = 2.$

Note that $f'(0)$ is computed using the chain rule:

23 $$f'(0) = \frac{du}{ds}\Big|_{s=1} \frac{ds}{dz}\Big|_{z=0}.$$

As in the proof of Theorem (13), an irrational unit h is constructed that satisfies

24 $h(0) = 1, h'(0) = -1, h(j) = 1+j2, h(1) = 2.$

Let $h = exp(g)$. In order for h to satisfy (24), g must satisfy

25 $g(0) = 0, g'(0) = -1$ (from (16))

$g(j) = \log(1+j2) \approx 0.8+j1.1$

$g(1) = \log 2 \approx 0.7$

Using Lagrange interpolation, one can readily find a polynomial g satisfying (25), namely

26 $$g(z) = 0.7\, z^2(z^2+1)/2 + z(z-1)(z^2+1)$$

$$+ (0.8+j1.1)\, z^2(z-1)(z-j)/(2+j2)$$

$$+ (0.8-j1.1)\, z^2(z-1)(z+j)/(2-2j)$$

$$= 2.3\, z^4 - 2.1\, z^3 + 1.5\, z^2 - z.$$

Now, $h(z) = exp(g(z))$ satisfies (24).

The construction of a *polynomial* unit satisfying (21) is more complicated.

Notes and References

The proof that the ring **S** is a Euclidean domain is given by Hung and Anderson [46]. Previously Morse [64] had shown that **S** is a principal ideal domain. For the related notion of Λ-generalized polynomialss, see Pernebo [70]. The interpolation theorem (2.3.3) is given by Youla, Bongiorno and Lu

[104], and the current proof is an adaptation of theirs. The more advanced proof of Theorem (2.3.3) given in Section 2.4 is found in Vidyasagar and Davidson [94].

CHAPTER 3. SCALAR SYSTEMS: AN INTRODUCTION

The objective of this chapter is to give a glimpse of the power of the factorization approach alluded to in the title of the book, by applying it to the special case of scalar systems (i.e. single-input-single-output systems). Attention is focused on two specific results, namely a parametrization of all compensators that stabilize a given plant, and the development of a necessary and sufficient condition for a plant to be stabilizable by a stable compensator. Proofs are kept to a minimum, especially in cases where they would be routine specializations of more general results for multivariable plants as developed in subsequent chapters. In the same vein, no discussion is given of control problems such as tracking, disturbance rejection, etc., which go beyond mere stabilization; these problems are studied directly in the multivariable case in Chapter 5.

Throughout the chapter, the symbol $\mathbf{S}$ is used to denote the set of proper stable rational functions, while $\mathbf{U}$ denotes the set of units of $\mathbf{S}$, i.e. functions in $\mathbf{S}$ whose reciprocal also belongs to $\mathbf{S}$.

3.1 PARAMETRIZATION OF ALL STABILIZING COMPENSATORS

Suppose $p \in \mathbb{R}(s)$, so that $p(s)$ is the transfer function of a scalar, lumped, linear, time-invariant system. Suppose a compensator $c \in \mathbb{R}(s)$ is connected to the plant p in the feedback configuration shown in Figure 3.1. The equations describing the closed-loop system are

1 $$\begin{bmatrix} e_1 \\ e_2 \end{bmatrix} = \begin{bmatrix} u_1 \\ u_2 \end{bmatrix} - \begin{bmatrix} 0 & p \\ -c & 0 \end{bmatrix} \begin{bmatrix} e_1 \\ e_2 \end{bmatrix},$$

which can be solved to yield

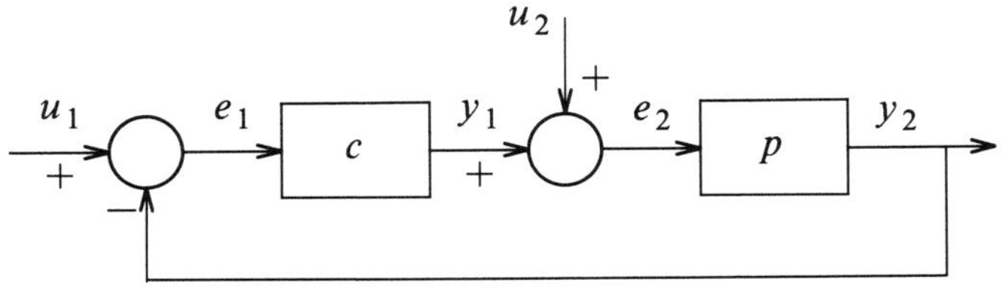

Figure 3.1. Feedback System.

2 $$\begin{bmatrix} e_1 \\ e_2 \end{bmatrix} = \begin{bmatrix} 1/(1+pc) & -p/(1+pc) \\ c/(1+pc) & 1/(1+pc) \end{bmatrix} \begin{bmatrix} u_1 \\ u_2 \end{bmatrix}$$

provided of course that $1+pc \neq 0$.[1] Let

3 $$H(p,c) = \begin{bmatrix} 1/(1+pc) & -p/(1+pc) \\ c/(1+pc) & 1/(1+pc) \end{bmatrix}$$

denote the 2×2 transfer matrix from (u_1,u_2) to (e_1,e_2). Then we say that *the pair* (p,c) *is stable*, and that c *stabilizes* p, if and only if $H(p,c)\in\mathbf{S}^{2\times 2}$. Thus the present notion of stability requires *each* of the four elements of $H(p,c)$ (of which two are the same) to represent a BIBO stable system: it is not enough to have just $p/(1+pc)\in\mathbf{S}$, which is an oft-stated definition of stability given in undergraduate textbooks. There two reasons for this: 1) If the compensator c is itself stable, then $H(p,c)\in\mathbf{S}^{2\times 2}$ if and only if $p/(1+pc)\in\mathbf{S}$. Hence, if we are using a stable compensator, then closed-loop stability as defined above reduces to the single (and familiar) requirement that $p/(1+pc)$ be a stable rational function. (See Lemma (5.1.11) for a multivariable version of this fact.) 2) In achieving closed-loop stability, we would like to be sure of having "internal" as well as "external" stability. In other words, it is desirable to insure that, whenever u_1, u_2 are any bounded inputs, *all* resulting signals in the closed-loop system are bounded. It turns out that $H(p,c)\in\mathbf{S}^{2\times 2}$ is a necessary and sufficient condition to ensure this.

[1] This means that $1+pc$ is not the zero function.

(See Lemmas (5.1.9) and (5.1.17) for multivariable versions of this fact).

Having thus defined closed-loop stability, we next give a necessary and sufficient condition for it.

4 *Lemma* Suppose $p,c \in \mathbb{R}(s)$, and let $p = n_p/d_p$, $c = n_c/d_c$, where $n_p,d_p,n_c,d_c \in \mathbf{S}$, and n_p,d_p are coprime, n_c,d_c are coprime. Define

5 $$\delta(p,c) = n_p n_c + d_p d_c.$$

Then the pair (p,c) is stable if and only if $\delta(p,c) \in \mathbf{U}$.

Proof "if" Suppose $\delta(p,c) \in \mathbf{U}$; then $1/\delta(p,c) \in \mathbf{S}$. It is now routine to verify that

6 $$H(p,c) = \frac{1}{\delta(p,c)} \begin{bmatrix} d_p d_c & -n_p d_c \\ d_p n_c & d_p d_c \end{bmatrix}$$

belongs to $\mathbf{S}^{2\times 2}$, so that the pair (p,c) is stable.

"only if" Suppose (p,c) is stable. Then certainly $1+pc \neq 0$. Also, $d_p \neq 0$, $d_c \neq 0$ since both are denominators of fractions. Hence $\delta(p,c) = d_p d_c(1+pc) \neq 0$, and the formula (6) is valid. Since $H(p,c) \in \mathbf{S}^{2\times 2}$, it follows from (6) that

7 $$\frac{d_p d_c}{\delta(p,c)} \in \mathbf{S}, \quad \frac{d_p n_c}{\delta(p,c)} \in \mathbf{S}, \quad \frac{n_p d_c}{\delta(p,c)} \in \mathbf{S}.$$

Also

8 $$1 - \frac{d_p d_c}{\delta(p,c)} = \frac{n_p n_c}{\delta(p,c)} \in \mathbf{S}.$$

A compact way of expressing (7) and (8) is

9 $$\begin{bmatrix} d_p \\ n_p \end{bmatrix} \frac{1}{\delta(p,c)} [d_c \;\; n_c] \in \mathbf{S}^{2\times 2}.$$

Since n_p,d_p are coprime and n_c,d_c are coprime, it follows from Fact (A.3.6) that there exist $x_p,y_p,x_c,y_c \in \mathbf{S}$ such that $x_p n_p + y_p d_p = 1$, $x_c n_c + y_c d_c = 1$. Now (9) implies that

10 $$[y_p \;\; x_p] \begin{bmatrix} d_p \\ n_p \end{bmatrix} \frac{1}{\delta(p,c)} [d_c \;\; n_c] \begin{bmatrix} y_c \\ x_c \end{bmatrix} = \frac{1}{\delta(p,c)} \in \mathbf{S},$$

which shows that $\delta(p,c) \in \mathbf{U}$. □

11 *Corollary* Suppose $p \in \mathbb{R}(s)$, and let $p = n_p/d_p$ where $n_p, d_p \in \mathbf{S}$ are coprime. Then $c \in \mathbb{R}(s)$ stabilizes p if and only if $c = n_c/d_c$ for some $n_c, d_c \in \mathbf{S}$ that sastisfy

12 $$n_p n_c + d_p d_c = 1.$$

Proof "if" If (12) holds, then $\delta(p,c) = 1$ which is certainly a unit of $\mathbf{S}$. Hence (p,c) is stable by Lemma (4).

"only if" Suppose c stabilizes p, and express c as n_1/d_1 where $n_1, d_1 \in \mathbf{S}$ are coprime. Then, by Lemma (4), it follows that $\delta = n_1 n_p + d_1 d_p \in \mathbf{U}$. Now define $n_c = n_1/\delta$, $d_c = d_1/\delta$. Then $n_c, d_c \in \mathbf{S}$, $c = n_c/d_c$, and n_c, d_c satisfy (12). □

The central result of this chapter, which provides a parametrization of *all* compensators that stabilize a given plant, now falls out almost routinely.

13 *Theorem* Suppose $p \in \mathbb{R}(s)$, and let $p = n_p/d_p$ where $n_p, d_p \in \mathbf{S}$ are coprime. Select $x, y \in \mathbf{S}$ such that

14 $$x n_p + y d_p = 1.$$

Then the set of all compensators that stabilize p, denoted by $S(p)$, is given by

15 $$S(p) = \{c = \frac{x + r d_p}{y - r n_p} : r \in \mathbf{S} \text{ and } y - r n_p \neq 0\}.$$

Proof Suppose that c is of the form

16 $$c = \frac{x + r d_p}{y - r n_p}$$

for some $r \in \mathbf{S}$. Then, since

17 $$(x+rd_p)\cdot n_p + (y-rn_p)\cdot d_p = xn_p + yd_p = 1,$$

it follows from Corollary (11) that c stabilizes p.

Conversely, suppose c stabilizes p. Then, from Corollary (11), $c = n_c/d_c$ where $n_c, d_c \in \mathbf{S}$ satisfy (12). Thus the proof is complete if it can be shown that every solution of (12) must be of the form

18 $$n_c = x + rd_p,\ d_c = y - rn_p$$

for some $r \in \mathbf{S}$. Subtracting (14) from (12) and rearranging gives

19 $$(n_c - x)n_p = (y - d_c)d_p.$$

Since d_p and n_p are coprime, it follows from Corollary (A.3.19) and Problem A.3.11 that d_p divides $(n_c - x)$ and that n_p divides $d_c - y$. Let r denote $(n_c - x)/d_p$. Then $n_c = x + rd_p$. Now (19) shows that $d_c = y - rn_p$. □

In order to apply Theorem (13) to determine the set of *all* compensators that stabilize a given plant p, one needs to do two things: (i) Express p as a ratio n_p/d_p where $n_p, d_p \in \mathbf{S}$ are coprime. (ii) Find a particular solution x, y of (14). Systematic procedures for achieving these two steps are given in Section 4.2. Note that, in view of Corollary (11), the second step is equivalent to finding *one* compensator that stabilizes p. For scalar systems, the first step is very easy, as shown in Fact (2.1.20).

Once these two steps are completed, (15) provides a parametrization of the set $S(p)$ of *all* stabilizing compensators for p. The condition $y - rn_p \neq 0$ is not very restrictive, as $y - rn_p$ can equal zero for at most one choice of r (see Problem 1).

The utility of Theorem (13) derives not merely from the fact that it provides a parametrization of $S(p)$ in terms of a "free" parameter r, but also from the simple manner in which this free parameter enters the resulting (stable) closed-loop transfer matrix.

20 *Corollary* Let the symbols p, n_p, d_p, x, y be as in Theorem (13). Suppose $c = (x + rd_p)/(y - rn_p)$ where $r \in \mathbf{S}$. Then

21 $$H(p,c) = \begin{bmatrix} d_p(y - rn_p) & -n_p(x + rd_p) \\ d_p(x + rd_p) & d_p(y - rn_p) \end{bmatrix}.$$

Proof Apply (6) with $n_c = x + rd_p$, $d_c = y - rn_p$. □

The application of Theorem (13) and Corollary (20) is now illustrated by means of a simple example.

22 *Example* Let

23 $$p(s) = \frac{s}{(s+1)(s-1)}.$$

Then $p = n_p / d_p$, where, as in Fact (2.1.20),

24 $$n_p(s) = \frac{s}{(s+1)^2},\ d_p(s) = \frac{s-1}{s+1}.$$

Suppose that, using some design procedure, we have found a stabilizing compensator for p, namely

25 $$c(s) = \frac{2(s+2)}{s-0.5}.$$

This leads to a particular solution x, y of (14), as follows: Let

26 $$n_1(s) = \frac{2(s+2)}{s+1},\ d_1(s) = \frac{s-0.5}{s+1}.$$

Then

27 $$\delta(p,c) = n_p n_1 + d_p d_1 = \frac{s^3 + 1.5s^2 + 3s + 0.5}{(s+1)^3}.$$

All zeros of the numerator polynomial of $\delta(p,c)$ are in the open left half-plane, so that $\delta(p,c)$ is a unit of $\mathbf{S}$. This is to be expected, since c is a stabilizing compensator. Now, as in the proof of Corollary (11), define

28 $$x(s) = \frac{n_1}{\delta(p,c)} = \frac{2(s+2)(s+1)^2}{\phi(s)},$$

29 $$y(s) = \frac{d_1}{\delta(p,c)} = \frac{(s-0.5)(s+1)^2}{\phi(s)},$$

where

30 $$\phi(s) = s^3 + 1.5s^2 + 3s + 0.5.$$

Then x,y satisfy (14). Using Theorem (13), it now follows that the set of *all* compensators that stabilize p is given, after clearing a few fractions, by those c that are of the form

31 $$c(s) = \frac{x+rd_p}{y-rn_p}$$

$$= \frac{2(s+2)(s+1)^4 + r(s)(s-1)(s+1)\phi(s)}{(s-0.5)(s+1)^4 - r(s)s\phi(s)}$$

where r is any function in $\mathbf{S}$.[2] The expression (31) can be simplified considerably. Dividing both numerator and denominator by $(s+1)^3$ gives

32 $$c(s) = \frac{2(s+2)(s+1) + r(s)\delta(s)(s-1)(s+1)}{(s-0.5)(s+1) - r(s)\delta(s)s},$$

where $\delta(s) = \delta(p,c)(s)$ as defined in (27). Now $\delta(p,c)$ is a *unit* of $\mathbf{S}$; hence, as r varies freely over $\mathbf{S}$, so does $r\delta$. In other words, the map $r \rightarrow r\delta$ is a one-to-one map of $\mathbf{S}$ onto itself. Therefore we may as well replace $r\delta$ by a new "free" parameter q, which is an arbitrary function in $\mathbf{S}$. Thus the set of compensators that stabilize p is given by

33 $$S(p) = \{c\colon c(s) = \frac{2(s+2)(s+1) + q(s)(s-1)(s+1)}{(s-0.5)(s+1) - sq(s)}, \, q \in \mathbf{S}\}.$$

From Corollary (20), the closed-loop transfer matrix $H(p,c)$ corresponding to a compensator c of the form (32) is given by (21). To illustrate the manipulations involved, consider the term $h_{11}(p,c)$; it equals

---

[2] Since $y(\infty) \neq 0$, $n_p(\infty) = 0$ and $r(\infty)$ is finite, we see that $(y - rn_p)(\infty) \neq 0 \; \forall r \in \mathbf{S}$, so that r is unrestricted.

34 $$d_p(y - rn_p) = \frac{s-1}{s+1}\left\{\frac{(s-0.5)(s+1)^2}{\phi(s)} - \frac{r(s)s}{(s+1)^2}\right\}$$

$$= \frac{(s-1)[(s-0.5)(s+1) - q(s)s}{\phi(s)},$$

where $q = r\delta$. After similar manipulations on the other three terms, the final result is

35 $$h_{21}(p,c)(s) = \frac{(s-1)(s+1)[2(s+2) + q(s)(s-1)]}{\phi(s)},$$

$$h_{12}(p,c)(s) = \frac{s[2(s+2) + q(s)(s-1)]}{\phi(s)},$$

and of course $h_{22}(p,c) = h_{11}(p,c)$. Thus as q varies over $\mathbf{S}$, the expression (33) generates *all* compensators that stabilize p, and (34-35) give the corresponding (stable) closed-loop transfer matrices.

Consider now the problem of stabilizing a given plant p with a general domain of stability $\mathbf{D}$ replacing the open left half-plane. Thus, given $p \in \mathbb{R}(s)$ and a domain of stability $\mathbf{D}$, the problem is to parametrize *all* compensators $c \in \mathbb{R}(s)$ such that the closed-loop transfer matrix $H(p,c)$ of (3) belongs to $\mathbf{S}_\mathbf{D}^{2\times 2}$. Looking back over the proof of Theorem (13) and all the preliminary results leading up to it, we see that the entire development depends on just two facts: (i) $\mathbf{S}$ is a proper Euclidean domain, and (ii) every $f \in \mathbb{R}(s)$ can be factorized as a/b where $a,b \in \mathbf{S}$ are coprime. Now, both of these statements remain valid if $\mathbf{S}$ is replaced by $\mathbf{S}_\mathbf{D}$. Hence Theorem (13) and Corollary (20) carry over *in toto* if $\mathbf{S}$ is replaced by $\mathbf{S}_\mathbf{D}$. This is illustrated by means of an example.

36 *Example* Consider again the plant p of (23), and let

37 $$\mathbf{D} = \{s : \text{Re}\, s < -2\}.$$

Suppose it is desired to parametrize all compensators c such that the closed-loop transfer matrix $H(p,c)$ is proper and has all of its poles in $\mathbf{D}$. The first step is to find $n_p, d_p \in \mathbf{S}_\mathbf{D}$ such that $p = n_p/d_p$ and n_p, d_p are coprime. This can be done using the procedure of Fact (2.1.20), except that the term $s+1$ is replaced by $s+a$ where $a > 2$. Let us take $a = 3$, so that

38 $$n_p(s) = \frac{s}{(s+3)^2},\ d_p(s) = \frac{(s+1)(s-1)}{(s+3)^2}.$$

In order to find a particular solution of (14), the bilinear transformation

39 $$z = \frac{s-3}{s+3},\ s = 3\frac{1+z}{1-z}$$

is introduced. This leads to

40 $$n_p = \frac{1-z^2}{12},\ d_p = \frac{2z^2+5z+2}{9}.$$

Now, since n_p and d_p are coprime *polynomials* in z, there exist *polynomials* x,y in z such that $xn_p + yd_p = 1$. One such pair x,y can be found by first performing the Euclidean division algorithm on n_p and d_p and then back-substituting. This gives

41 $$x = \frac{10z+17}{9} = \frac{9s-7}{3(s+3)},$$

42 $$y = 5z-4 = \frac{s-27}{s+3}.$$

The corresponding nominal (or starting) stabilizing compensator[3] is

43 $$c = \frac{x}{y} = \frac{9s-7}{3(s-27)}.$$

The set of *all* compensators that stabilize p, in the sense that $H(p,c) \in \mathbf{S}_{\mathbf{D}}^{2\times 2}$, is given by

44 $$S_{\mathbf{D}}(p) = \{c : c = \frac{x+rd_p}{y-rn_p} \text{ for some } r \in \mathbf{S}_{\mathbf{D}}\},$$

and the corresponding $H(p,c)$ is given by (21).

---

[3] Recall that the phrase "stabilize" has here the meaning "place all closed-loop poles in the prescribed domain of stability."

Now suppose the performance specifications on the closed-loop system are made more stringent, as follows: Define

45 $$\mathbf{D}_1 = \{s: \text{Re } s < -2, \ |\text{Im } s| \leq |\text{Re } s|\}.$$

The region $\mathbf{D}_1$ is shown in Figure 3.2.

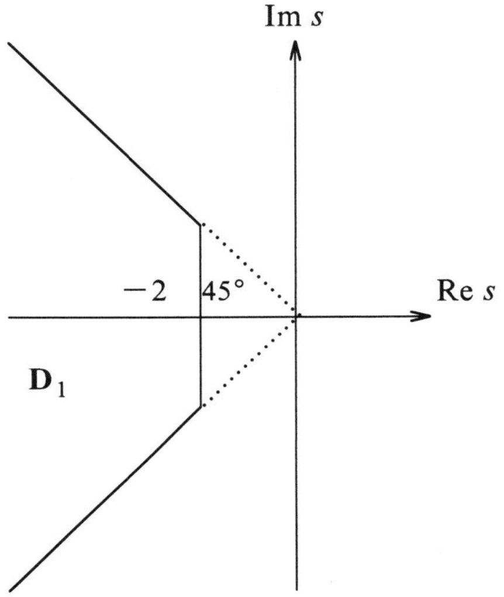

Figure 3.2. Generalized Region of Stability.

Suppose it is desired to parametrize all compensators c such that $H(p,c)$ is proper and has all of its poles inside the region $\mathbf{D}_1$ (rather than just $\mathbf{D}$ as in the previous paragraph). To obtain such a parametrization, it is necessary first to obtain a coprime pair $n_p, d_p \in \mathbf{S}_{\mathbf{D}_1}$ and then to find a pair $x, y \in \mathbf{S}_{\mathbf{D}_1}$ such that $xn_p + yd_p = 1$. Fortunately, the functions n_p, d_p, x, y in (38), (41), (42) already satisfy this requirement. Hence the desired parametrization is

46 $$\mathbf{S}_{\mathbf{D}_1}(p) = \{c: c = \frac{x + rd_p}{y - rn_p} \text{ for some } r \in \mathbf{S}_{\mathbf{D}_1}\}.$$

Note that the expressions in (44) and (45) are the same; the only difference is that the "free" parameter r in (45) is only permitted to range over the smaller set $\mathbf{S}_{\mathbf{D}_1}$, whereas in (44) it is permitted to range over the larger set $\mathbf{S}_{\mathbf{D}}$.

Problems

1. Given that $a,b,c,d \in \mathbf{S}$ satisfy $ab + cd = 1$, show that $a - rc$ can equal zero for atmost one $r \in \mathbf{S}$. (Hint: If $c \neq 0$, then $a - rc = 0$ implies that $a - qc \neq 0$ for all $q \neq r$. If $c = 0$ then $a \neq 0$.) Construct $a,b,c,d \in \mathbf{S}$ such that $ab + cd = 1$ and c divides a (which is equivalent to saying that $a - rc = 0$ for some r). Describe *all* such c.

2. Suppose $a,b,c,d \in \mathbf{S}$ satisfy $ab + cd = 1$. Show that if c is not a unit then c does not divide a (i.e.,$a - rc \neq 0 \; \forall r$). Apply this result to Theorem (13) to show that if the plant p is strictly proper, then $y - rn_p \neq 0 \; \forall r \in \mathbf{S}$.

3. In Theorem (13), show that distinct choices of the parameter r lead to distinct compensators.

4. Find all compensators that stabilize the plant

$$p(s) = \frac{s-1}{(s-2)(s^2+1)}.$$

5. Given the BIBO stable plant

$$p(s) = \frac{s^2+1}{(s+1)^2},$$

find all compensators c such that $H(p,c) \in \mathbf{S}_{\mathbf{D}_1}^{2 \times 2}$, where $\mathbf{D}_1$ is as in (45).

3.2 STABILIZATION USING A STABLE COMPENSATOR

A plant $p \in \mathbb{R}(s)$ is said to be *strongly stabilizable* if there exists a $c \in \mathbf{S}$ such that (p,c) is stable, i.e. if p can be stabilized using a stable compensator. A study of strong stabilizability is important for at least two reasons: (i) Suppose p is strongly stabilizable. Let $c \in \mathbf{S}$ stabilize p and let $p_1 = p/(1+pc)$ denote the resulting plant transfer function. Since c is stable, a factorization of c is given by $n_c = c$, $d_c = 1$. It now follows from (3.1.6) that $p_1 = n_p$ where n_p is the numerator of a factorization of p. In other words, if p is stabilized using a *stable* compensator, then the resulting plant transfer function $p_1 = p/(1+pc)$ has the same C_{+e}-zeros as p *and no others*. The argument is reversible: If p is *not* strongly stabilizable, then d_c in (3.1.6) can never be a unit, so that p_1 always has some additional C_{+e}-zeros beyond those of p. Since the C_{+e}-zeros of p_1 affect its sensitivity to disturbances and its ability to track reference inputs (see Section 5.6), it is desirable not to introduce, in the process of stabilization, additional C_{+e}-zeros into the closed-loop transfer function beyond those of p. But this is possible if and only if p is strongly stabilizable. (ii) Several problems in compensator design, such as simultaneous stabilization (Section 5.4) and two-step stabilization (Section 5.2) are closely related to strong stabilizability. Accordingly, in this section necessary and sufficient conditions are derived for a plant to be strongly stabilizable. These conditions turn out to be very easily stated in terms of the locations of the real C_{+e}-zeros and poles of the plant under study.

1 *Theorem* Given $p \in \mathbb{R}(s)$, let $\sigma_1, \cdots, \sigma_l$ denote the real C_{+e}-zeros of p (including ∞ if p is strictly proper), arranged in ascending order. Let η_i denote the number of poles of p (counting multiplicity) in the interval (σ_i, σ_{i+1}), and let η denote the number of odd integers in the sequence $\eta_1, \cdots, \eta_{l-1}$. Then every c that stabilizes p has at least η poles in C_{+e}. Moreover, this lower bound is exact in that there is a c that stabilizes p that has exactly η C_{+e}-poles.

2 *Corollary* A plant $p \in \mathbb{R}(s)$ is strongly stabilizable if and only if the number of poles of p between every pair of real C_{+e}-zeros of p is even.

The property described in Corollary (2) is called the *parity interlacing property (p.i.p.)*. Thus a plant p is strongly stabilizable if and only if it has the p.i.p.

Proof of Theorem (1) By Theorem (3.1.13), every c that stabilizes p has the form $(x+rd_p)/(y-rn_p)$ for some $r \in \mathbf{S}$, where all symbols are defined in Theorem (3.1.13). Further, $x+rd_p$ and $y-rn_p$ are coprime, by (3.1.17). Hence, by Fact (2.1.20), the number of C_{+e}-poles of c equals $\delta(y-rn_p)$, i.e.

the degree of $y - rn_p$ in the Euclidean domain **S**. Now a lower bound on this degree as r varies can be computed using Theorem (2.3.5). Again by Fact (2.1.20), the C_{+e}-zeros of n_p are precisely $\sigma_1, \cdots, \sigma_l$. Thus by Theorem (2.3.5), $\min_{r \in \mathbf{S}} \delta(y - rn_p)$ equals the number of sign changes in the sequence $\{y(\sigma_1), \cdots, y(\sigma_l)\}$. Next, observe that, since $xn_p + yd_p = 1$, it follows that $y(\sigma_i)d_p(\sigma_i) = 1 \ \forall i$ (since $n_p(\sigma_i) = 0$). Hence $y(\sigma_i)$ and $d_p(\sigma_i)$ have the same sign for all i. Thus $\min_{r \in \mathbf{S}} \delta(y - rn_p)$ also equals the number of sign changes in the sequence $\{d_p(\sigma_1), \cdots, d_p(\sigma_l)\}$. To conclude the proof, note that $d_p(\sigma_{i+1})$ has the same (resp. opposite) sign as $d_p(\sigma_i)$ if and only if the number of zeros of d_p in the interval (σ_i, σ_{i+1}) is even (resp. odd). Since the C_{+e}-zeros of d_p are same as the poles of p (Fact (2.1.20)), the result follows. □

The proof of the corollary is omitted as it is obvious.

The following alternate form of Corollary (2) is sometimes useful.

3 *Corollary* Suppose $p \in \mathbb{R}(s)$ equals n_p/d_p where $n_p, d_p \in \mathbf{S}$ are coprime. Then p is strongly stabilizable if and only if $d_p(\cdot)$ has the same sign at all real C_{+e}-zeros of p (or equivalently, n_p).

Once the strong stabilizability of a given plant p is established using Corollary (2) or Corollary (3), the construction of a stable stabilizing compensator for p is straight-forward using the results of Sections 2.3 or 2.4. Specifically, factor p as n_p/d_p where $n_p, d_p \in \mathbf{S}$ are coprime. Then the conditions of either corollary assure the existence of a unit u that interpolates the values of d_p at the C_{+e}-zeros of n_p; moreover, such a unit can be constructed using the procedure described in the proof of Theorem (2.3.3) or Theorem (2.4.13). Once such a u is constructed, the stable stabilizing compensator is given by $c = (u - d_p)/n_p$.

4 *Example* Consider again the plant

5 $$p(s) = \frac{s}{(s+1)(s-1)}.$$

This plant has two real C_{+e}-zeros, namely $\sigma_1 = 0$, $\sigma_2 = \infty$. Since p has one pole in the interval $(0,\infty)$, it is *not* strongly stabilizable.

6 *Example* Consider the plant

7 $$p(s) = \frac{s(s-0.4)}{(s+1)(s-1)(s-2)}.$$

By applying the criterion of Corollary (2), we see that p is strongly stabilizable. To construct a stable c that stabilizes p, we proceed as follows: First, factorize p as a ratio n_p/d_p where $n_p, d_p \in \mathbf{S}$ are coprime. This is most easily done using Fact (2.1.20), which leads to

8 $$n_p(s) = \frac{s(s-0.4)}{(s+1)^3},\ d_p(s) = \frac{(s-1)(s-2)}{(s+1)^2}.$$

The problem of constructing a stable stabilizing compensator is equivalent to constructing a unit $u \in \mathbf{S}$ that interpolates the values of d_p at the C_{+e}-zeros of n_p. Thus the problem is one of constructing a unit $u \in \mathbf{S}$ that satisfies

9 $$u(0) = 2,\ u(0.4) = \frac{24}{49},\ u(\infty) = 1.$$

Let us find such a unit using the iterative procedure contained in the proof of Theorem (2.3.1). First, let $u_0(s) \equiv 1$; this unit interpolates the required value at infinity. Next, to meet the constraint at 0.4, define $f_1(s) = 1/(s+1)$, and let

10 $$u_1 = (1+bf_1)^r u_0,$$

where the constant b and the integer r are to be selected such that $|b| < \|f_1\|^{-1}$ and $u_1(0.4) = 24/49$. Routine calculations show that $\|f_1\| = 1$, so that $b = 5/7$, $r = 1$ are suitable, leading to

11 $$u_1(s) = \frac{7s+2}{7s+7}.$$

This unit meets the interpolation constraints at $s = 0.4$ and $s = \infty$. To meet the last interpolation constraint, define

12 $$f_2(s) = \frac{s-0.4}{(s+1)^2},$$

$$u_2 = (1+bf_2)^r u_1.$$

where b must satisfy $|b| < \|f_2\|^{-1}$ and r is an integer. In the present instance this translates to $|b| < 2.0$. Again it is routine to verify that $r = 4$ is a suitable choice, leading to $b = (1-7^{1/4})/0.4 \approx -1.5165$, and

13 $$1+bf_2 = \frac{s^2+0.4835s+1.6266}{s^2+2s+1},$$

$$u_2 = u(s) = \left\{ \frac{s^2+0.4835s+1.6266}{s^2+2s+1} \right\}^4 \frac{7s+2}{7s+7}.$$

Finally, the compensator is given by $c = (u-d_p)/n_p$.

Next, consider the problem of strong stabilizability with a general domain of stability $\mathbf{D}$. Thus the problem is whether a given plant $p \in \mathbb{R}(s)$ can be "stabilized" by a compensator $c \in \mathbf{S}_{\mathbf{D}}$ such that $H(p,c) \in \mathbf{S}_{\mathbf{D}}^{2\times 2}$. If the complement of the region $\mathbf{D}$ is simply connected, then both Theorem (1) and Corollary (2) apply, the only modification being that C_{+e} is replaced by "the union of $\{\infty\}$ and the complement of $\mathbf{D}$."

14 *Example* Suppose

15 $$p(s) = \frac{(s+1)(s-1)(s^2+1)}{s(s-2)^2[(s-1)^2+1]}.$$

Then p has two *real* C_{+e}-zeros, namely $\sigma_1 = 1$ and $\sigma_2 = \infty$. Since p has two poles in the interval $(1,\infty)$, namely the double pole at $s = 2$, p is strongly stabilizable.

Now let

16 $$\mathbf{D} = \{s: \text{Re } s < -2, \ |\text{Im } s| \leq 2 |\text{Re } s| \}.$$

Then p has three real zeros in the extended complement of $\mathbf{D}$, namely $\sigma_1 = -1, \sigma_2 = 1, \sigma_3 = \infty$. In the interval $(-1,1)$ p has one pole. Hence no $c \in \mathbf{S_D}$ exists such that $H(p,c) \in \mathbf{S}_{\mathbf{D}}^{2\times 2}$, i.e. p is *not* strongly stabilizable with respect to the domain of stability $\mathbf{D}$.

Finally, consider the following generalization of the strong stabilization problem: Given a plant p and a general domain of stability $\mathbf{D}$, find a compensator $c \in \mathbf{S}$ such that the closed-loop transfer matrix $H(p,c) \in \mathbf{S}_{\mathbf{D}}^{2\times 2}$. Thus the objective is to design a stable compensator such that the closed-loop performance meets more stringent requirements than just stability with respect to the left half-plane. Note that the compensator is merely required to be stable in the conventional sense. The solution to this problem is given without proof, and is this: Such a compensator exists if and only if the plant p satisfies the parity interlacing property with respect to the region C_{+e}, i.e. p is strongly stabilizable in the conventional sense. Applying this test to the plant p of Example (14), we see that all closed-loop poles can be placed in the region $\mathbf{D}$ using a stable compensator, even though p is not strongly stabilizable with respect to the region $\mathbf{D}$.

The case of discrete-time systems can be handled in a straight-forward manner. Indeed, the contents of Section 2.4 make clear that the strong stabilization problem is most naturally solved by employing a bilinear transformation to map the extended right half-plane into the closed unit disc. Thus a discrete-time system p is strongly stabilizable if and only if it satisfies the parity interlacing property with respect to the interval [-1,1].

Problems

1. Determine whether each of the strictly proper plants below satisfies the parity interlacing property:

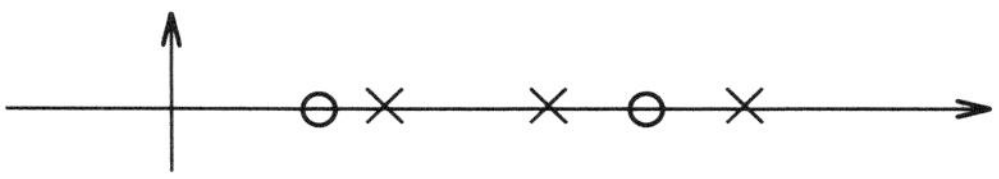

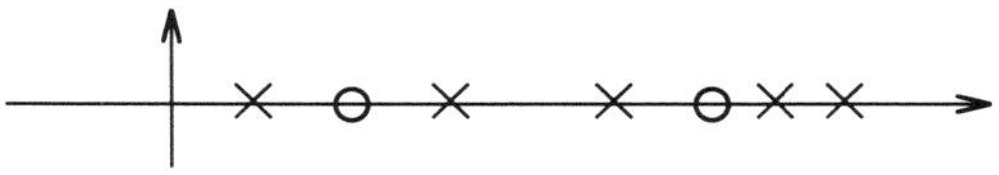

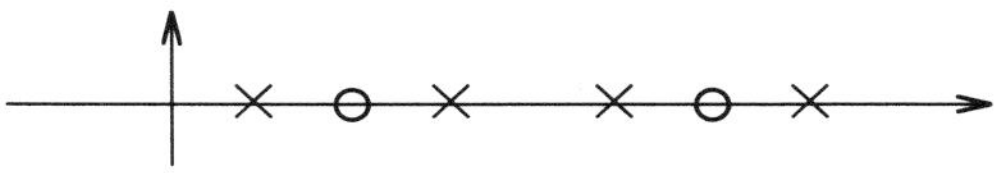

2. Find a stable stabilizing compensator for the plant

$$p(s) = \frac{s(s-1)}{(s^2+1)(s-2)(s-3)}.$$

3.3 PARAMETRIZATION OF ALL STABLE STABILIZING COMPENSATORS

Given a plant $p \in \mathbb{R}(s)$, the results of the previous section can be used to determine whether or not p can be stabilized using a stable compensator, i.e. whether or not p is strongly stabilizable. If p is strongly stabilizable, the methods of Sections 2.3 and 2.4 can be used to construct *a* stable stabilizing compensator for p. The objective of this section is to parametrize *all* such compensators.

After the bilinear transformation introduced in Section 2.4, the problem under study can be formulated as follows: Given rational functions d and n in the Banach algebra A_s, find all rational $c \in A_s$ such that $d+cn$ is a unit of A_s. Clearly this is equivalent to finding *all* units $u \in A_s$ such that $d-u$ is a multiple of n. Let $z_1, \cdots, z_l$ be the zeros of n in the unit disc $\mathbf{D}$, with multiplicities $m_1, \cdots, m_l$. Then the problem is one of finding all units u in A_s that satisfy

1 $$u^{(j)}(z_i) = d^{(j)}(z_i), 0 \le j \le m_i - 1; 1 \le i \le l.$$

Suppose u_1 and u_2 are two units that each satisfy (1), and let v denote the unit $u_2 u_1^{-1}$. Then a routine calculation shows that

2 $$v(z_i) = 1, 1 \le i \le l;$$

3 $$v^{(j)}(z_i) = 0, 1 \le j \le m_i - 1; 1 \le i \le l.$$

Equivalently, v is of the form $1 + fn$ for some $f \in A_s$, since v interpolates the constant function "1" and its derivatives at the zeros of n. This leads to the next result.

4 *Proposition* Given $d, n \in A_s$, let u be any unit in A_s satisfying (1). Then the set of *all* units satisfying (1) is given by $\{uv: v \in A_s$ is a unit of the form $1 + fn$ for some $f \in A_s\}$.

For a given $g \in A_s$, let $\mathbf{U}(g)$ denote the set of all units of the form $1 + fg$ where $f \in A_s$. If we can parametrize $\mathbf{U}(g)$, then Proposition (4) enables us to find all units that satisfy (1). Then the set of all $c \in A_s$ that stabilize the plant n/d is just $\{(u - d)/c: u$ satisfies $(1)\}$.

The next two results give an explicit description of $\mathbf{U}(g)$ given a $g \in A_s$. It is necessary to treat separately the cases where g has real zeros in $\mathbf{D}$ and where it does not. To aid in the presentation of the results, some notation is introduced. Suppose $g \in A_s$ has only a finite number of zeros in $\mathbf{D}$. Let $z_1, \cdots, z_s, \bar{z}_1, \cdots, \bar{z}_s, \sigma_1, \cdots, \sigma_r$ denote the distinct zeros of g in $\mathbf{D}$, where $z_1, \cdots, z_s$ are nonreal and $\sigma_1, \cdots, \sigma_r$ are real. For convenience, let $z_1, \cdots, z_{2s+r}$ denote the same sequence, and let μ_i denote the multiplicity of z_i as a zero of g. Select polynomials $p_1(z), \cdots, p_{2s}(z)$ such that

5 $$p_i(z_i) = 1, p_i^{(j)}(z_i) = 0 \text{ for } j = 1, \cdots, \mu_i - 1,$$

$$p_i^{(j)}(z_k) = 0 \text{ for } j = 0, \cdots, \mu_i - 1 \text{ if } k \ne i.$$

Since the zeros of g occur in complex conjugate pairs, we may suppose that $p_i(z) = \bar{p}_{i+s}(\bar{z})$ for $i = 1, \cdots, s$. Finally, define the polynomials

6 $$\phi_i(z) = j[p_i(z) - p_{i+s}(z)],\ i = 1, \cdots, s,$$

and observe that $\phi_i \in A_s\ \forall i$.

7 *Proposition* Suppose $g \in A_s$ has only a finite number of zeros in **D**, and that g has at least one real zero in **D**. Then every unit in A_s of the form $1+fg$, $f \in A_s$, can be expressed as $\exp(v)$, where $v \in A_s$ has the form

8 $$v = hg + \sum_{i=1}^{s} 2\pi m_i \phi_i,$$

where $h \in A_s$, $m_1, \cdots, m_s$ are arbitrary integers, and ϕ_i is defined in (6). Conversely, every unit $\exp(v)$ where v is of the form (8) can also be written as $1+fg$ for some $f \in A_s$. In summary

9 $$\mathbf{U}(g) = \{\exp(v): v \text{ is of the form (8)}\}.$$

Proof Suppose $u = 1+fg$ is a unit. Since g has at least one real zero in **D**, $u(z)$ equals one for some real z in **D**. Hence $u(z) > 0$ for $z \in [-1,1]$, and by Lemma (2.4.8), u has a logarithm v in A_s. Since $g(z_i) = 0\ \forall i$, it follows that

10 $$v(z_i) = j2\pi m_i\ \forall i,$$

where m_i is some integer. Since $v(z) = \bar{v}(\bar{z})\ \forall z$, it is immediate from (10) that $m_i = -m_{i+s}$ for $i = 1, \cdots, s$, and that $m_i = 0$ for $i = 2s+1, \cdots, 2s+r$. At multiple zeros of g, successive higher derivatives of g vanish, which implies that the corresponding derivatives of u and v also vanish. Thus

11 $$v^{(j)}(z_i) = 0 \text{ for } j = 1, \cdots, \mu_i - 1;\ i = 1, \cdots, 2s+r.$$

Now (10) and (11) lead to the conclusion that g divides the function $v - \sum_{i=1}^{2s} j2\pi m_i p_i$, which equals $v - \sum_{i=1}^{s} 2\pi m_i \phi_i$. This is precisely (8).

Conversely, suppose ν is of the form (8). Then ν satisfies (10) and (11), which in turn implies that exp(ν)-1 is divisible by g. Hence exp(ν) is of the form $1+fg$ for some $f \in A_s$. □

In practice, the assumption that g has at least one real zero in **D** does not pose a major restriction, because if g is the "numerator" of a strictly proper plant, then $g(1) = 0$. Also, note that g is not required to be rational - merely to have only a finite number of zeros in **D**.

We state the next result in the interests of completeness. The proof is omitted, as it closely follows that of Proposition (7).

12 *Proposition* Suppose $g \in A_s$ has only a finite number of zeros in **D**, all of them nonreal. Then every unit $u \in A_s$ of the form $1+fg, f \in A_s$, can either be written as exp(ν) where

13 $$\nu = hg + \sum_{i=1}^{s} 2\pi m_i \phi_i,$$

or else as -exp(ν) where

14 $$\nu = hg + \sum_{i=1}^{s} (2m_i+1)\pi\phi_i,$$

where $h \in A_s$, $m_1, \cdots, m_s$ are arbitrary integers, and ϕ_i is defined in (6). Conversely, every unit exp(ν) where ν is of the form (13) and every unit -exp(ν) where ν is of the form (14) can be written as $1+fg$ for some $f \in A_s$.

15 *Example* Consider the problem of determining all stable stabilizing compensators for the plant whose transfer function is

$$p(s) = \frac{s^2+1}{(s+2)^2(s-3)}.$$

After substituting $s = (1+z)/(1-z)$, this becomes

$$\frac{n(z)}{d(z)} = \frac{(z^2+1)(1-z)}{(z-3)^2(2z-1)}$$

where n and d denote the numerator and denominator polynomials respectively. Thus the problem is one of finding all $c \in A_s$ such that $d+cn$ is a unit of A_s. This can be solved using Propositions (4) and (7). First, we construct a unit u_0 such that $u_0 - d$ is a multiple of n. Since the zeros of n inside the unit disc are at $1, \pm j$, we require $u_0(z)$ to equal $d(z)$ at these points. Thus we must have

$$u_0(1) = 4, \; u_0(j) = 4+j22.$$

If $u_0 = \exp(f_0)$, then f_0 must satisfy

$$f_0(1) = \log 4 \approx 1.4, f_0(j) = \log(4+j22) \approx 3.1+j1.4.$$

Using Lagrange interpolation, one can find

$$f_0(z) = -1.55z^2+1.4z+1.55.$$

Therefore, *one* stabilizing compensator is given by

$$c_0(z) = [d(z)-\exp(f_0(z))]/n(z).$$

To find *all* stabilizing compensators, we compute the set $\mathbf{U}(n)$. Again using Lagrange interpolation, one can find polynomials to satisfy (5), namely

$$p_1(z) = \frac{(z-1)(z+j)}{2+j2}, p_2(z) = \bar{p}_1(\bar{z}),$$

$$\phi_1(z) = j[p_1(z)-p_2(z)] = -\frac{\pi}{2}(z-1)^2.$$

Thus $\mathbf{U}(n) = \{\exp(v)\}$ where $v = hn + \pi m(z-1)^2$, m an arbitrary integer and $h \in A_s$. Therefore the set of all stable stabilizing compensators is given by

$$\{\exp[(f(z))-d(z)]/n(z) \text{ where}$$

$$f(z) = f_0(z) + h(z)n(z) + \pi m(z-1)^2, m \in \mathbf{Z}, h \in A_s\}.$$

Notes and References

The parametrization of all compensators that stabilize a given plant is given in [105], and in the present form, in [25]. The necessity and sufficiency of the parity interlacing property for strong stabilizability is shown in [104], and its extension as stated in Theorem (3.2.1) is contained in [98]. Anderson [2] gives a method for testing the p.i.p. without computing the poles and zeros of a plant. The results of Section 3.3 are from [94].

Chapter 4. MATRIX RINGS

In the preceding chapter, we have seen the solution of some important problems in controller design for *scalar* systems. The objective of this chapter is to develop the machinery that will enable us to tackle *multivariable* systems. Accordingly, several problems are formulated and solved in the set of *matrices* over a commutative ring. These problems are posed in a fairly general setting so as to unify the controller synthesis techniques for continuous-time as well as discrete-time systems within a single framework.

Throughout this chapter, **R** denotes a commutative domain with identity and **F** denotes the field of fractions associated with **R**. Through most of the chapter, **R** is merely assumed to be a principal ideal domain. At various places, **R** is further restricted to be a proper Euclidean domain.

Throughout this chapter we shall be dealing with matrices whose elements belong to either **R** or **F**. Much of the time, the actual orders of the matrices under study are unimportant to the concepts being put forward. For this reason, $\mathbf{M}(\mathbf{R})$ is used as a generic symbol to denote the set of matrices with elements in **R**, of whatever order. Thus "$A,B \in \mathbf{M}(\mathbf{R})$" means that A,B are matrices whose elements are in **R**, but does *not* mean that A,B have the same order. In cases where it is necessary to display explicitly the order of a matrix, a notation of the form "$A \in \mathbf{R}^{n \times m}$" is used to indicate that A is an $n \times m$ matrix with elements in **R**. The symbols $\mathbf{M}(\mathbf{F})$ and $\mathbf{F}^{n \times m}$ are defined analogously. Finally, **U** denotes the set of units in **R**, and $\mathbf{U}(\mathbf{R})$ denotes the set of unimodular matrices (of whatever order) in $\mathbf{M}(\mathbf{R})$. The symbol $\mathbf{U}(\mathbf{F})$ is never used.

Throughout, I denotes the identity matrix. If appropriate, I_m is used to denote the identity matrix of order $m \times m$.

4.1 COPRIME FACTORIZATIONS OVER A PRINCIPAL IDEAL DOMAIN

Suppose **R** is a principal ideal domain (p.i.d.), **F** is the field of fractions associated with **R**, and that a/b is a fraction in **F**. Then it is always possible to express a/b as an equivalent fraction f/g where f,g are coprime (i.e. their greatest common divisor is 1; see Fact (A.3.11)). The main objective of this section is to develop analogous results for matrices in $\mathbf{M}(\mathbf{F})$.

We begin by defining the concepts of a multiple, common divisor, and coprime pair for the set $\mathbf{M}(\mathbf{R})$. Since matrix multiplication is in general noncommutative, it is necessary to make a distinction between *left* multiple and *right* multiple, etc.

1 *Definition* Suppose $A \in \mathbf{M}(\mathbf{R})$; then a *square* matrix $D \in \mathbf{M}(\mathbf{R})$ is a *right divisor* of A, and A is a *left multiple* of D, if there is a $C \in \mathbf{M}(\mathbf{R})$ such that $A = CD$. Suppose $A, B \in \mathbf{M}(\mathbf{R})$ and have the same number of columns. Then a *square matrix* $D \in \mathbf{M}(\mathbf{R})$ is a *greatest common right divisor (g.c.r.d.)* of A, B if

(GCRD1) D is a right divisor of both A and B, and

(GCRD2) D is a left multiple of every common right divisor of A, B.

Two matrices $A, B \in \mathbf{M}(\mathbf{R})$ having the same number of columns are *right-coprime* if every g.c.r.d. of A and B is unimodular.

The definitions of left divisor, g.c.l.d., and left-coprimeness are entirely analogous.

The next result is a matrix version of Fact (A.3.2).

2 *Lemma* Suppose $A, B \in \mathbf{M}(\mathbf{R})$ and D is a g.c.r.d. of A, B; then so is UD whenever $U \in \mathbf{U}(\mathbf{R})$. If D_1 is another g.c.r.d. of A, B, then there exists a $U \in \mathbf{U}(\mathbf{R})$ such that $D_1 = UD$. Thus if D is any g.c.r.d. of A, B, then the set of all g.c.r.d.'s of A, B is given by $\{UD: U \in \mathbf{U}(\mathbf{R})\}$.

Proof[1] The first sentence is easily proved. To prove the second sentence,

[1] Proof due to Chris Ma.

observe that since D and D_1 are both g.c.r.d.'s of A,B, there exist square matrices $V,W \in \mathbf{M}(\mathbf{R})$ such that $D_1 = VD$, $D = WD_1$. If $|D| \neq 0$, the rest of the proof is easy. Since $D = WVD$, it follows that $|V| \cdot |W| = 1$, whence both V and W are unimodular. To complete the proof, we must address the case where $|D| = 0$.

If $|D| = 0$, then the relation $D = WVD$ does not determine V and W uniquely. Thus the claim is that for some unimodular U, we have $D_1 = UD$ (though D_1 could also equal VD for some nonunimodular matrix V; see Problem 1). The first step in the proof is to observe that since D_1 is a left multiple of D, we have rank $(D_1) \leq$ rank (D). By symmetry, rank $(D_1) \leq$ rank (D), which implies that D_1 and D have the same rank. To be specific, suppose D, D_1 have rank r and are of order $n \times n$. By Theorem (B.2.6) there exist $Y,Y_1 \in \mathbf{R}^{n \times n}$, $H,H_1 \in \mathbf{R}^{r \times n}$ such that

3 $$YD = \begin{bmatrix} H \\ 0 \end{bmatrix} =: \bar{H},\; Y_1 D_1 = \begin{bmatrix} H_1 \\ 0 \end{bmatrix} =: \bar{H}_1,$$

where Y,Y_1 are unimodular and H,H_1 have rank r . Since D is a left multiple of D_1 and vice versa, it follows from (3) that $\bar{H}$ is a left multiple of $\bar{H}_1$ and vice versa. Suppose $\bar{H} = R\bar{H}_1$, $\bar{H}_1 = S\bar{H}$, and partition R,S appropriately. Then $\bar{H} = R\bar{H}_1$ implies that

4 $$\begin{bmatrix} H \\ 0 \end{bmatrix} = \begin{bmatrix} R_{11} & R_{12} \\ R_{21} & R_{22} \end{bmatrix} \begin{bmatrix} H_1 \\ 0 \end{bmatrix}.$$

Now, since H_1 has full rank, $R_{21}H_1 = 0$ implies that $R_{21} = 0$. Also, R_{12} and R_{22} are multiplied by the zero matrix and can therefore be replaced by any arbitrary matrices without affecting the validity of (4). In particular,

5 $$\begin{bmatrix} H \\ 0 \end{bmatrix} = \begin{bmatrix} R_{11} & 0 \\ 0 & I_{n-r} \end{bmatrix} \begin{bmatrix} H_1 \\ 0 \end{bmatrix} =: \bar{R}\bar{H}_1.$$

Similarly,

6 $$\begin{bmatrix} H_1 \\ 0 \end{bmatrix} = \begin{bmatrix} S_{11} & 0 \\ 0 & I_{n-r} \end{bmatrix} \begin{bmatrix} H \\ 0 \end{bmatrix} =: \overline{S}\overline{H}.$$

Now (5) and (6) together imply that $\overline{H} = \overline{R}\overline{S}\overline{H}$, and in particular, that $H = R_{11}S_{11}H$. Since H has full row rank, it follows that $R_{11}S_{11} = I_r$. Thus $R_{11},S_{11} \in \mathbf{U}(\mathbf{R})$, whence $\overline{R},\overline{S} \in \mathbf{U}(\mathbf{R})$. Finally, it follows from (3) that $D_1 = Y_1^{-1}\overline{H}_1 = Y_1^{-1}\overline{S}Y^{-1}D = UD$, where $U := Y_1^{-1}\overline{S}Y^{-1} \in \mathbf{U}(\mathbf{R})$. □

7 *Theorem* Suppose $A,B \in \mathbf{M}(\mathbf{R})$ have the same number of columns. Then A,B have a g.c.r.d. Moreover, if D is any g.c.r.d. of A,B, then there exist $X,Y \in \mathbf{M}(\mathbf{R})$ such that

8 $$XA + YB = D.$$

Proof Define $F \in \mathbf{M}(\mathbf{R})$ by

9 $$F = \begin{bmatrix} A \\ B \end{bmatrix},$$

and consider first the case where F has at least as many rows as columns. By the existence of the Hermite form (Corollary (B.2.15)), there is a matrix $T \in \mathbf{U}(\mathbf{R})$ such that

10 $$TF = \begin{bmatrix} D \\ 0 \end{bmatrix},$$

where D is upper traingular (though this is not needed in the proof). Partitioning T appropriately and recalling the definition of F gives

11 $$\begin{bmatrix} T_{11} & T_{12} \\ T_{21} & T_{22} \end{bmatrix} \begin{bmatrix} A \\ B \end{bmatrix} = \begin{bmatrix} D \\ 0 \end{bmatrix}.$$

Now (11) implies that

12 $$D = T_{11}A + T_{12}B.$$

On the other hand, if we let $S = T^{-1}$ and partition S appropriately, then

(11) implies that

13 $$\begin{bmatrix} A \\ B \end{bmatrix} = \begin{bmatrix} S_{11} & S_{12} \\ S_{21} & S_{22} \end{bmatrix} \begin{bmatrix} D \\ 0 \end{bmatrix} = \begin{bmatrix} S_{11}D \\ S_{21}D \end{bmatrix}.$$

Now (13) shows that D is a right divisor of both A and B, so that D satisfies (GCRD1). Further, if C is also a right divisor of A and B, then $A = A_1C, B = B_1C$ for some $A_1, B_1 \in \mathbf{M}(\mathbf{R})$. Then (12) shows that $D = (T_{11}A_1 + T_{12}B_1)C$. Hence D is a left multiple of C, so that D satisfies (GCRD2). Thus D is a g.c.r.d. of A,B and is of the form (8). Finally, suppose D_1 is any other g.c.r.d. of A,B. Then by Lemma (2) there exists a $U \in \mathbf{U}(\mathbf{R})$ such that $D_1 = UD$. Thus $D_1 = UD = UT_{11}A + UT_{12}B$ is also of the form (8).

To complete the proof, we examine the case where F in (9) has fewer rows than columns. To be specifc, suppose $A \in \mathbf{R}^{n \times m}$, $B \in \mathbf{R}^{l \times m}$ and that $n+l < m$. In this case, let $k = m-(n+l)$ and define

14 $$D = \begin{bmatrix} A \\ B \\ 0_{k \times m} \end{bmatrix}.$$

Then D is a right divisor of both A and B, since

15 $$A = [I_n \; 0 \; 0] \, D, B = [0 \; I_l \; 0] \, D.$$

Moreover, D is of the form (8) since

16 $$D = \begin{bmatrix} I_n \\ 0 \\ 0 \end{bmatrix} A + \begin{bmatrix} 0 \\ I_l \\ 0 \end{bmatrix} B.$$

Now (15) and (16) are enough to show that D is a g.c.r.d. of A and B; the details are left to the reader. □

17 *Corollary* Suppose $A, B \in \mathbf{M}(\mathbf{R})$ have the same number of columns. Then A and B are right-coprime if and only if there exist $X, Y \in \mathbf{M}(\mathbf{R})$ such that

18 $$XA + YB = I.$$

Proof A and B are right-coprime if and only if I is a g.c.r.d. of A and B. Now apply the theorem. □

A useful way of interpreting Corollary (17) is the following: A, B are right-coprime if and only if the matrix $F = [A' \ \ B']'$ has a left inverse in $\mathbf{M}(\mathbf{R})$.

The relation (18) is called the ***right Bezout identity***, or the ***right Diophantine identity***. Though we have derived (18) as a corollary to Theorem (7), it turns out that (18) is the most useful characterization of right-coprimeness in subsequent applications.

19 *Corollary* Two matrices $A \in \mathbf{R}^{n \times m}$, $B \in \mathbf{R}^{l \times m}$ are right-coprime if and only if 1 is a g.c.d. of all $m \times m$ minors of the matrix $F = [A' \ \ B']'$.

Remark It is implicit in the above corollary that A, B cannot be right-coprime unless $n + l \geq m$.

Proof "if" Suppose 1 is a g.c.d. of all $m \times m$ minors of F. Then all invariant factors of F are units (see Theorem (B.2.17)). Putting F into its Smith form, we see that there exist $U, V \in \mathbf{M}(\mathbf{R})$ such that

20 $$UFV = \begin{bmatrix} I \\ 0 \end{bmatrix}.$$

If U is partitioned in the obvious way, (20) implies that

21 $$(U_{11}A + U_{12}B)V = I,$$

22 $$U_{11}A + U_{12}B = V^{-1},$$

23 $$VU_{11}A + VU_{12}B = I.$$

Since V is unimodular, (23) is of the form (18), with $X = VU_{11}$, $Y = VU_{12}$. Hence A,B are right-coprime.

"only if" Suppose A,B are right-coprime. By (10) there exists a $T \in \mathbf{U}(\mathbf{R})$ such that

24 $$TF = \begin{bmatrix} D \\ 0 \end{bmatrix},$$

where D is upper triangular and is a g.c.r.d. of A,B. Since A,B are right-coprime, D is unimodular. Hence

25 $$TFD^{-1} = \begin{bmatrix} I \\ 0 \end{bmatrix}.$$

Since T and D^{-1} are both unimodular, the right side of (20) is a Smith form of F. By Theorem (B.2.17), it follows that 1 is a g.c.d. of all $m \times m$ minors of F. □

26 *Corollary* Suppose $A,B \in \mathbf{M}(\mathbf{R})$ have the same number of columns, and let $F = [A' \ \ B']'$. Then A,B are right-coprime if and only if there exists a $U \in \mathbf{U}(\mathbf{R})$ of the form $U = [F \ \ G]$.

Remarks To put it into words, Corollary (26) states that A and B are right-coprime if and only if there exists a unimodular matrix containing $[A' \ \ B']$ among its columns, or equivalently, if and only if the matrix F can be "complemented" by appropriate columns to form a unimodular matrix. For a generalization of this property to more general situations, see Section 8.1.

Proof "if" Suppose there exists a $U \in \mathbf{U}(\mathbf{R})$ of the form $[F \ \ G]$, and let $V = U^{-1}$. Partitioning V appropriately and expanding $VU = I$ gives

27 $$V_{11}A + V_{12}B = I.$$

Hence A,B are right-coprime.

"only if" Suppose A,B are right-coprime. By Corollary (19), there exist $T,V \in \mathbf{U}(\mathbf{R})$ such that

28 $$T \begin{bmatrix} A \\ B \end{bmatrix} V = \begin{bmatrix} I \\ 0 \end{bmatrix},$$

or equivalently,

29 $$T \begin{bmatrix} A \\ B \end{bmatrix} = \begin{bmatrix} V^{-1} \\ 0 \end{bmatrix}.$$

Let $W = T^{-1}$ and partition T,W appropriately. Then

30 $$\begin{bmatrix} T_{11} & T_{12} \\ T_{21} & T_{22} \end{bmatrix} \begin{bmatrix} A & W_{12} \\ B & W_{22} \end{bmatrix} = \begin{bmatrix} V^{-1} & 0 \\ 0 & I \end{bmatrix},$$

where we use the fact that $TW = I$. As the matrix on the right side of (30) is unimodular, so are both the matrices on the left side. If we define

31 $$U = \begin{bmatrix} A & W_{12} \\ B & W_{22} \end{bmatrix},$$

then the proof is complete. □

Suppose $A,B \in \mathbf{M}(\mathbf{R})$ are right-coprime. Then, by Corollary (17), there exist $X,Y \in \mathbf{M}(\mathbf{R})$ such that $XA + YB = I$. Lemma (32) below characterizes *all* solutions of the equation (18). It turns out that this is an important part of the controller synthesis theory to be developed in Chapter 5.

32 *Lemma* Supppose $A,B \in \mathbf{M}(\mathbf{R})$ are right-coprime, and select $U \in \mathbf{U}(\mathbf{R})$ of the form in Corollary (26) (i.e. such that U contains $[A' \;\; B']'$ as its first columns). Then the set of $X,Y \in \mathbf{M}(\mathbf{R})$ that satisfy

33 $$XA + YB = I$$

is given by

34 $$[X \;\; Y] = [I \;\; R]\, U^{-1},\; R \in \mathbf{M}(\mathbf{R}).$$

Proof It is necessary to show that (i) every X, Y of the form (34) satisfies (33), and (ii) every X, Y that satisfies (33) is of the form (34) for some $R \in \mathbf{M}(\mathbf{R})$.

To prove (i), observe that $U^{-1}U = I$. Hence

35 $$XA + YB = [X \;\; Y] \begin{bmatrix} A \\ B \end{bmatrix}$$

$$= [I \;\; R]\, U^{-1} \begin{bmatrix} A \\ B \end{bmatrix}$$

$$= [I \;\; R] \begin{bmatrix} I \\ 0 \end{bmatrix} = I,$$

so that X, Y satisfy (33). To prove (ii), suppose X, Y sastisfy (33) and define $R = [X \;\; Y]G$. Then

36 $$[X \;\; Y]U = [X \;\; Y] \begin{bmatrix} A & \\ & G \\ B & \end{bmatrix} = [I \;\; R]$$

from which (34) follows readily. □

The concept of right-coprimeness can be readily extended to more than two matrices. Suppose $A_1, \cdots, A_p \in \mathbf{M}(\mathbf{R})$ all have the same number of columns. Then a square matrix D is a *g.c.r.d.* of $A_1, \cdots, A_p$ if

(i) D is a right divisor of A_i for all i, and

(ii) If C is a right divisor of A_i for all i, then D is a left multiple of C.

The matrices $A_1, \cdots, A_p$ are *right-coprime* if every g.c.r.d. of $A_1, \cdots, A_p$ is unimodular.

The following facts are straight-forward extensions of the corresponding results for the case of two matrices; the proofs are left to the reader.

37 *Fact* Suppose $A_1, \cdots ,A_p \in \mathbf{M(R)}$ all have the same number of columns. Then they have a g.c.r.d. Moreover, if D is *any* g.c.r.d. of $A_1, \cdots ,A_p$, then there exist $X_1, \cdots ,X_p \in \mathbf{M(R)}$ such that

38 $$D = \sum_{i=1}^{p} X_i A_i.$$

39 *Fact* Suppose $A_1, \cdots ,A_p \in \mathbf{M(R)}$ all have the same number of columns, and define

40 $$F = \begin{bmatrix} A_1 \\ A_2 \\ \vdots \\ A_p \end{bmatrix}.$$

Then the following statements are equivalent:

(i) $A_1, \cdots ,A_p$ are right-coprime.

(ii) F has a left inverse in $\mathbf{M(R)}$.

(iii) F has at least as many rows as columns, and 1 is a g.c.d. of all the largest-size minors of F.

(iv) There exists a $U \in \mathbf{U(R)}$ of the form $U = [F \;\; G]$. Now suppose $A_1, \cdots ,A_p$ are right-coprime. Then the set of all left inverses in $\mathbf{M(R)}$ of F is given by

41 $$\{X : X = [I \;\; R]U^{-1}, R \in \mathbf{M(R)}\},$$

where U is any unimodular matrix of the form $[F \;\; G]$.

The next definition introduces a concept that is central to this book.

42 *Definition* Suppose $P \in \mathbf{M(F)}$. An ordered pair (N,D) where $N,D \in \mathbf{M(R)}$ is a *right-coprime factorization (r.c.f.) of* P if

(RCF1) D is square and $|D| \neq 0$,

(RCF2) $P = ND^{-1}$, and

(RCF3) N and D are right-coprime.

43 *Theorem* Suppose $P \in \mathbf{M}(\mathbf{F})$. Then it has an r.c.f. (N,D). Further (NU,DU) is also an r.c.f. of P for every $U \in \mathbf{U}(\mathbf{R})$. Finally, suppose (N_1,D_1) is another r.c.f. of P. Then there exists a $U \in \mathbf{U}(\mathbf{R})$ such that $N_1 = NU$, $D_1 = DU$.

Remarks Theorem (43) states that every $P \in \mathbf{M}(\mathbf{F})$ has an r.c.f., and that the r.c.f. is unique except for the possibility of multiplying the "numerator" and "denominator" matrices on the right by a unimodular matrix.

Proof (There are many ways to prove that every $P \in \mathbf{M}(\mathbf{F})$ has an r.c.f. One way is given below. See the proof of Theorem (49) for another method of constructing r.c.f.'s.) Write every element p_{ij} as a_{ij}/b_{ij}, where $a_{ij} \in \mathbf{R}$, $b_{ij} \in \mathbf{R} - 0$. Let b denote a least common multiple of all the b_{ij}'s, and let $F = bP$. Then $F \in \mathbf{M}(\mathbf{R})$ and $P = F(bI)^{-1}$. In accordance with Theorem (7), let E be a g.c.r.d. of F and bI, and define $N,D \in \mathbf{M}(\mathbf{R})$ by $F = NE$, $bI = DE$. Then, as in (8), there exist $X,Y \in \mathbf{M}(\mathbf{R})$ such that

44 $$XF + Yb = E.$$

Since $|D| \cdot |E| = |bI| \neq 0$, it follows that $|D| \neq 0$, $|E| \neq 0$. Hence we can multiply both sides of (44) by E^{-1}, which gives

45 $$XN + YD = I$$

showing that N and D are right-coprime. Finally, it is immediate that $ND^{-1} = (NE)(DE)^{-1} = P$. Hence (N,D) is an r.c.f. of P.

It is left to the reader to verify that (NU,DU) is also an r.c.f. of P whenever $U \in \mathbf{U}(\mathbf{R})$.

To prove the final assertion, let (N_1,D_1) be another r.c.f. of P. Then, by definition, $|D_1| \neq 0$, $P = N_1 D_1^{-1}$, and there exist $X_1,Y_1 \in \mathbf{M}(\mathbf{R})$ such that

46 $X_1N_1 + Y_1D_1 = I.$

Multiplying both sides by D_1^{-1} gives

47 $X_1N_1D_1^{-1} + Y_1 = D_1^{-1}.$

Multiplying both sides by D, and noting that $N_1D_1^{-1} = ND^{-1}$, gives

48 $D_1^{-1}D = X_1N + Y_1D \in \mathbf{M}(\mathbf{R}).$

By symmetrical reasoning, we also get $D^{-1}D_1 := U \in \mathbf{M}(\mathbf{R})$. Since U has an inverse in $\mathbf{M}(\mathbf{R})$, it is unimodular. Further, $D_1 = DU$ and $N_1 = PD_1 = PDU = NU$. $\square$

It is left to the reader to state and prove the "left" analogs of the results presented up to now. These results are used freely in the sequel, without comment. As a mnemonic aid, a tilde "~" is used to denote "left" quantities. For example, (N,D) is used for an r.c.f. of $P \in \mathbf{M}(\mathbf{F})$ and $(\tilde{D},\tilde{N})$ for an l.c.f. of P. Note that the order of the "denominator" and "numerator" matrices is interchanged in the latter case. This is to reinforce the point that if (N,D) is an r.c.f. of P, then $P = ND^{-1}$, whereas if $(\tilde{D},\tilde{N})$ is an l.c.f. of P, then $P = \tilde{D}^{-1}\tilde{N}$.

Up to now, it has been shown that every $P \in \mathbf{M}(\mathbf{F})$ has both an r.c.f. and an l.c.f. The next question to be answered is whether there is any relationship between the two.

49 *Theorem* Suppose $P \in \mathbf{M}(\mathbf{F})$ and that (N,D), $(\tilde{D},\tilde{N})$ are an r.c.f. and an l.c.f. of P. Then N, $\tilde{N}$ are equivalent, and the nonunit invariant factors of D, $\tilde{D}$ are the same. In particular, $|D|$ and $|\tilde{D}|$ are associates.[2]

Proof Suppose to be specific that P is of order $n \times m$ and has rank r. The discussion below applies to the case where $n \le m$; the case $n > m$ is entirely similar and is left to the reader. For convenience, let l denote $m-n$.

---

[2] See Section B.2 for the definition of equivalence.

Putting P in Smith-McMillan form (Theorem (B.2.29)), we see that there exist $U,V \in \mathbf{U}(\mathbf{R})$ such that

$$50 \qquad UPV = \begin{bmatrix} a_1/b_1 & & & \\ & \cdot & 0 & \\ & & \cdot & 0_{n\times l} \\ & & \cdot\ a_r/b_r & \\ & 0 & & \end{bmatrix},$$

where a_i, b_i are coprime for all i. Now define $A = \text{Diag}\ \{a_1, \cdots, a_r\}$, $B = \text{Diag}\ \{b_1, \cdots, b_r\}$, and let

$$51 \qquad N_1 = U^{-1} \begin{bmatrix} A & 0 & \\ & & 0_{n\times l} \\ 0 & 0 & \end{bmatrix}$$

$$52 \qquad D_1 = V \begin{bmatrix} B & 0 \\ 0 & I \end{bmatrix}$$

Then $|D_1| = \prod_{i=1}^{r} b_i \neq 0$, and $P = N_1 D_1^{-1}$. Also, since a_i, b_i are coprime for all i, there exist elements $x_i, y_i \in \mathbf{R}$ such that $x_i a_i + y_i b_i = 1$ for all i. Now define the matrices $X, Y \in \mathbf{R}^{n\times n}$ by $X = \text{Diag}\ \{x_1, \cdots, x_r, 0, \cdots, 0\}$, $Y = \text{Diag}\ \{y_1, \cdots, y_r, 1, \cdots, 1\}$, and let

$$54 \qquad S = \begin{bmatrix} X \\ 0_{l\times n} \end{bmatrix},$$

then it follows that

$$55 \qquad SN_1 + YD_1 = I.$$

Thus (N_1, D_1) is an r.c.f. of P.

Using entirely parallel reasoning, one can show that $(\tilde{D}_1,\tilde{N}_1)$ is an l.c.f. of P, where

56 $$\tilde{N}_1 = \begin{bmatrix} A & 0 & \\ & & 0_{n \times l} \\ 0 & 0 & \end{bmatrix} V,$$

57 $$\tilde{D}_1 = \begin{bmatrix} B & 0 \\ 0 & I \end{bmatrix} U^{-1}.$$

It is clear from (51) and (56) that N_1 and $\tilde{N}_1$ are equivalent matrices, as they both have the same invariant factors (see Corollary (B.2.28)). Similarly, (52) and (57) show that D_1 and $\tilde{D}_1$ have the same nonunit invariant factors. Thus we have established the theorem for the *particular* r.c.f. (N_1,D_1) and l.c.f. $(\tilde{D}_1,\tilde{N}_1)$ of P.

To complete the proof, let (N,D), $(\tilde{D},\tilde{N})$ be *any* r.c.f. and l.c.f. of P. Then by Theorem (43) and its left analog, there exist unimodular matrices $W \in \mathbf{R}^{m \times m}$, $Z \in \mathbf{R}^{n \times n}$ such that

58 $$N = N_1 W, \tilde{N} = Z\tilde{N}_1,$$

59 $$D = D_1 W, \tilde{D} = Z\tilde{D}_1.$$

Now (58) shows that N and N_1 are equivalent, and that $\tilde{N}$ and $\tilde{N}_1$ are equivalent. Since equivalence of matrices is transitive, it follows that N and $\tilde{N}$ are also equivalent. Similarly, (59) shows that D and $\tilde{D}$ have the same nonunit invariant factors. Finally, the fact that $|D|$ and $|\tilde{D}|$ are associates is a direct consequence of the preceding sentence. □

We conclude this section by introducing yet another type of useful factorization.

60 *Theorem* Suppose $P \in \mathbf{M}(\mathbf{F})$, and let (N,D), $(\tilde{D},\tilde{N})$ be any r.c.f. and l.c.f. of P. Suppose $X,Y \in \mathbf{M}(\mathbf{R})$ satisfy

61 $$XN + YD = I.$$

Then there exist $\tilde{X},\tilde{Y} \in \mathbf{M}(\mathbf{R})$ such that

62 $$\begin{bmatrix} Y & X \\ -\tilde{N} & \tilde{D} \end{bmatrix} \begin{bmatrix} D & -\tilde{X} \\ N & \tilde{Y} \end{bmatrix} = I.$$

Remark The ordered pair of matrices in (61) is referred to as a *doubly coprime factorization* of P.

Proof Given $N,D,\tilde{N},\tilde{D}$, there exist $\tilde{X}_1,\tilde{Y}_1 \in \mathbf{M}(\mathbf{R})$ such that

63 $$\tilde{N}\tilde{X}_1 + \tilde{D}\tilde{Y}_1 = I,$$

since $\tilde{N},\tilde{D}$ are left-coprime. Define

64 $$E = \begin{bmatrix} Y & X \\ -\tilde{N} & \tilde{D} \end{bmatrix}.$$

Then

65 $$E \begin{bmatrix} D & -\tilde{X}_1 \\ N & \tilde{Y}_1 \end{bmatrix} = \begin{bmatrix} I & \Delta \\ 0 & I \end{bmatrix},$$

where $\Delta = -Y\tilde{X}_1 + X\tilde{Y}_1$. (In (65) we use the fact that $\tilde{N}D = \tilde{D}N$, since $\tilde{D}^{-1}\tilde{N} = ND^{-1}$.) Since the matrix on the right side of (65) is unimodular, so is E. Moreover, from (65),

66 $$E^{-1} = \begin{bmatrix} D & -\tilde{X}_1 \\ N & \tilde{Y}_1 \end{bmatrix} \begin{bmatrix} I & \Delta \\ 0 & I \end{bmatrix}^{-1} = \begin{bmatrix} D & -(\tilde{X}_1 + D\Delta) \\ N & \tilde{Y}_1 + N\Delta \end{bmatrix}.$$

Thus (62) is satisfied with $\tilde{X} = \tilde{X}_1 + D\Delta$, $\tilde{Y} = \tilde{Y}_1 + N\Delta$. □

The above proof makes it clear that, once $N,D,\tilde{N},\tilde{D}$ have been selected, every X,Y such that $XN + YD = I$ determines a unique $\tilde{X},\tilde{Y}$ such that (61) holds, and vice-versa. The proof also demonstrates the following result.

67 *Corollary* Suppose $P \in \mathbf{M}(\mathbf{F})$, and that (N,D), $(\tilde{D},\tilde{N})$ are a r.c.f. and l.c.f. of P. Then for every $X,Y,\tilde{X},\tilde{Y} \in \mathbf{M}(\mathbf{R})$ such that $XN + YD = I$, $\tilde{N}\tilde{X} + \tilde{D}\tilde{Y} = I$, the matrices

68 $$U_1 = \begin{bmatrix} Y & X \\ -\tilde{N} & \tilde{D} \end{bmatrix},\ U_2 = \begin{bmatrix} D & -\tilde{X} \\ N & \tilde{Y} \end{bmatrix}$$

are unimodular. Moreover, U_1^{-1} is a complementation of $[D' \ \ N']'$, in that U_1^{-1} is of the form

69 $$U_1^{-1} = \begin{bmatrix} D & \\ & G \\ N & \end{bmatrix}$$

for some $G \in \mathbf{M}(\mathbf{R})$. Similarly, U_2^{-1} is a complementation of the matrix $[-\tilde{N} \ \ \tilde{D}]$ in that U_2^{-1} is of the form

70 $$U_2^{-1} = \begin{bmatrix} & H & \\ -\tilde{N} & & \tilde{D} \end{bmatrix}$$

for some $H \in \mathbf{M}(\mathbf{R})$.

Problems

1. As an illustration of Lemma (2), let $\mathbf{R}$ be the ring $\mathbb{R}[s]$ of polynomials with real coefficients. Suppose

$$A = \begin{bmatrix} 1 & s \\ s^2 & s^3 \end{bmatrix},\ B = [s+1 \ \ s^2+s]$$

(i) Show that both

$$D = \begin{bmatrix} 1 & s \\ 0 & 0 \end{bmatrix},\ D_1 = \begin{bmatrix} 1 & s \\ s & s^2 \end{bmatrix}$$

are g.c.r.d.'s of A and B.

(ii) Verify that $D_1 = VD$, where

$$V = \begin{bmatrix} 1 & s \\ s & 1 \end{bmatrix},$$

and that V is *not* unimodular over $\mathbb{R}[s]$.

(iii) Find a unimodular matrix U such that $D_1 = UD$.

2. Suppose $A, B \in \mathbf{M}(\mathbf{R})$ have the same number of columns, and let $F = [A' \;\; B']'$. Show that every g.c.r.d. of A, B is nonsingular if and only if F has full column rank. (Hint: See (10))

3. Suppose $A, B \in \mathbf{M}(\mathbf{R})$ are right-coprime. Suppose A is square and let $a = |A|$. Finally, let b denote the smallest invariant factor of B. Show that a and b are coprime.

4. Prove the following generalization of Problem 3: Suppose $A \in \mathbf{R}^{n \times n}$, $B \in \mathbf{R}^{m \times n}$ are right-coprime. Let $a_1, \cdots, a_n$ denote the invariant factors of A, and let $b_1, \cdots, b_n$ ($b_1, \cdots, b_m$ if $m < n$) denote the invariant factors of B. Then a_n, b_1 are coprime, a_{n-1}, b_2 are coprime, ..., a_1, b_n are coprime. (If $m < n$, then $a_m, a_{m-1}, \cdots, a_1$ are all units.)

5. Prove, or disprove by means of an example: Suppose A and B are square and right-coprime. Then $|A|$ and $|B|$ are coprime.

6. Prove, or disprove by means of an example: Suppose $A, B \in \mathbf{M}(\mathbf{R})$ are square and $|A|$, $|B|$ are coprime. Then A, B are right- as well as left-coprime.

7. Show that $A, B \in \mathbf{M}(\mathbf{R})$ are right-coprime if and only if A', B' are left-coprime.

8. Prove Fact (32).

9. Prove Fact (33).

10. Suppose $A,B \in \mathbf{M}(\mathbf{R})$ are right-coprime. Show that $A+RB,B$ are right-coprime for all $R \in \mathbf{M}(\mathbf{R})$. State and prove the left analog of this result.

11. Suppose $G \in \mathbf{M}(\mathbf{F})$, $H \in \mathbf{M}(\mathbf{R})$ have the same order, and let (N,D) be an r.c.f. of G. Show that $(N+HD,D)$ is an r.c.f. of $G+H$. State and prove the left analog of this result. (Hint: See Problem 10)

12. This problem is a generalization of Problem 11. Suppose $G \in \mathbf{M}(\mathbf{F})$, $H \in \mathbf{M}(\mathbf{R})$ and let $N,D,\tilde{N},\tilde{D},X,Y,\tilde{X},\tilde{Y} \in \mathbf{M}(\mathbf{R})$ be a doubly coprime factorization of G. Thus (62) holds, and in addition $G = ND^{-1} = \tilde{D}^{-1}\tilde{N}$. Show that a doubly coprime factorization of $G+H$ can be obtained by the following replacements:

$$N \leftarrow N+HD,\ Y \leftarrow Y-XH,\ \tilde{N} \leftarrow \tilde{N}+\tilde{D}H,\ \tilde{Y} \leftarrow \tilde{Y}-H\tilde{X}.$$

4.2 COPRIME FACTORIZATIONS OVER S

In this section, a method is presented for obtaining left- and right-coprime factorizations of a system transfer matrix from its state-space description. This method makes it easy to apply the subsequently developed synthesis theory to systems described in state-space form.

Consider a system described by the equations

1 $$\dot{x}(t) = Ax(t) + Bu(t),$$

2 $$y(t) = Cx(t) + Eu(t).$$

where A,B,C,E are constant matrices of compatible dimensions. The transfer matrix of this system is

3 $$P(s) = C(sI-A)^{-1}B + E.$$

The objective is to derive a doubly coprime factorization of P. One such factorization is given in Theorem (4) below.

4 *Theorem* Given the system (1)-(2), suppose the pairs (A,B), (A,C) are stabilizable and detectable, respectively. Select constant matrices K and F such that the matrices $A_0 := A - BK$, $\tilde{A}_0 := A - FC$ are both Hurwitz.[3] Then $P = N_p D_p^{-1} = \tilde{D}_p^{-1} \tilde{N}_p$ and

5 $$\begin{bmatrix} Y & X \\ -\tilde{N}_p & \tilde{D}_p \end{bmatrix} \begin{bmatrix} D_p & -\tilde{X} \\ N_p & \tilde{Y} \end{bmatrix} = I,$$

where the various matrices are defined as follows:

6 $$\tilde{N}_p = C(sI - \tilde{A}_0)^{-1}(B - FE) + E,$$

$$\tilde{D}_p = I - C(sI - \tilde{A}_0)^{-1}F,$$

$$N_p = (C - EK)(sI - A_0)^{-1}B + E,$$

$$D_p = I - K(sI - A_0)^{-1}B,$$

$$X = K(sI - \tilde{A}_0)^{-1}F,$$

$$Y = I + K(sI - \tilde{A}_0)^{-1}(B - FE),$$

$$\tilde{X} = K(sI - A_0)^{-1}F,$$

$$\tilde{Y} = I + (C - EK)(sI - A_0)^{-1}F.$$

As shown in Problem 4.1.12, if we can find a doubly coprime factorization for the transfer matrix

7 $$G(s) = P(s) - E = C(sI - A)^{-1}B,$$

then it is a simple matter to find a corresponding doubly coprime factorization for $P = G + E$. Hence the proof of Theorem (4) is straight-forward once the following lemma is established.

---

[3] A matrix is *Hurwitz* if all its eigenvalues have negative real parts.

8 *Lemma* Let all symbols be as in Theorem (4), and define

9 $$G(s) = C(sI-A)^{-1}B.$$

Then a doubly coprime factorization of G is given by the matrices

10 $$\tilde{N}_g = C(sI-\tilde{A}_0)^{-1}B,$$

$$\tilde{D}_g = I - C(sI-\tilde{A}_0)^{-1}F,$$

$$N_g = C(sI-A_0)^{-1}B,$$

$$D_g = I - K(sI-A_0)^{-1}B,$$

$$X_g = K(sI-\tilde{A}_0)^{-1}F,$$

$$Y_g = I + K(sI-\tilde{A}_0)^{-1}B,$$

$$\tilde{X}_g = K(sI-A_0)^{-1}F,$$

$$\tilde{Y}_g = I + C(sI-A_0)^{-1}F.$$

Proof It is first shown that $G = N_g D_g^{-1} = \tilde{D}_g^{-1}\tilde{N}_g$. This part of the proof uses the matrix identity $T(I+UT)^{-1} = (I+TU)^{-1}T$. Now

11 $$G = C(sI-A)^{-1}B$$

$$= C(sI-A_0-BK)^{-1}B$$

$$= C[(I-BK(sI-A_0)^{-1})(sI-A_0)]^{-1}B$$

$$= C(sI-A_0)^{-1}[I-BK(sI-A_0)^{-1}]^{-1}B$$

$$= C(sI-A_0)^{-1}B[I-K(sI-A_0)^{-1}B]^{-1}$$

$$= N_g D_g^{-1}.$$

The proof that $G = \tilde{D}_g^{-1}\tilde{N}_g$ is entirely similar.

The next part of the proof shows that the Bezout identities are satisfied. We have

$$12 \quad \tilde{N}_g\tilde{X}_g + \tilde{D}_g\tilde{Y}_g = C(sI-\tilde{A}_0)^{-1}BK(sI-A_0)^{-1}F$$
$$+ [I-C(sI-\tilde{A}_0)^{-1}F][I+C(sI-A_0)^{-1}F]$$
$$= C(sI-\tilde{A}_0)^{-1}BK(sI-A_0)^{-1}F$$
$$+ I - C(sI-\tilde{A}_0)^{-1}F + C(sI-A_0)^{-1}F$$
$$- C(sI-\tilde{A}_0)^{-1}FC(sI-A_0)^{-1}F.$$

To show that $\tilde{N}_g\tilde{X}_g + \tilde{D}_g\tilde{Y}_g = I$, it suffices to show that

$$13 \quad C(sI-\tilde{A}_0)^{-1}[BK-(sI-A_0)+sI-\tilde{A}_0-FC](sI-A_0)^{-1}F = 0.$$

But this relation is immediate, since the matrix inside the brackets is zero. The remaining equations follow in the same manner. □

The proof of Theorem (4) based on Lemma (8) is left to the reader.

4.3 BICOPRIME FACTORIZATIONS, CHARACTERISTIC DETERMINANTS

The objectives of this section are two-fold: (i) to introduce the concept of bicoprime factorizations, and (ii) to define characteristic determinants and McMillan degree, and to relate these to bicoprime factorizations. Throughout this section, $\mathbf{R}$ denotes a principal ideal domain and $\mathbf{F}$ the field of fractions associated with $\mathbf{R}$.

1 *Definition* Suppose $F \in \mathbf{M}(\mathbf{F})$. A quadruple $(N_r, D, N_l, K) \in \mathbf{M}(\mathbf{R})$ is a *bicoprime factorization of* F if (i) $|D| \neq 0$ and $F = N_r D^{-1} N_l + K$, (ii) N_r and D are right-coprime, and (iii) N_l and D are left-coprime.

Note that the term right-left-coprime factorization is also used by other authors.

There are at least two good reasons for studying bicoprime factorizations: 1) In the study of feedback systems, such as in Figure 1.2, it turns out that, starting from coprime factorizations for the plant and compensator, one is led naturally to a *bicoprime* (and not coprime) factorization of the resulting closed-loop transfer matrix (see Section 5.1). 2) In the case where the ring **R** is $\mathbb{R}[s]$, if a system is described in state-space form by

2 $$\dot{x}(t) = Ax(t) + Bu(t), y(t) = Cx(t) + Eu(t),$$

then its transfer matrix $H(s)$ equals $C(sI-A)^{-1}B + E$; further, if the description (2) is *minimal*, then $(C,sI-A,B,E)$ is a bicoprime factorization of H.

It is easy to see that if $N_l = I, K = 0$, (resp. $N_r = I, K = 0$), then a bicoprime factorization becomes an r.c.f. (resp. an l.c.f.).

3 *Definition* Suppose $F \in \mathbf{M}(\mathbf{F})$. A *characteristic determinant* of F is an element $\phi \in \mathbf{R}$ such that

(i) the product of ϕ and every minor of F belongs to **R**, and

(ii) if $\psi \in \mathbf{R}$ also satisfies (i) above (i.e. the product of ψ and every minor of F belongs to **R**), then ϕ divides ψ.

An equivalent, and more transparent, definition of a characteristic determinant is the following: Consider all minors of F, and express them as reduced-form fractions. Then ϕ is a least common multiple of all the denominators of the various minors of F. Since a least common multiple is only unique to within multipication by a unit of **R**, the same is true of the characteristic determinant.

Note that in the case where $\mathbf{R} = \mathbb{R}[s]$ and $\mathbf{F} = \mathbb{R}(s)$, it is customary to refer to the characteristic determinant as the characteristic polynomial.

4 *Example* Let $\mathbf{R} = \mathbb{R}[s]$, $\mathbf{F} = \mathbb{R}(s)$, and suppose

5 $$F = \begin{bmatrix} \dfrac{1}{s-1} & \dfrac{2}{s} \\ \dfrac{-s}{s+1} & \dfrac{s}{s^2-1} \end{bmatrix}.$$

Then F has five minors (four 1×1 and one 2×2). The monic l.c.m. of all the denominators of these minors is $\phi(s) = s(s+1)(s-1)^2$. Hence the set of *all* characteristic polynomials of F is $a\phi(s)$, where a is any nonzero real number.

It is obvious from the definition that $F \in \mathbf{M(F)}$ actually belongs to $\mathbf{M(R)}$ if and only if its characteristic determinant is a unit.

The next theorem is the main result of this section.

6 *Theorem* Suppose $F \in \mathbf{M(F)}$, and that (N_r, D, N_l, K) is a bicoprime factorization of F. Then $|D|$ is a characteristic determinant of F.

The proof requires a few lemmas, which are presented next. In the interests of brevity, let $\phi(A)$ denote a characteristic determinant of $A \in \mathbf{M(F)}$.

7 *Lemma* Suppose $A, B \in \mathbf{M(F)}$ and $A - B \in \mathbf{M(R)}$. Then $\phi(A)$ and $\phi(B)$ are associates.

Proof Suppose $A = B + C$ where $C \in \mathbf{M(R)}$. It is first shown that $\phi(A)$ divides $\phi(B)$. Consider a typical (say $n \times n$) minor of A, of the form

8 $$\begin{vmatrix} a_{i_1 j_1} & \cdots & a_{i_1 j_n} \\ \vdots & & \vdots \\ a_{i_n j_1} & \cdots & a_{i_n j_n} \end{vmatrix}$$

$$= \begin{vmatrix} b_{i_1 j_1} + c_{i_1 j_1} & \cdots & b_{i_1 j_n} + c_{i_1 j_n} \\ \vdots & & \vdots \\ b_{i_n j_1} + c_{i_n j_1} & \cdots & b_{i_n j_n} \end{vmatrix}.$$

Since the determinant is a multilinear function, the minor in (8) can be expanded as a sum of 2^n determinants, namely

9 $$\begin{vmatrix} b_{i_1 j_1} & \cdots & b_{i_1 j_n} \\ \vdots & & \vdots \\ b_{i_n j_1} & \cdots & b_{i_n j_n} \end{vmatrix} + \begin{vmatrix} c_{i_1 j_1} & b_{i_1 j_2} & \cdots & b_{i_1 j_n} \\ \vdots & \vdots & & \vdots \\ c_{i_n j_1} & b_{i_n j_2} & \cdots & b_{i_n j_n} \end{vmatrix}$$

$$+ \cdots + \begin{vmatrix} c_{i_1j_1} & \cdots & c_{i_1j_n} \\ \vdots & & \vdots \\ c_{i_nj_1} & \cdots & c_{i_nj_n} \end{vmatrix}.$$

Moreover, each of the 2^n-1 minors involving some elements of C can be further expanded, using Laplace's expansion, into a sum of products of minors of B and of C. Since $C \in \mathbf{M}(\mathbf{R})$, every minor of C belongs to $\mathbf{R}$. Since the product of $\phi(B)$ and every minor of B belongs to $\mathbf{R}$, we conclude that the product of $\phi(B)$ and the sum in (9) belongs to $\mathbf{R}$. Since this is true of *every* minor of B, it follows that $\phi(A)$ divides $\phi(B)$, by condition (ii) of Definition (3).

Next, write $B = A + (-C)$ and note that $-C \in \mathbf{M}(\mathbf{R})$. By the same reasoning as in the preceding paragraph, $\phi(B)$ divides $\phi(A)$. Hence $\phi(A)$ and $\phi(B)$ are associates. □

10 *Lemma* Suppose $X, Y \in \mathbf{M}(\mathbf{R})$, $B \in \mathbf{M}(\mathbf{F})$, and let $A = XBY$. Then $\phi(A)$ divides $\phi(B)$.

Proof To be specific, suppose X, B, Y have dimensions $k \times l$, $l \times m$, $m \times n$, respectively. Let $r \leq \min\{k, l, m, n\}$ and suppose $J \in S(r,k)$, $K \in S(r,n)$. By the Binet-Cauchy formula (Fact (B.1.19)), the minor a_{JK} is given by

11 $$a_{JK} = \sum_{K_1 \in S(r,l)} \sum_{K_2 \in S(r,m)} x_{JK_1} b_{K_1K_2} y_{K_2K}.$$

Hence

12 $$a_{JK} \cdot \phi(B) = \sum_{K_1} \sum_{K_2} x_{JK_1} \cdot [b_{K_1K_2} \phi(B)] \cdot y_{K_2K}.$$

Since every term in the summation on the right side of (12) belongs to $\mathbf{R}$, so does $a_{JK} \cdot \phi(B)$ for all J, K. Hence by condition (ii) of Definition (3), $\phi(A)$ divides $\phi(B)$. □

13 *Lemma* Suppose $P,Q,R \in \mathbf{M}(\mathbf{R})$, $|Q| \neq 0$, and let $A = PQ^{-1}R \in \mathbf{M}(\mathbf{F})$. Then $\phi(A)$ divides $|Q|$.

Proof By Fact (B.1.28), $(Q^{-1})_{JK} \cdot |Q|$ equals $\pm$ an appropriate minor of Q, and hence belongs to $\mathbf{R}$, for all J,K. Hence $\phi(Q^{-1})$ divides $|Q|$. By Lemma (10), $\phi(A)$ divides $\phi(Q^{-1})$. The conclusion follows. □

14 *Lemma* Suppose $Q \in \mathbf{M}(\mathbf{R})$ is square and $|Q| \neq 0$. Then $\phi(Q^{-1}) = |Q|$.

Proof As in Lemma (13), it is immediate that $\phi(Q^{-1})$ divides $|Q|$. To prove the converse, note that $|Q^{-1}| = 1/|Q|$. Hence any l.c.m. of the denominators of all minors of Q^{-1} must, in particular, be a multiple of $|Q|$.
□

Proof of Theorem (6) It is first shown that $\phi(F)$ divides $|D|$. We know that $F = N_r D^{-1} N_l + K$, and since $K \in \mathbf{M}(\mathbf{R})$ it follows from Lemma (7) that $\phi(F) = \phi(F-K) = \phi(N_r D^{-1} N_l)$. By Lemma (13), the latter divides $|D|$.

Next it is shown that $|D|$ divides $\phi(F)$. For convenience, let $\bar{F}$ denote $F-K$. Then $N_r D^{-1} N_l = \bar{F}$. Since N_r and D are right-coprime, there exist $X,Y \in \mathbf{M}(\mathbf{R})$ such that $XN_r + YD = I$. Similarly, since D and N_l are left-coprime, there exist $\tilde{X},\tilde{Y} \in \mathbf{M}(\mathbf{R})$ such that $N_l\tilde{X} + D\tilde{Y} = I$. Now

15 $$D^{-1} = D^{-1}(N_l\tilde{X} + D\tilde{Y}) = D^{-1}N_l\tilde{X} + \tilde{Y},$$

16 $$N_r D^{-1} = N_r D^{-1} N_l \tilde{X} + N_r \tilde{Y} = \bar{F}\tilde{X} + N_r\tilde{Y}.$$

Similarly,

17 $$\begin{aligned} D^{-1} &= (XN_r + YD)D^{-1} \\ &= XN_r D^{-1} + Y \\ &= X\bar{F}\tilde{X} + XN_r\tilde{Y} + Y. \end{aligned}$$

18 $$\begin{aligned} |D| &= \phi(D^{-1}) \text{ by Lemma (14)} \\ &= \phi(X\bar{F}\tilde{X} + XN_r\tilde{Y} + Y) \text{ by (17)} \\ &= \phi(X\bar{F}\tilde{X}) \text{ by Lemma (7)} \end{aligned}$$

since $XN_r\tilde{Y} + Y \in \mathbf{M}(\mathbf{R})$. However, by Lemma (10), $\phi(X\bar{F}\tilde{X})$ divides $\phi(\bar{F}) = \phi(F-K) = \bar{F}$, using Lemma (7) again. Hence $|D|$ divides $\phi(F)$. □

19 *Corollary* Suppose $F \in \mathbf{M}(\mathbf{F})$, and that (N_r, D, N_l, K) is a bicoprime factorization of F. Then $F \in \mathbf{M}(\mathbf{R})$ if and only if $D \in \mathbf{U}(\mathbf{R})$.

20 *Corollary* Suppose $F \in \mathbf{M}(\mathbf{F})$, and let (N, D), $(\tilde{D}, \tilde{N})$ be any r.c.f. and any l.c.f. of F. Then $|D|$ and $|\tilde{D}|$ are both chracteristic determinants of F.

Proof Note that $(N, D, I, 0)$ and $(I, \tilde{D}, \tilde{N}, 0)$ are both bicoprime factorizations of F, and apply Theorem (6). □

The rest of this section is devoted to extending Fact (2.1.20) to rational matrices, by introducing the notion of McMillan degree. Suppose $F \in \mathbf{M}(\mathbb{R}(s))$ and that $p \in C$ is a pole of $F(\cdot)$; the *McMillan degree* of p as a pole of F is the highest order it has as a pole of any minor of F.

21 *Example* Consider again the rational matrix F of (5). This matrix has three poles, namely 0,1, and -1. The McMillan degrees of 0 and -1 are 1, while the McMillan degree of 1 is 2.

The following important result is now obtained almost routinely from the preceding discussion.

22 *Theorem* Suppose $F \in \mathbf{M}(\mathbb{R}(s))$, and suppose (N, D), $(\tilde{D}, \tilde{N})$, (C, A, B, K) are any r.c.f., any l.c.f. and any bicoprime factorization of F over the ring $\mathbf{S}$. Then

(i) $|D|$, $|\tilde{D}|$, $|A|$ are associates.

(ii) An element $p \in C_{+e}$ is a pole of F if and only if it is a zero of a characteristic determinant of F, and its McMillan degree as a pole of F equals its multiplicity as a zero of $\phi(F)$.

(iii) An element $p \in C_{+e}$ is a pole of F if and only if $|D(p)| = 0$ (or equivalently $|\tilde{D}(p)| = 0$ or $|A(p)| = 0$). If $p \in C_{+e}$ is a pole of F, its McMillan degree equals its multiplicity as a zero of $|D|$ (or $|\tilde{D}|$ or $|A|$).

4.4 MATRIX EUCLIDEAN DIVISION

Throughout this section, $\mathbf{R}$ denotes a *proper* Euclidean domain. Recall that a domain $\mathbf{R}$ is *Euclidean* if there is a "degree" function $\delta : \mathbf{R} - 0 \rightarrow \mathbf{Z}_+$ such that

(i) Given any $f, g \in \mathbf{R}$ with $g \neq 0$, there exists a $q \in \mathbf{R}$ such that either $r := f - gq$ is zero or else $\delta(r) < \delta(g)$.

(ii) If x divides y, then $\delta(x) \leq \delta(y)$.

A Euclidean domain $\mathbf{R}$ is *proper* if it is not a field, and $\delta(fg) = \delta(f) + \delta(g)$ whenever f, g are nonzero. However, even in a proper Euclidean domain, the "quotient" q and "remainder" r corresponding to a given pair f, g need not be unique: this is so in the ring $\mathbf{S}$, for example. Accordingly, given $f, g \in \mathbf{R} - 0$, define

1 $$I(f,g) = \min_{q \in \mathbf{R}} \delta(f - gq),$$

where $\delta(0)$ is taken as $-\infty$. Clearly $I(f,g) < \delta(g)$, and it is either nonnegative or $-\infty$. In the case $\mathbf{R} = \mathbf{S}$, Theorem (2.3.5) provides an expression for $I(\cdot,\cdot)$. The objective of this section is to answer the following question: Given $A, B \in \mathbf{M}(\mathbf{R})$ with A square and A, B right-coprime, over what elements of $\mathbf{R}$ does $|A + RB|$ vary? This question arises in the study of the strong stabilizability of multivariable plants. It turns out that the answer to this question is quite simple, though its derivation is rather involved.

2 *Theorem* Suppose $A, B \in \mathbf{M}(\mathbf{R})$ are right-coprime and A is square. Let $a = |A|$ and let b denote the smallest invariant factor of B. Then the sets

3 $\{a+rb\colon r\in\mathbf{R}\}$,

4 $\{\,|A+RB|\colon R\in\mathbf{M}(\mathbf{R})\}$

are equal. As a consequence,

5 $$\min_{R\in\mathbf{M}(\mathbf{R})} \delta(|A+RB|) = \min_{r\in\mathbf{R}} \delta(a+rb) = I(a,b).$$

Remarks The first part of the theorem means that, if any element $f\in\mathbf{R}$ can be expressed as $a+rb$ for some $r\in\mathbf{R}$, then there exists an $R\in\mathbf{M}(\mathbf{R})$ such that $f = |A+RB|$, and conversely. The relation (5) follows readily once the equality of the sets in (3) and (4) is established.

Proof The proof is divided into two parts: First, it is shown that, corresponding to every $R\in\mathbf{M}(\mathbf{R})$, there exists an $r\in\mathbf{R}$ such that $|A+RB| = a+rb$. Then it is shown that, corresponding to every $r\in\mathbf{R}$, there exists an $R\in\mathbf{M}(\mathbf{R})$ such that $a+rb = |A+RB|$.

To prove the first assertion, suppose $R\in\mathbf{M}(\mathbf{R})$ is selected arbitrarily. Note that

6 $$A + \overline{R} = [I \quad R]\begin{bmatrix} A \\ B \end{bmatrix}.$$

Using (6) and the Binet-Cauchy formula (Fact (B.1.19)), one can obtain an expansion for $|A+RB|$ as a sum involving products of minors of $[I \quad R]$ and of $F = [A' \; B']'$. Now, except for $|A|$, every other minor of F contains at least one row from B and is hence a multiple of b. In the Binet-Cauchy expansion, $|A|$ is multiplied by $|I|$, which is one. Hence $|A+RB| = a+rb$ for an appropriate $r\in\mathbf{R}$.

To prove the second assertion, suppose $r\in\mathbf{R}$ is selected arbitrarily; it is then necessary to find an $R\in\mathbf{M}(\mathbf{R})$ such that $a+rb = |A+RB|$. This is first proved under the assumption that $|A| \neq 0$; this assumption is removed later on. Let A^{adj} denote the adjoint matrix of A, and let $c_1, \cdots, c_k$ denote the invariant factors of BA^{adj} (where $k = \text{rank } BA^{adj} = \text{rank } BA^{-1} = \text{rank } B$). The major part of the proof consists of demonstrating the existence of an $R_0 \in \mathbf{M}(\mathbf{R})$ such that $|aI + R_0 C| = (a+rb)a^{n-1}$, where $n \times n$ is the size of A, and

7 $$C = \begin{bmatrix} c_1 & & 0 & \\ & \ddots & & \\ 0 & & c_k & \\ & 0 & & 0 \end{bmatrix}$$

is a Smith form of BA^{adj}. If this can be shown, then the existence of an $R \in \mathbf{M}(\mathbf{R})$ such that $|A+RB| = a+rb$ follows readily: Select $U, V \in \mathbf{U}(\mathbf{R})$ such that $UBA^{adj}V = C$, and define

8 $$R = VR_0U.$$

To show that $|A+RB| = a+rb$, it is enough to show that $|A+RB| \cdot |A^{adj}| = (a+rb)a^{n-1}$, since $|A^{adj}| = a^{n-1}$. But

9 $$\begin{aligned} |A+RB| \cdot |A^{adj}| &= |V^{-1}(A+RB)A^{adj}V| \\ &= |aI + R_0UBA^{adj}V| \\ &= |aI + R_0C| \\ &= a^{n-1}(a+rb) \end{aligned}$$

provided R_0 satisfies $|aI + R_0C| = (a+rb)a^{n-1}$.

To demonstrate the existence of such an R_0, let

10 $$D = \begin{bmatrix} s_1/t_1 & & & 0 & \\ & \cdot & & & \\ & & \cdot & & \\ & & & \cdot & \\ & & & s_k/t_k & \\ & 0 & & & 0 \end{bmatrix}$$

be a Smith-McMillan form of BA^{-1}. Since B,A are right-coprime, (B,A) is an r.c.f. of BA^{-1}. Hence, from the proof of Theorem (4.1.49), it follows that

11 $$B \sim \begin{bmatrix} S & 0 \\ 0 & 0 \end{bmatrix}, A \sim \begin{bmatrix} T & 0 \\ 0 & I_{n-k} \end{bmatrix},$$

where $S = \text{Diag}\ \{s_1, \cdots, s_k\}$, $T = \text{Diag}\ \{t_1, \cdots, t_k\}$. In particular, we have

12 $$b \sim s_1, a \sim \prod_{i=1}^{k} t_i.$$

Without loss of generality, it is assumed that the equivalences in (12) are actually equalities. Now BA^{adj} is equivalent to C defined in (7), so that Ca^{-1} is a Smith-McMillan form for BA^{-1}. Hence

13 $$\frac{s_i}{t_i} = \frac{c_i}{a} \text{ for } i = 1, \cdots, k,$$

where again equivalence has been replaced by equality. Thus

14 $$c_1 = \frac{s_1}{t_1} \cdot a = s_1 t_2 \cdots t_k,$$

$$c_1 c_2 = \frac{s_1 s_2}{t_1 t_2} \cdot a^2 = (s_1 s_2 t_3 \cdots t_k) a$$

$$\vdots$$

$$c_1 c_2 \cdots c_k = (s_1 \cdots s_k) a^{k-1}.$$

Since s_i, t_i are coprime for all i, it is straight-forward to show that

15 $$\text{g.c.d. } \{t_2t_3 \cdots t_k, s_2t_3 \cdots t_k, s_2s_3 \cdots t_k, \cdots, s_2s_3 \cdots s_k\} = 1.$$

Hence from (14) and (15),

16 $$\text{g.c.d. } \{a^{k-1}c_1, a^{k-2}c_1c_2, \cdots, c_1c_2 \cdots c_k\} = a^{k-1}s_1 = a^{k-1}b.$$

In view of (16), there exist elements $q_1, \cdots, q_k \in \mathbf{R}$ such that

17 $$q_1a^{k-1}c_1 + q_2a^{k-2}c_1c_2 + \cdots + q_kc_1c_2 \cdots c_k = a^{k-1}br.$$

Now define R_0 as the bordered companion matrix

18 $$R_0 = \begin{bmatrix} q_1 & q_2 & \cdots & q_k & 0 \cdots 0 \\ -1 & 0 & & \vdots & \\ 0 & -1 & & 0 & \\ & & & & 0 \end{bmatrix}.$$

It is claimed that $|aI + R_0C| = (a+rb)a^{n-1}$. To see this, note that $aI + R_0C$ is the sum of a diagonal matrix and a nondiagonal matrix. Using Fact (B.1.22) it is possible to expand $|aI + R_0C|$ as a sum of products of principal minors of aI and of R_0C. This gives

19 $$|aI + R_0C| = a^n + \sum_{i=1}^{k} a^{n-i}$$
$$(\textstyle\sum i \times i \text{ principal minors of } R_0C),$$

because every $i \times i$ minor of aI equals a^i, and all minors of R_0C larger than $k \times k$ equal zero (recall that C has rank k). Now, since R_0 is a companion matrix, the only nonzero $i \times i$ principal minor of R_0 is the one corresponding to rows (and columns) 1 to i; it equals q_i. All other $i \times i$ principal minors of R_0 equal zero. Hence there is only one nonzero $i \times i$ principal minor of R_0C, and it equals $q_ic_1 \cdots c_i$. Finally, from (19) and (17),

20 $$|aI + R_0C| = a^n + a^{n-k}(a^{k-1}q_1c_1 + a^{k-2}q_2c_1c_2$$
$$+ \cdots + q_kc_1 \cdots c_k)$$

$$= a^n + a^{n-k} \cdot a^{k-1} br = a^{n-1}(a + rb).$$

It only remains to address the case where $|A| = 0$ in order to complete the proof. In this case, since A,B are right-coprime, it follows from Problem 4.1.3 that b is a unit; so let $b = 1$, without loss of generality. Then $\{a + rb: r \in \mathbf{R}\} = \mathbf{R}$. The conclusion of the proof requires a result of independent interest, so it is displayed separately.

21 *Lemma* Suppose $P,Q \in \mathbf{M}(\mathbf{R})$ have the same number of columns, and that $T = [P' \ \ Q']'$ has full column rank. Then there exists an $R \in \mathbf{M}(\mathbf{R})$ such that $|P + RQ| \neq 0$.

Proof If $|P| \neq 0$ let $R = 0$. Otherwise select a nonzero full-size minor of T having as few rows from Q as possible. Such a minor must contain at least one row of Q since $|P| = 0$. Let t denote this minor, and suppose it is obtained by excluding rows $i_1, \cdots, i_k$ of P and including rows $j_1, \cdots, j_k$ of Q. By assumption, *every* minor of T containing fewer than k rows of Q equals zero. Now define $R \in \mathbf{M}(\mathbf{R})$ by

22 $$r_{i_1 j_1} = \cdots = r_{i_k j_k} = 1; \ r_{ij} = 0 \text{ for all other } i,j.$$

Then every $k \times k$ minor of R equals zero, except for the minor consisting of rows $i_1, \cdots, i_k$ and columns $j_1, \cdots, j_k$, which equals one. Every larger minor of R equals zero. Next, observe that

23 $$P + RQ = [I \ \ R] \begin{bmatrix} P \\ Q \end{bmatrix}.$$

Hence, using the Binet-Cauchy formula (Fact (B.1.19)), one can expand $|P + RQ|$ in terms of the minors of $[I \ \ R]$ and of T. Further, by Laplace's expansion (Fact (B.1.16)), every minor of $[I \ \ R]$ equals $\pm$ a corresponding minor of R. In particular, every minor of $[I \ \ R]$ containing more than k columns of R equals zero. There is only one nonzero minor of $[I \ \ R]$ containing exactly k columns of R, and it is obtained by omitting columns $i_1, \cdots, i_k$ of the identity matrix and including columns $j_1, \cdots, j_k$ of R; it equals ± 1. Since every minor of T containing fewer than k rows of Q equals zero, we finally get

24 $$|P+RQ| = \pm t$$

which is nonzero. □.

Proof of Theorem (2) (conclusion) Suppose $|A| = 0$. Then $b = 1$, as shown previously. Suppose $r \in \mathbf{R}$ is specified. First select $R_0 \in \mathbf{M}(\mathbf{R})$ such that $|A+R_0B| \neq 0$; such an R_0 exists, by Lemma (21). Define $\bar{A} = A+R_0B$, $\bar{a} = |\bar{A}|$. By previous discussion, there exists an $R_1 \in \mathbf{M}(\mathbf{R})$ such that $|\bar{A}+R_1B| = \bar{a}+(r-\bar{a})b = r$. Now let $R = R_0+R_1$. □

25 *Corollary* Suppose $A,B \in \mathbf{M}(\mathbf{R})$ are right-coprime and A is square. Let b denote the smallest invariant factor of B. If b is not a unit, then $|A+RB| \neq 0$ for all $R \in \mathbf{M}(\mathbf{R})$; moreover, $|A+RB|$ and b are coprime for all $R \in \mathbf{M}(\mathbf{R})$.

Proof Let $a = |A|$. In view of Theorem (2), it is enough to show that $a+rb \neq 0$ for all $r \in \mathbf{R}$ and that $a+rb$, b are coprime for all $r \in \mathbf{R}$. By Problem 2.1.3, a and b are coprime, since A and B are right-coprime. The desired conclusions now follow readily. □

26 *Corollary* Suppose $A,B \in \mathbf{M}(\mathbf{R})$ are both square, with $|B| \neq 0$. Then there exists an $R \in \mathbf{M}(\mathbf{R})$ such that

27 $$\delta(|A+RB|) < \delta(|B|).$$

Proof If $|A| = 0$, (27) is satisfied with $R = 0$, so suppose $|A| \neq 0$. Let F be a g.c.r.d. of A,B, and let $A = A_1F$, $B = B_1F$. Let $a_1 = |A_1|$ and let b_1 denote the smallest invariant factor of B_1. Then Theorem (2) implies that, for some $R \in \mathbf{M}(\mathbf{R})$,

28 $$\delta(|A+RB|) = \delta(|F|) + \delta(|A_1+RB_1|)$$
$$= \delta(|F|) + I(a_1,b_1)$$

$$< \delta(|F|) + \delta(b_1)$$

$$\leq \delta(|F|) + \delta(|B_1|) = \delta(|B|).$$

This completes the proof. □

Notes and References

Most of the material in Section 4.1 is standard. Kailath [49] introduces many of these results for matrices with elements in the polynomial ring $\mathbb{R}[s]$ or the field $\mathbb{R}(s)$, but his methods largely carry over to the general situation studied here. See also [102]. The factorization in Section 4.2 is from [68], which extends earlier results from [92]. The results concerning matrix Euclidean division in Section 4.4 are taken from [98].

Chapter 5. STABILIZATION

This is one of the central chapters in the book, whose theme is the stabilization of one or more plants using one or more compensators. The first step is a study of the problem of stabilizing a (not necessarily stable) plant using an appropriate compensator, and the main result is a parametrization of *all* compensators that stabilize a given plant. This parametrization enables one to answer questions such as the following: (i) When can a plant be stabilized using a stable compensator? (ii) When can a single compensator stabilize each of two or more plants? (iii) When can a plant be stabilized using a multi-compensator configuration? (iv) What are the advantages of a "two degrees of freedom" compensator? (v) Once a plant has been stabilized, what type of performance can be expected from it in tasks such as tracking, disturbance rejection, etc.?

Throughout the chapter, the symbol $\mathbf{S}$, introduced in Chapter 2, is used to denote the set of proper stable rational functions. The symbol $\mathbf{M}(\mathbf{S})$ (resp. $\mathbf{U}(\mathbf{S})$) denotes the set of matrices (resp. unimodular matrices) whose elements belong to $\mathbf{S}$.

5.1 CLOSED-LOOP STABILITY

Consider the feedback system shown in Figure 5.1, where P represents the plant and C the compensator; u_1,u_2 denote the externally applied inputs, e_1,e_2 denote the inputs to the compensator and plant respectively, and y_1,y_2 denote the outputs of the compensator and plant respectively. This model is versatile enough to accomodate several control problems. For instance, in a problem of tracking, u_1 would be a reference signal to be tracked by the plant output y_2. In a problem of disturbance rejection or desensitization to noise, u_1 would be the disturbance/noise. Depending on whether u_1 or u_2 is the externally applied control signal (as opposed to noise etc.), the configuration of Figure 5.1 can represent either feedback or cascade compensation. For convenience, we will refer to this set-up as a feedback system.

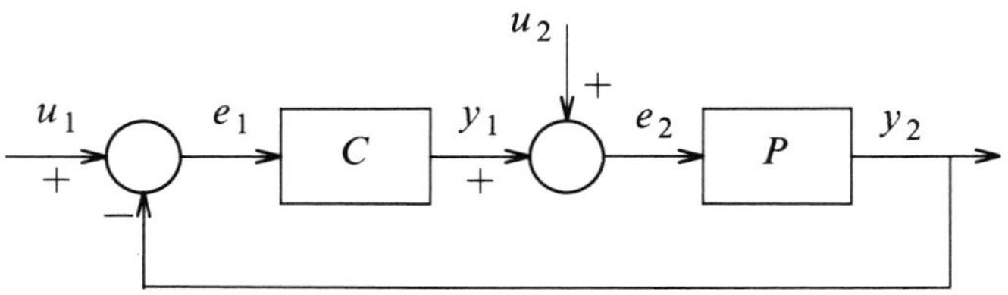

Figure 5.1. Feedback System.

Suppose $P, C \in \mathbf{M}(\mathbb{R}(s))$ so that they represent lumped linear time-invariant (continuous-time) systems, and are of compatible dimensions so that the interconnection in Figure 5.1 makes sense. The system under study is then described by

1 $$\begin{bmatrix} e_1 \\ e_2 \end{bmatrix} = \begin{bmatrix} u_1 \\ u_2 \end{bmatrix} - \begin{bmatrix} 0 & P \\ -C & 0 \end{bmatrix} \begin{bmatrix} e_1 \\ e_2 \end{bmatrix}; \quad \begin{bmatrix} y_1 \\ y_2 \end{bmatrix} = \begin{bmatrix} C & 0 \\ 0 & P \end{bmatrix} \begin{bmatrix} e_1 \\ e_2 \end{bmatrix}.$$

This system of equations can be rewritten as

2 $$e = u - FGe; \quad y = Ge,$$

where

3 $$e = \begin{bmatrix} e_1 \\ e_2 \end{bmatrix}, u = \begin{bmatrix} u_1 \\ u_2 \end{bmatrix}, y = \begin{bmatrix} y_1 \\ y_2 \end{bmatrix}, F = \begin{bmatrix} 0 & I \\ -I & 0 \end{bmatrix}, G = \begin{bmatrix} C & 0 \\ 0 & P \end{bmatrix}.$$

It is easy to verify that $|I+FG| = |I+PC| = |I+CP|$. The system (1) is *well-posed* if this determinant is nonzero as an element of $\mathbb{R}(s)$, i.e. if $|I+(FG)(s)|$ is not identically zero for all s. This condition is necessary and sufficient to ensure that (1) that has a unique solution in $\mathbf{M}(\mathbb{R}(s))$ for e_1, e_2 corresponding to every $u_1, u_2 \in \mathbf{M}(\mathbb{R}(s))$ of appropriate dimension.

If the system (1) is well-posed, then (1) can be solved for e_1, e_2; this gives

4 $$e = (I+FG)^{-1}u =: H(P,C)u.$$

It is possible to obtain several equivalent expressions for $H(P,C)$. Perhaps the most easily proved is

5 $$H(P,C) = \begin{bmatrix} (I+PC)^{-1} & -P(I+CP)^{-1} \\ C(I+PC)^{-1} & (I+CP)^{-1} \end{bmatrix}$$

In (5) both $(I+PC)^{-1}$ and $(I+CP)^{-1}$ occur, which is not very convenient. To get around this, one can use the following well-known matrix identities [26]:

6 $$(I+PC)^{-1} = I-P(I+CP)^{-1}C,\ C(I+PC)^{-1} = (I+CP)^{-1}C.$$

Of course (6) also holds with P and C interchanged throughout. Using these identities gives two other expressions for $H(P,C)$, namely

7 $$H(P,C) = \begin{bmatrix} I-P(I+CP)^{-1}C & -P(I+CP)^{-1} \\ (I+CP)^{-1}C & (I+CP)^{-1} \end{bmatrix}$$

$$= \begin{bmatrix} (I+PC)^{-1} & -(I+PC)^{-1}P \\ C(I+PC)^{-1} & I-C(I+PC)^{-1}P \end{bmatrix}.$$

Of these, the first involves only $(I+CP)^{-1}$ and the second only $(I+PC)^{-1}$.

The pair (P,C) is *stable* if $|I+PC| = |I+CP| \neq 0$ and $H(P,C) \in \mathbf{M}(\mathbf{S})$. Under these conditions, we also say that the feedback system (1) is stable. Put into other words, the system (1) is stable if it is well-posed, and if the transfer matrix from u to e belongs to $\mathbf{M}(\mathbf{S})$. The reader can verify that the conditions for stability are symmetric in P and C; thus (P,C) is stable if and only if (C,P) is stable.

A natural question at this stage is: Why should stability be defined in terms of the transfer matrix from u to e? Why not in terms of the transfer matrix from u to y? The answer is that both notions of stability are equivalent.

9 *Lemma* Suppose the system (1) is well-posed, and define $W(P,C) = G(I+FG)^{-1}$, where G and F are as in (3). Then $W(P,C) \in \mathbf{M}(\mathbf{S})$ if and only if $H(P,C) \in \mathbf{M}(\mathbf{S})$.

Proof Observe that F is nonsingular and that $F^{-1} = -F$. Hence

10 $$H = I - FW, \quad W = F^{-1}(I-H) = F(H-I),$$

where H, W are shorthand for $H(P,C)$, $W(P,C)$ respectively. It is immediate from (10) that $W \in \mathbf{M}(\mathbf{S})$ if and only if $H \in \mathbf{M}(\mathbf{S})$. □

In some undergraduate textbooks the feedback system (1) is defined to be "stable" if $P(I+CP)^{-1} \in \mathbf{M}(\mathbf{S})$. Thus one may ask why the current definition of stability requires *four different* transfer matrices to belong to $\mathbf{M}(\mathbf{S})$. There are two answers: (i) If $C \in \mathbf{M}(\mathbf{S})$, then $H(P,C) \in \mathbf{M}(\mathbf{S})$ if and only if $P(I+CP)^{-1} \in \mathbf{M}(\mathbf{S})$. Thus if one uses a stable compensator, then the stability of (1) can be ascertained by examining only $P(I+CP)^{-1}$. (ii) In the general case where no assumptions are made about C, "$H(P,C) \in \mathbf{M}(\mathbf{S})$" is a necessary and sufficient condition for "internal" as well as "external" stability. Each of these points is elaborated below.

11 *Lemma* Suppose $C \in \mathbf{M}(\mathbf{S})$. Then $H(P,C) \in \mathbf{M}(\mathbf{S})$ if and only if $P(I+CP)^{-1} \in \mathbf{M}(\mathbf{S})$.

Proof "only if" Obvious.

"if" Suppose $P(I+CP)^{-1} \in \mathbf{M}(\mathbf{S})$, and label the four matrices in (7) as H_{11} through H_{22}. Thus the hypothesis is that $-H_{12} \in \mathbf{M}(\mathbf{S})$. Clearly this implies that $H_{12} \in \mathbf{M}(\mathbf{S})$. Next,

12 $$H_{11} = I - P(I+CP)^{-1}C = I + H_{12}C \in \mathbf{M}(\mathbf{S})$$

since both H_{12} and C belong to $\mathbf{M}(\mathbf{S})$. This in turn implies that

13 $$H_{21} = CH_{11} \in \mathbf{M}(\mathbf{S}).$$

Finally,

14 $$H_{22} = I - C(I+PC)^{-1}P = I - CP(I+PC)^{-1}$$

$$= I + CH_{12} \in \mathbf{M}(\mathbf{S}).$$

Thus $H(P,C) \in \mathbf{M}(\mathbf{S})$. □

Suppose now that no assumptions are made regarding the stability of C. In order to discuss the internal stability of the feedback system (1), suppose C and P are described by state-space equations, of the form

15 $$\dot{x}_c = Q_c x_c + R_c u_c; \; y_c = S_c x_c + T_c u_c,$$

16 $$\dot{x}_p = Q_p x_p + R_p u_p; \; y_p = S_p x_p + T_p u_p.$$

17 *Lemma* Suppose the triples (S_c, Q_c, R_c), (S_p, Q_p, R_p) are both stabilizable and detectable, and that $|I + C(\infty)P(\infty)| = |I + T_c T_p| \neq 0$. Under these conditions, the system of Figure 5.1 is asymptotically stable if and only if $H(P,C) \in \mathbf{M}(\mathbf{S})$.

Remarks Lemma (17) shows that internal stability and external stability are equivalent if the state-space representations of C and P are stabilizable and detectable. Note that the representations of C and P need not be minimal; but any "hidden modes" must be stable.

The proof of Lemma (17) uses the following well-known result, which can be found in [49].

18 *Lemma* Suppose a triple $(\bar{S}, \bar{Q}, \bar{R})$ is stabilizable and detectable, and that T is a matrix of compatible dimensions. Under these conditions, the system

19 $$\dot{x}(t) = \bar{Q}x(t)$$

is asymptotically stable if and only if

20 $$W(s) = \bar{S}(sI - \bar{Q})^{-1}\bar{R} + \bar{T}$$

belongs to $\mathbf{M}(\mathbf{S})$.

Proof of Lemma (17) Define block-diagonal matrices Q, R, S, T in the obvious manner, and let F be as in (3). Then it is routine to verify that the interconnected system is described by the equations

21 $$\dot{x} = \bar{Q}x + \bar{R}u, \; y = \bar{S}x + \bar{T}u,$$

where

22 $$\bar{Q} = Q - R(I+TF)^{-1}S,$$

$$\bar{R} = R[I - F(I+TF)^{-1}T] = R(I+FT)^{-1},$$

$$\bar{S} = (I+TF)^{-1}S,$$

$$\bar{T} = (I+TF)^{-1}T.$$

It is now shown that the triple $(\bar{S},\bar{Q},\bar{R})$ is stabilizable and detectable. Recall that $(\bar{Q},\bar{R})$ is stabilizable if and only if there exists a constant matrix $\bar{K}$ such that $\bar{Q}-\bar{R}\bar{K}$ is Hurwitz [74].[1] By assumption, the pair (Q,R) is stabilizable, so that there exists a matrix K such that $Q-RK$ is Hurwitz. Now define

23 $$\bar{K} = (I+FT)(K-\bar{S}).$$

Then

24 $$\bar{Q}-\bar{R}\bar{K} = Q-R\bar{S}-RK+R\bar{S} = Q-RK,$$

which is Hurwitz by construction. This shows that the pair $(\bar{Q},\bar{R})$ is stabilizable. The detectability is shown in the same manner.

To conclude the proof, observe that, by Lemma (18), the overall system is asymptotically stable (or "internally" stable) if and only if its transfer matrix $W(P,C) \in \mathbf{M}(\mathbf{S})$. However, by Lemma (9) the latter condition is equivalent to the requirement that $H(P,C) \in \mathbf{M}(\mathbf{S})$ (i.e. the system is "externally" stable). □

Having defined the notion of stability for feedback systems and examined its implications, we now move on to the main objective of this section, which is to characterize the stability of the system (1) in terms of coprime

[1] A matrix is *Hurwitz* if all of its eigenvalues have negative real parts.

factorizations over **S** of the plant and compensator. This is done next in Theorem (25), which is the principal result of this section.

25 *Theorem* Suppose $P,C \in \mathbf{M}(\mathbb{R}(s))$. Let (N_p,D_p), $(\tilde{D}_p,\tilde{N}_p)$ be any r.c.f. and any l.c.f. of P, and let (N_c,D_c), $(\tilde{D}_c,\tilde{N}_c)$ be any r.c.f. and l.c.f. of C. Under these conditions, the following are equivalent:

(i) The pair (P,C) is stable.

(ii) The matrix $\tilde{N}_c N_p + \tilde{D}_c D_p$ is unimodular.

(iii) The matrix $\tilde{N}_p N_c + \tilde{D}_p D_c$ is unimodular.

The proof is based on the following lemma, which is of independent interest.

26 *Lemma* Let all symbols be as in Theorem (25), and suppose the pair (P,C) is well-posed. Define

27 $$\Delta(P,C) = \tilde{N}_c N_p + \tilde{D}_c D_p, \; \tilde{\Delta}(P,C) = \tilde{N}_p N_c + \tilde{D}_p D_c.$$

Then both $|\Delta(P,C)|$ and $|\tilde{\Delta}(P,C)|$ are characteristic determinants of $H(P,C)$ (in the sense of Definition (4.3.3)).

Proof It is only shown that $|\Delta(P,C)|$ is a characteristic determinant of $H(P,C)$. The parallel assertion for $|\tilde{\Delta}(P,C)|$ is proved in an entirely analogous fashion.

Note that $\Delta = \Delta(P,C) = \tilde{D}_c(I+CP)D_p$. If the pair (P,C) is well-posed, then $|I+CP| \neq 0$, and by definition $|\tilde{D}_c| \neq 0$, $|D_p| \neq 0$. Hence $|\Delta| \neq 0$, and from (7),

28 $$H(P,C) = \begin{bmatrix} I - N_p\Delta^{-1}\tilde{N}_c & -N_p\Delta^{-1}\tilde{D}_c \\ D_p\Delta^{-1}\tilde{N}_c & D_p\Delta^{-1}\tilde{D}_c \end{bmatrix}$$

$$= \begin{bmatrix} I & 0 \\ 0 & 0 \end{bmatrix} + \begin{bmatrix} -N_p \\ D_p \end{bmatrix} \Delta^{-1} [\tilde{N}_c \;\; \tilde{D}_c].$$

Let K denote the first (constant) matrix on the right side of (28). It is claimed that the quadruple $([-N_p' \;\; D_p']',\Delta,[\tilde{N}_c \;\; \tilde{D}_c],K)$ is a bicoprime factorization of $H(P,C)$. To show this, it is only necessary to show that

29 $$\begin{bmatrix} -N_p \\ D_p \\ \Delta \end{bmatrix} \sim \begin{bmatrix} I \\ \\ 0 \end{bmatrix}, \; [\tilde{N}_c \;\; \tilde{D}_c \;\; \Delta] \sim [I \;\; 0].$$

But this is immediate from the right-coprimeness of N_p,D_p and the left-coprimeness of $\tilde{N}_c,\tilde{D}_c$: Any common right divisor of $-N_p,D_p,\Delta$ must, in particular, be a common right divisor of $-N_p,D_p$ and hence must be unimodular; the second part of (29) is proved in the same manner. Now by Theorem (4.3.6) we conclude that $|\Delta(P,C)|$ is a characteristic determinant of $H(P,C)$. □

Proof of Theorem (25) (i) $\Rightarrow$ (ii) Suppose the pair (P,C) is stable. Then in particular it is well-posed. By Lemma (26) and Corollary (4.3.19), it follows that $\Delta(P,C) \in \mathbf{U}(\mathbf{S})$.

(ii) $\Rightarrow$ (i) Suppose $\Delta(P,C) \in \mathbf{U}(\mathbf{S})$. Since $\Delta(P,C) = \tilde{D}_c(I+CP)D_p$, it follows that $|I+CP| \neq 0$ so that the pair (P,C) is well-posed. Further, (28) shows that $H(P,C) \in \mathbf{M}(\mathbf{S})$, since $[\Delta(P,C)]^{-1} \in \mathbf{M}(\mathbf{S})$.

The proof that (i) and (iii) are equivalent is entirely similar. □

Let us say that a compensator C *stabilizes* the plant P if the pair (P,C) is stable. Then Theorem (25) leads to the result below.

30 *Corollary* Suppose $P \in \mathbf{M}(\mathbb{R}(s))$, and let (N_p,D_p), $(\tilde{D}_p,\tilde{N}_p)$ be any r.c.f. and any l.c.f. of P. Then the following are equivalent:

(i) C stabilizes P.

(ii) C has an l.c.f. $(\tilde{D}_c,\tilde{N}_c)$ such that $\tilde{D}_c D_p + \tilde{N}_c N_p = I$.

(iii) C has an r.c.f. (N_c,D_c) such that $\tilde{D}_p D_c + \tilde{N}_p N_c = I$.

Proof Once again, only the equivalence of (i) and (ii) is demonstrated; the equivalence of (i) and (iii) follows along similar lines.

(i)$\Longrightarrow$(ii) Since I is unimodular, it follows from Theorem (25) that (P,C) is stable, i.e. that C stabilizes P.

(ii)$\Longrightarrow$(i) Suppose C stabilizes P, and let $(\tilde{D},\tilde{N})$ be any l.c.f. of C. By Theorem (25), the matrix $\Delta := \tilde{D}D_p + \tilde{N}N_p$ is unimodular. Now let $\tilde{D}_c = \Delta^{-1}\tilde{D}$, $\tilde{N}_c = \Delta^{-1}\tilde{N}$. Then by the left analog of Theorem (4.1.43), $(\tilde{D}_c,\tilde{N}_c)$ is also an l.c.f. of C, and moreover, $\tilde{D}_c D_p + \tilde{N}_c N_p = \Delta^{-1}(\tilde{D}D_p + \tilde{N}N_p) = I$. □

31 *Remarks* Let P be a plant, and let (N_p,D_p), $(\tilde{D}_p,\tilde{N}_p)$ be any r.c.f. and any l.c.f. of P. Then every C that stabilizes P has a *unique* l.c.f. $(\tilde{D}_c,\tilde{N}_c)$ such that $\tilde{D}_c D_p + \tilde{N}_c N_p = I$. Moreover, in terms of this *particular* r.c.f. (N_p,D_p) of P and l.c.f. $(\tilde{D}_c,\tilde{N}_c)$ of C, we have

32 $$H(P,C) = \begin{bmatrix} I - N_p\tilde{N}_c & -N_p\tilde{D}_c \\ D_p\tilde{N}_c & D_p\tilde{D}_c \end{bmatrix}.$$

This follows readily from (26). Similarly,

33 $$W(P,C) = F^{-1}(I - H(P,C)) = \begin{bmatrix} D_p\tilde{N}_c & D_p\tilde{D}_c - I \\ N_p\tilde{N}_c & N_p\tilde{N}_c \end{bmatrix}.$$

The "right" analogs of (32) and (33) are easy to derive and are left to the reader.

Problems

1. (a) Suppose P,C are given, and let (N_p,D_p), (N_c,D_c) be any r.c.f.'s of P and C respectively. Show that the pair (P,C) is stable if and only if

$$U = \begin{bmatrix} D_c & -N_p \\ N_c & D_p \end{bmatrix} \in \mathbf{U}(\mathbf{S}).$$

(b) State and prove the "left" analog of (a).

2. The objective of this problem is to show that in a stable feedback system there can be no unstable pole-zero cancellations between P and C.

(a) Suppose P,C are scalar, and that (n_p,d_p), (n_c,d_c) are coprime factorizations of P and C. Suppose the pair (P,C) is stable. Show that no C_{+e}-pole of P can be a zero of C and vice versa. (Hint: Show that n_p,d_c are coprime and that n_c,d_p are coprime.)

(b) Prove the following multivariable generalization of (a): If (P,C) is stable, then N_p,D_c are left-coprime and N_c,D_p are left-coprime. (Hint: See Problem 1.)

(c) State and prove the left analog of (b).

5.2 PARAMETRIZATION OF ALL STABILIZING COMPENSATORS

Let $P \in \mathbf{M}(\mathbb{R}(s))$ and let $S(P)$ denote the set of all $C \in \mathbf{M}(\mathbb{R}(s))$ that stabilize P, i.e. the set of all $C \in \mathbf{M}(\mathbb{R}(s))$ such that (P,C) is stable. The main objective of this section is to parametrize the set $S(P)$, as well as the set $\{H(P,C): C \in S(P)\}$. Thus we wish to parametrize the set of all compensators that stabilize a given plant P, together with the set of all stable closed-loop transfer matrices that can be generated from P by using an appropriate stabilizing compensator.

1 *Theorem* Suppose $P \in \mathbf{M}(\mathbb{R}(s))$, and let (N_p,D_p), $(\tilde{D}_p,\tilde{N}_p)$ be any r.c.f. and any l.c.f. of P. Select matrices $X,Y,\tilde{X},\tilde{Y} \in \mathbf{M}(\mathbf{S})$ such that $XN_p + YD_p = I$, $\tilde{N}_p\tilde{X} + \tilde{D}_p\tilde{Y} = I$. Then

2 $$S(P) = \{(Y-R\tilde{N}_p)^{-1}(X+R\tilde{D}_p): R \in \mathbf{M}(\mathbf{S}),\ |Y-R\tilde{N}_p| \neq 0\}$$
$$= \{(\tilde{X}+D_pS)(\tilde{Y}-N_pS)^{-1}: S \in \mathbf{M}(\mathbf{S}),\ |\tilde{Y}-N_pS| \neq 0\}.$$

Remarks Equation (2) gives two representations for the set $S(P)$ of compensators that stabilize the plant P. The first of these states the following: Let R be any element of $\mathbf{M}(\mathbf{S})$ of appropriate dimensions such that $|Y-R\tilde{N}_p| \neq 0$; then the corresponding compensator $C = (Y-R\tilde{N}_p)^{-1}(X+R\tilde{D}_p)$ stabilizes P. Conversely, suppose

$C \in \mathbf{M}(\mathbb{R}(s))$ stabilizes P; then C is of the form $(Y - R\tilde{N}_p)^{-1}(X + R\tilde{D}_p)$ for some $R \in \mathbf{M}(\mathbf{S})$ (and it is implicit that $|Y - R\tilde{N}_p| \neq 0$). The second representation has a similar interpretation.

Proof Only the first representation is proved below. The proof of the second is entirely similar and is left to the reader.

It is shown in Corollary (5.1.30) that C stabilizes P if and only if C has an l.c.f. $(\tilde{D}_c, \tilde{N}_c)$ such that $\tilde{D}_c D_p + \tilde{N}_c N_p = I$. Now consider the equation

3 $$\bar{X} N_p + \bar{Y} D_p = I$$

in the unknowns $\bar{X}$ and $\bar{Y}$. Then C stabilizes P if and only if C is of the form $\bar{Y}^{-1}\bar{X}$ for some $\bar{X}, \bar{Y} \in \mathbf{M}(\mathbf{S})$ such that (3) holds and $|\bar{Y}| \neq 0$. From Corollary (4.1.67), the matrix

4 $$U_1 = \begin{bmatrix} Y & X \\ -\tilde{N}_p & \tilde{D}_p \end{bmatrix}$$

is unimodular, and moreover U_1^{-1} is of the form

5 $$U_1^{-1} = \begin{bmatrix} D_p & \\ & G \\ N_p & \end{bmatrix}$$

where $G \in \mathbf{M}(\mathbf{S})$. By Corollary (4.1.32), it now follows that all solutions for $(\bar{Y}, \bar{X})$ of (3) are of the form

6 $$[\bar{Y} \;\; \bar{X}] = [I \;\; R]U_1 = [Y - R\tilde{N}_p \;\; X + R\tilde{D}_p] \text{ for some } R \in \mathbf{M}(\mathbf{S}).$$

The first representation in (2) is implied by (6). □

Theorem (1) characterizes the set of all compensators that stabilize a given plant in terms of a certain "free" parameter. The correspondence between the parameter and the compensator is one-to-one in the following sense: Suppose P is a given plant, and that we have chosen a *particular* r.c.f. (N_p, D_p), l.c.f. $(\tilde{D}_p, \tilde{N}_p)$, as well as particular matrices $X, Y, \tilde{X}, \tilde{Y} \in \mathbf{M}(\mathbf{S})$ such that $XN_p + YD_p = I$, $\tilde{N}_p\tilde{X} + \tilde{D}_p\tilde{Y} = I$; then, corresponding to each $C \in S(P)$, there is a *unique* $R \in \mathbf{M}(\mathbf{S})$ such that $C = (Y - R\tilde{N}_p)^{-1}(X + R\tilde{D}_p)$, as well as a *unique* $S \in \mathbf{M}(\mathbf{S})$ such that $C = (\tilde{X} + D_p S)(\tilde{Y} - \tilde{N} S)^{-1}$. This fact is easy to prove and is left to the reader as an exercise (see Problem 1).

Theorem (1) shows that there is a simple formula that generates all the stabilizing compensators for a given plant. By substituting this formula into the expression for $H(P,C)$, we can obtain a parametrization of all stable closed-loop transfer matrices that can be obtained by stabilizing a given plant. The bonus is that this parametrization is *affine* in the free parameters R or S.

7 *Corollary* Suppose $P \in \mathbf{M}(\mathbb{R}(s))$ and let $N_p, D_p, \tilde{N}_p, \tilde{D}_p, X, Y, \tilde{X}, \tilde{Y}$ be as in Theorem (1). Suppose $C \in S(P)$. Then

8 $$H(P,C) = \begin{bmatrix} I - N_p(X + R\tilde{D}_p) & -N_p(Y - R\tilde{N}_p) \\ D_p(X + R\tilde{D}_p) & D_p(Y - R\tilde{N}_p) \end{bmatrix},$$

where R is the unique element of $\mathbf{M}(\mathbf{S})$ such that $C = (Y - R\tilde{N}_p)^{-1}(X + R\tilde{D}_p)$; alternatively,

9 $$H(P,C) = \begin{bmatrix} (\tilde{Y} - N_p S)\tilde{D}_p & -(\tilde{Y} - N_p S)\tilde{N}_p \\ (\tilde{X} + D_p S)\tilde{D}_p & I - (\tilde{X} + D_p S)\tilde{N}_p \end{bmatrix},$$

where S is the unique element of $\mathbf{M}(\mathbf{S})$ such that $C = (\tilde{X} + D_p S)(\tilde{Y} - N_p S)^{-1}$. Similarly,

10 $$W(P,C) = \begin{bmatrix} D_p(X + R\tilde{D}_p) & D_p(Y - R\tilde{N}_p) - I \\ N_p(X + R\tilde{D}_p) & N_p(Y - R\tilde{N}_p) \end{bmatrix}$$

$$= \begin{bmatrix} (\tilde{X} + D_p S)\tilde{D}_p & (\tilde{X} + D_p S)\tilde{N}_p \\ (\tilde{Y} - N_p S)\tilde{D}_p - I & (\tilde{Y} - N_p S)\tilde{N}_p \end{bmatrix}.$$

The proof of the corollary is left to the reader; it is a ready consequence of Theorem (1) and Remark (5.1.31).

If $C \in S(P)$, it follows from (10) that the transfer matrices from u_1 to y_2 and from u_2 to y_2 are both right multiples of N_p, no matter which stabilizing compensator is used. This is a multivariable generalization of a fact well-known in the scalar case, namely that if an unstable plant is stabilized by feedback compensation, then the nonminimum phase zeros of the plant continue to be zeros of the resulting stable closed-loop transfer function, irrespective of which stabilizing compensator is used.

In the characterization (2) of $S(P)$, the parameters R and S are not entirely free, since the nonsingularity constraint in (2) must be respected. It is shown next that "almost all" R and S satisfy this constraint.

Lemma (11) below can be thought of as an extension of Lemma (4.4.21) to topological rings.

11 *Lemma* Suppose $A, B \in \mathbf{M}(\mathbf{S})$ have the same number of columns, that A is square, and that $F = [A' \;\; B']'$ has full (column) rank. Define

12 $$\mathbf{V} = \{R \in \mathbf{M}(\mathbf{S}) : |A + RB| \neq 0\}$$

Then $\mathbf{V}$ is an open dense subset of $\mathbf{M}(\mathbf{S})$.

Proof It is necessary to prove two things: (i) If $R \in \mathbf{V}$, then there is a neighborhood $\mathbf{N}(R)$ of R such that $\mathbf{N}(R) \subseteq \mathbf{V}$ (this shows that $\mathbf{V}$ is open). (ii) If $R \notin \mathbf{V}$, then there is a sequence $\{R_i\}$ in $\mathbf{V}$ such that $R_i \rightarrow R$ (this shows that $\mathbf{V}$ is dense).

Accordingly, suppose first that $R \in \mathbf{V}$, so that $|A + RB| =: d \neq 0$. Observe that if $\varepsilon < \| d \|$, then 0 does not belong to the ball $\mathbf{B}(d, \varepsilon)$ defined by

13 $$\mathbf{B}(d, \varepsilon) = \{x \in \mathbf{S} : \| x - d \| < \varepsilon\}$$

Also note that the function $R \rightarrow f(R) = |A + RB|$ is a continuous map from $\mathbf{M}(\mathbf{S})$ into $\mathbf{S}$. Thus the preimage $f^{-1}(\mathbf{B}(d, \varepsilon))$ is an open neighborhood of R, and is a subset of $\mathbf{V}$. This shows that $\mathbf{V}$ is open.

Next, suppose that $R \notin \mathbf{V}$ so that $|A+RB| = 0$. Define $\bar{A} = A+RB$ and observe that $\bar{F} := [\bar{A}' \ \ B']'$ has full column rank since

14 $$\begin{bmatrix} \bar{A} \\ B \end{bmatrix} = \begin{bmatrix} I & R \\ 0 & I \end{bmatrix} \begin{bmatrix} A \\ B \end{bmatrix}.$$

It is shown that there exists a sequence $\{C_i\}$ in $\mathbf{M}(\mathbf{S})$ converging to zero such that $|\bar{A}+C_iB| \neq 0 \ \forall i$. Setting $R_i = R+C_i$ completes the proof that $\mathbf{V}$ is dense.

The rest of the proof is very reminiscent of that of Lemma (4.4.21). Select a nonzero full-size minor of $\bar{F}$ containing as few rows of B as possible. Let $\bar{f}$ denote this minor and suppose it is obtained by omitting rows $i_1, \cdots, i_k$ of A and including rows $j_1, \cdots, j_k$ of B. Define $C^\varepsilon \in \mathbf{M}(\mathbf{S})$ by

15 $$c^\varepsilon_{i_1 j_1} = \cdots = c^\varepsilon_{i_k j_k} = \varepsilon, \ c^\varepsilon_{ij} = 0 \text{ for all other } i,j.$$

Then, as in the proof of Lemma (4.4.21),

16 $$|\bar{A} + C^\varepsilon B| = \pm \varepsilon^k \bar{f} \text{ if } \varepsilon \neq 0.$$

Now let $\{\varepsilon_i\}$ be any sequence of nonzero numbers converging to zero, and let $C_i = C^{\varepsilon_i}$. Then $C_i \rightarrow 0$ but $|\bar{A} + C_iB| \neq 0$ for all i. □

17 *Corollary* With all symbols as in Theorem (1), the set of $R \in \mathbf{M}(\mathbf{S})$ such that $|Y - R\tilde{N}_p| \neq 0$ is an open dense subset of $\mathbf{M}(\mathbf{S})$, and the set of $S \in \mathbf{M}(\mathbf{S})$ such that $|\tilde{Y} - N_pS| \neq 0$ is an open dense subset of $\mathbf{M}(\mathbf{S})$.

Proof Since the matrix U_1 of (4) is unimodular, it follows in particular that Y and $\tilde{N}_p$ are right-coprime. Hence the matrix $F = [Y' \ \ \tilde{N}_p']'$ has full column rank. Now apply Lemma (11). □

18 *Lemma* Let all symbols be as in Theorem (1), and let n_p denote the smallest invariant factor of $\tilde{N}_p$. Suppose n_p is not a unit. Then

19 $$S(P) = \{(Y - R\tilde{N}_p)^{-1}(X + R\tilde{D}_p) \colon R \in \mathbf{M}(\mathbf{S})\}$$

$$= \{(\tilde{X}+D_pS)(\tilde{Y}-N_pS)^{-1}: S\in\mathbf{M}(\mathbf{S})\}.$$

Moreover, every $C\in S(P)$ is analytic at every C_{+e}-zero of n_p.

Proof By Corollary (4.4.25), if n_p is not a unit, then $|Y-R\tilde{N}_p| \neq 0$ for *all* $R\in\mathbf{M}(\mathbf{S})$, and $|Y-R\tilde{N}_p|$, n_p are coprime. □

Lemma (18) can be interpreted as follows: An $s\in C_{+e}$ where $P(s) = 0$ is called a *blocking zero of P*. It is easily established that the C_{+e}-blocking zeros of P are precisely the C_{+e}-zeros of n_p (see Problem 3). The lemma states that if P has any C_{+e}-blocking zeros, then the "nonsingularity constraint" in (2) is automatically satisfied and can therefore be omitted (compare (2) and (19)). Moreover, every C that stabilizes P is analytic at the C_{+e}-blocking zeros of P.

If P is strictly proper, then infinity is a blocking zero of P. In this case Lemma (18) leads to the following result:

20 *Corollary* Suppose P is strictly proper. Then every $C\in S(P)$ is proper.

Lemma (18) assumes a particularly simple form if P is stable.

21 *Corollary* Let all symbols be as in Lemma (18). Suppose $P\in\mathbf{M}(\mathbf{S})$ and that n_p is not a unit. Then

22 $$S(P) = \{(I-RP)^{-1}R: R\in\mathbf{M}(\mathbf{S})\}$$

$$= \{R(I-PR)^{-1}: R\in\mathbf{M}(\mathbf{S})\}.$$

In this case

23 $$H(P,C) = \begin{bmatrix} I-PR & -P(I-RP) \\ R & I-RP \end{bmatrix},$$

24 $$W(P,C) = \begin{bmatrix} R & -RP \\ PR & P(I-RP) \end{bmatrix}.$$

Proof Since P is stable, select $N_p = \tilde{N}_p = P$, $D_p = I$, $\tilde{D}_p = I$, $Y = I$, $\tilde{Y} = I$, $X = \tilde{X} = 0$, and apply Lemma (18) together with the formula (8). This establishes (23). Similarly (24) follows from Lemma (18) and (10). □

In particular, if P is stable and has some C_{+e}-blocking zeros, then the set of transfer matrices from u_1 to y_2 that can be obtained by applying a stabilizing compensator is precisely the set of right multiples of P.

The next lemma shows that, given any plant P and any *finite* set of points in C_{+e} that is disjoint from the set of poles of P, there exists a stabilizing compensator for P that has a blocking zero at each of these points. The main motivation for this lemma is to show that every proper plant can be stabilized by a strictly proper compensator. However, the general result follows with very little more effort.

25 *Lemma* Given a plant P and points $s_1, \cdots, s_n \in C_{+e}$ such that none is a pole of P, there exists a $C \in S(P)$ such that $C(s_i) = 0$ for all i.

Proof Select an $f \in \mathbf{S}$ such that $f(s_i) = 0 \ \forall i$. Let (N_p, D_p), $(\tilde{D}_p, \tilde{N}_p)$ be any r.c.f. and any l.c.f. of P, and select any $X, Y \in \mathbf{M}(\mathbf{S})$ such that $XN_p + YD_p = I$. Since none of the s_i is a pole of P, none is a zero of $|\tilde{D}_p|$ (see Theorem (4.3.22iii)). Hence f can be selected so that $|\tilde{D}_p|, f$ are coprime. Let d denote $|\tilde{D}_p|$ for convenience. Since d, f are coprime, for every $x \in \mathbf{S}$ there exist $a, b \in \mathbf{S}$ such that $ad + bf = x$; equivalently, for every $x \in \mathbf{S}$ there exists an $a \in \mathbf{S}$ such that $x + ad$ is divisible by f. Accordingly, select $A \in \mathbf{M}(\mathbf{S})$ such that every element of $X + Ad$ is divisible by f, and let $R = A\tilde{D}_p^{adj}$. Then $X + R\tilde{D}_p = X + Ad$ is a multiple of f and vanishes at each s_i. Moreover,

26 $$(Y - R\tilde{N}_p)D_p + (X + R\tilde{D}_p)N_p = I.$$

Finally, $|Y - R\tilde{N}_p| \neq 0$ since $[(Y - R\tilde{N}_p)D_p](s_i) = I$ for all i. Thus $(Y - R\tilde{N}_p)^{-1}$ is well-defined, and $C = (Y - R\tilde{N}_p)^{-1}(X + R\tilde{D}_p)$ is in $S(P)$ and satisfies $C(s_i) = 0$ for all i. □

Proposition (27) below rounds out the contents of this section; it characterizes *all* compensators that satisfy the conditions of Lemma (25).

27 *Proposition* Given a plant P and points $s_1, \cdots, s_n \in C_{+e}$ such that none of the s_i is a pole of P, let (N_p, D_p), $(\tilde{D}_p, \tilde{N}_p)$ be any r.c.f. and l.c.f. of P. Let $\mathbf{I}$ denote the ideal in $\mathbf{S}$ consisting of all functions that vanish at each of the s_i, and let f denote a generator of this ideal. In accordance with Lemma (25), select $X, Y, \tilde{X}, \tilde{Y} \in \mathbf{M}(\mathbf{S})$ such that

28 $$XN + YD = I, \tilde{N}\tilde{X} + \tilde{D}\tilde{Y} = I,$$

29 $$X, \tilde{X} \in \mathbf{M}(\mathbf{I}),$$

where the last equation merely states that both X and $\tilde{X}$ vanish at each s_i. Then the set of all $C \in S(P)$ that vanish at each s_i is given by

30 $$\{(Y - R\tilde{N}_p)^{-1}(X + R\tilde{D}_p): R \in \mathbf{M}(\mathbf{I})\}$$
$$= \{(\tilde{X} + D_p S)(\tilde{Y} - N_p S)^{-1}: S \in \mathbf{M}(\mathbf{I})\}.$$

Proof Suppose $C = (Y - R\tilde{N}_p)^{-1}(X + R\tilde{D}_p)$ where $R \in \mathbf{M}(\mathbf{I})$. Then $X + R\tilde{D}_p \in \mathbf{M}(\mathbf{I})$, so that $C(s_i) = 0 \; \forall i$.

Conversely, suppose $C \in S(P)$ and $C(s_i) = 0 \; \forall i$. Then, by Theorem (1), C equals $(Y - R\tilde{N}_p)^{-1}(X + R\tilde{D}_p)$ for some $R \in \mathbf{M}(\mathbf{S})$, and all that remains to be shown is that this R actually belongs to $\mathbf{M}(\mathbf{I})$. Now, since C vanishes at each s_i, it follows that $X + R\tilde{D}_p \in \mathbf{M}(\mathbf{I})$. Since $X \in \mathbf{M}(\mathbf{I})$, this implies successively that $R\tilde{D}_p \in \mathbf{M}(\mathbf{I})$, and that $R \mid \tilde{D}_p \mid = R\tilde{D}_p \tilde{D}_p^{adj} \in \mathbf{M}(\mathbf{I})$. Since $\mid \tilde{D}_p \mid$ is coprime to f, this implies that $R \in \mathbf{M}(\mathbf{I})$. □

As a final observation, suppose it is desired to determine all compensators C such that $H(P,C) \in \mathbf{M}(\mathbf{S_D})$, where $\mathbf{D}$ is a prespecified subset of the complex plane. Then all of the results of this section carry over completely, provided $\mathbf{S}$ is replaced by $\mathbf{S_D}$, and C_{+e} is replaced by the extended complement of $\mathbf{D}$, i.e. the union of $\{\infty\}$ and the complement of $\mathbf{D}$. An illustration of this is provided in Section 3.1. In the same manner, the results of this section can be extended in a completely transparent way to discrete-time systems.

Problems

1. In the parametrization (2) of $S(P)$, show that distinct choices of R lead to distinct compensators. (Hint: Suppose $(Y-R_1\tilde{N}_p)^{-1}(X+R_1\tilde{D}_p) = (Y-R_2\tilde{N}_p)^{-1}(X+R_2\tilde{D}_p)$ for some $R_1, R_2 \in \mathbf{M}(\mathbf{S})$. Show first the existence of a $U \in \mathbf{U}(\mathbf{S})$ such that

$$[Y-R_1\tilde{N}_p \quad X+R_1\tilde{D}_p] = U[Y-R_2\tilde{N}_p \quad X+R_2\tilde{D}_p].$$

Write this in the form

$$[I \quad R_1]\begin{bmatrix} Y & X \\ -\tilde{N}_p & \tilde{D}_p \end{bmatrix} = U[I \quad R_2]\begin{bmatrix} Y & X \\ -\tilde{N}_p & \tilde{D}_p \end{bmatrix}.$$

Now use Corollary (4.1.67) to show that $U = I$, $R_1 = R_2$.)

2. In (2), suppose the matrices $X, Y, \tilde{X}, \tilde{Y}$ also satisfy $X\tilde{Y} = Y\tilde{X}$ (such matrices can always be found, by Theorem (4.1.60)). Show that $C_1 = (Y-R\tilde{N}_p)^{-1}(X+R\tilde{D}_p)$ equals $C_2 = (\tilde{X}+D_pS)(\tilde{Y}-N_pS)^{-1}$ if and only if $R = S$.

3. Suppose $P \in \mathbf{M}(\mathbb{R}(s))$ and let (N_p, D_p), $(\tilde{D}_p, \tilde{N}_p)$ be any r.c.f. and l.c.f. of P. Let n_p denote the smallest invariant factor of N_p (and of $\tilde{N}_p$; see Theorem (4.1.49)). Show that the C_{+e}-blocking zeros of P are precisely the C_{+e}-zeros of n_p.

5.3 STRONG STABILIZATION

The question studied in this section is the following: Given an unstable plant P, when does there exist a *stable* compensator C that stabilizes P? More generally, what is the minimum number of C_{+e}-poles (counted according to their McMillan degrees) that any stabilizing compensator for P must have? A plant P is *strongly stabilizable* if $S(P)$ contains a stable compensator. A study of strong stabilizability is of interest for its own sake; some consequences of strong stabilizability are discussed towards the end of this section. In addition, several other problems in reliable stabilization are intimately related to strong stabilizability (see Sections 5.4 and 5.5). It turns out that there exist very simple necessary and sufficient conditions for strong stabilizability, based on the numbers and locations of the real C_{+e}-poles and zeros of the plant.

1 *Theorem* Given $P \in \mathbf{M}(\mathbb{R}(s))$, let $\sigma_1, \cdots, \sigma_l$ denote the real C_{+e}-blocking zeros of P (including ∞ if P is strictly proper), arranged in ascending order. Let η_i denote the number of poles of P in the interval (σ_i, σ_{i+1}), counted according to their McMillan degrees, and let η denote the number of odd integers in the sequence $\{\eta_1, \cdots, \eta_{l-1}\}$. Then every $C \in S(P)$ has at least η poles in C_{+e}. Moreover, this lower bound is exact in that there is a $C \in S(P)$ with exactly η poles in C_{+e}.

2 *Corollary* P is strongly stabilizable if and only if the number of poles of P (counted according to their McMillan degrees) between any pair of real C_{+e}-blocking zeros of P is even.

Remarks Both Theorem (1) and Corollary (2) are natural generalizations of the corresponding scalar results (Theorem (3.2.1) and Corollary (3.2.2), respectively). As in the scalar case, the condition for strong stabilizability given in Corollary (2) is referred to as the *parity interlacing property (p.i.p).*

Proof of Theorem (1) By Theorem (5.2.1), every $C \in S(P)$ is of the form $(Y - R\tilde{N}_p)^{-1}(X + R\tilde{D}_p)$ for some $R \in \mathbf{M}(\mathbf{S})$ such that $|Y - R\tilde{N}_p| \neq 0$. By Theorem (4.2.22), the sum of the McMillan degrees of the C_{+e}-poles of C equals $\delta(Y - R\tilde{N}_p)$. Thus the theorem is proved if it can be shown that

3 $$\min_{R \in \mathbf{M}(\mathbf{S}),\ |Y - R\tilde{N}_p| \neq 0} \delta(Y - R\tilde{N}_p) = \eta.$$

Let n_p denote the smallest invariant factor of $\tilde{N}_p$. There are two cases to consider, namely where n_p is a unit and where it is not. If n_p is a unit then $\eta = 0$. Also, by Theorem (4.4.2), the quantity $|Y - R\tilde{N}_p|$ can made to equal every element of $\mathbf{S}$ by an appropriate choice of R. In particular, there exists an $R \in \mathbf{M}(\mathbf{S})$ such that $\delta(Y - R\tilde{N}_p) = 0$. Since $\delta(f) \geq 0\ \forall f \in \mathbf{S}$, (3) holds in the case where n_p is a unit.

If n_p is not a unit, then by Corollary (4.4.25), $|Y-R\tilde{N}_p| \neq 0$ for all $R \in \mathbf{M}(\mathbf{S})$. Hence the nonsingularity constraint in (3) can be dropped, and (3) simplifies to

4 $$\min_{R \in \mathbf{M}(\mathbf{S})} \delta(|Y-R\tilde{N}_p|) = \eta.$$

Now by Theorem (4.4.2),

5 $$\min_{R \in \mathbf{M}(\mathbf{S})} \delta(|Y-R\tilde{N}_p|) = \min_{r \in \mathbf{M}(\mathbf{S})} \delta(y-rn_p)$$

where $y = |Y|$. Now note that the C_{+e}-zeros of n_p are precisely the C_{+e}-blocking zeros of P (see Problem 5.2.3). Hence, by Theorem (2.3.5), the minimum on the right side of (5) equals the number of sign changes in the sequence $\{y(\sigma_1), \cdots, y(\sigma_l)\}$. Next, observe that, since $XN_p + YD_p = I$ and $N_p(\sigma_i) = 0$, we have $(YD_p)(\sigma_i) = I$ for all i. This shows that $y(\sigma_i)$ and $|D_p(\sigma_i)|$ have the same sign for all i. Thus the minimum in (5) equals the number of sign changes in the sequence $\{|D_p(\sigma_1)|, \cdots, |D_p(\sigma_l)|\}$. Finally, note that $|D_p(\sigma_i)|$ and $|D_p(\sigma_{i+1})|$ have opposite signs if and only if the number of zeros of $|D_p(\cdot)|$ in the interval (σ_i, σ_{i+1}) is odd. By Theorem (4.2.22), the zeros of $|D_p(\cdot)|$ are in one-to-one correspondence with the poles of P, and the multiplicity of a zero of $|D_p(\cdot)|$ is the same as its McMillan degree as a pole of P. This completes the proof. □

The proof of the corollary is straight-forward and is left to the reader.

Note that the condition of Corollary (2) is vacuously satisfied if P has at most one real C_{+e}-blocking zero, of if P has no real C_{+e}-poles.

The following alternate form of Corollary (2) is sometimes useful.

6 *Corollary* Suppose $P \in \mathbf{M}(\mathbb{R}(s))$, and let (N_p, D_p) be any r.c.f. of P. Then P is strongly stabilizable if and only if $|D_p(\cdot)|$ has the same sign at all real C_{+e}-blocking zeros of P.

The next result gives a canonical form for a strongly stabilizable plant. This form is not very useful as a test for strong stabilizability (such a test is provided by Corollary (2)), but is very useful for proving theorems about strongly stabilizable systems.

7 *Lemma* A plant $P \in \mathbf{M}(\mathbb{R}(s))$ is strongly stabilizable if and only if there exist $C, V \in \mathbf{M}(\mathbf{S})$ such that $(V, I - CV)$, $(I - VC, V)$ are respectively an r.c.f. and an l.c.f. of P.

Proof "if" Suppose $(V, I - CV)$ is an r.c.f. of P. Then, since

8 $$C \cdot V + I \cdot (I - CV) = I,$$

it follows that C stabilizes P. Hence P is strongly stabilizable.

"only if" Suppose P is strongly stabilizable. Select $C \in \mathbf{M}(\mathbf{S})$ such that $P(I + CP)^{-1} =: V \in \mathbf{M}(\mathbf{S})$. It is now easy to verify that $P = (I - VC)^{-1} V = V(I - CV)^{-1}$, and that $(I - VC, V)$ $(V, I - CV)$ are respectively left-coprime and right-coprime. □

It is now possible to explain one of the essential differences between a strongly stabilizable plant and one that is not. Refer to (5.1.33), which gives an expression for all possible $W(P, C)$ that can be generated by a stabilizing compensator C. If P is strongly stabilizable and if $C \in S(P)$ is chosen to be stable, then $\tilde{D}_c$ is unimodular. As a result, N_p and $W_{22}(P, C) = N_p \tilde{D}_c$ have the same invariant factors. To put it another way, if P is stabilized using a *stable* compensator, then the resulting stable transfer matrix has the same C_{+e}-zeros as the original plant P. On the other hand, if P is stabilized using an *unstable* compensator (and this is all one can do if P is not strongly stabilizable), then $|\tilde{D}_c|$ is not a unit, so that at least one invariant factor of $N_p \tilde{D}_c$ is a strict multiple of the corresponding invariant factor of N_p. In other words, stabilization using an unstable compensator always introduces additional C_{+e}-zeros into the closed-loop transfer matrix beyond those of the original plant. As we will see in Chapter 6, the RHP zeros of a plant affect its ability to track reference signals and/or to reject disturbances. Hence it is preferrable to use a stable stabilizing compensator in such situations.

In some design problems, one uses a so-called *two-stage procedure* for selecting an appropriate compensator. Given a plant P, the first stage consists of selecting a stabilizing compensator for P. Let $C \in S(P)$ denote this compensator and define $P_1 = P(I+CP)^{-1}$. The second stage consists of selecting a stabilizing compensator for P_1 that also achieves some other design objectives such as decoupling, sensitivity minimization, etc. The rationale behind this procedure is that the design problems are often easier to solve when the plant is stable. The resulting configuration with its inner and outer loops is shown in Figure 5.2. (Note that the system of Figure 5.1 has been slightly rearranged.)

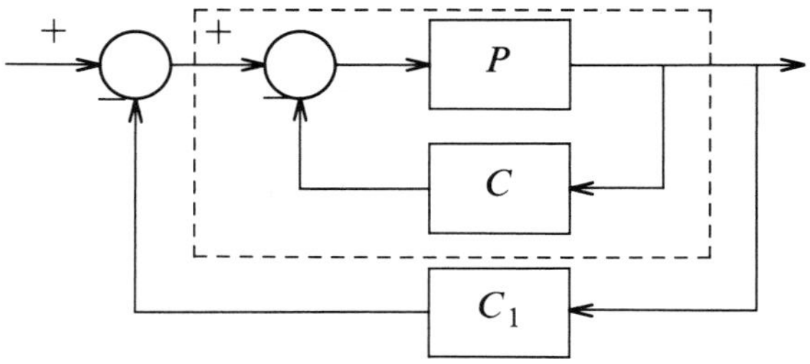

Figure 5.2. Two-Stage Compensator Design.

However, in order for this two-stage procedure to be valid, it must be true that

9 $$S(P) = C + S(P_1),$$

or, in other words, C_1 stabilizes P_1 if and only if $C+C_1$ stabilizes P. This is not always true in general.

10 *Theorem* Suppose $P \in \mathbf{M}(\mathbb{R}(s))$, $C \in S(P)$, and let P_1 denote $P(I+CP)^{-1}$. Then

11 $$C+S(P_1) \subseteq S(P),$$

with equality holding if and only if $C \in \mathbf{M}(\mathbf{S})$.

Remarks Theorem (10) states that, in general, if $C \in S(P)$ and $C_1 \in S(P_1)$, then $C+C_1 \in S(P)$. However, unless $C \in \mathbf{M}(\mathbf{S})$ (which is possible only if P is strongly stabilizable), there exists a C_1 such that $C+C_1 \in S(P)$ but $C_1 \notin S(P_1)$. To put it another way, the two-stage technique always yields a compensator that stabilizes the original plant P.

However, not all compensators in $S(P)$ can be realized in this fashion; a necessary and sufficient condition for such a decomposition to be possible is that the inner-loop compensator C is stable (which implies, *a fortiori*, that P is strongly stabilizable).

Proof In the first part of the proof it is shown that if $C_1 \in S(P_1)$ then $C+C_1 \in S(P)$. By assumption, $H(P,C) \in \mathbf{M}(\mathbf{S})$ and $H(P_1,C_1) \in \mathbf{M}(\mathbf{S})$. Now

12 $$I+P(C+C_1) = (I+PC+PC_1) = (I+PC)[I+(I+PC)^{-1}PC_1]$$
$$= (I+PC)(I+P_1C_1)$$

Thus

13 $$[I+P(C+C_1)]^{-1} = (I+P_1C_1)^{-1}(I+PC)^{-1} \in \mathbf{M}(\mathbf{S}).$$

Similarly,

14 $$[I+P(C+C_1)]^{-1}P = (I+P_1C_1)^{-1}(I+PC)^{-1}P \in \mathbf{M}(\mathbf{S}).$$

15 $$[I+(C+C_1)P]^{-1} = (I+CP)^{-1}(I+C_1P_1)^{-1} \in \mathbf{M}(\mathbf{S}).$$

In order to complete the proof that $H(P_1,C_1) \in \mathbf{M}(\mathbf{S})$, it only remains to show that

16 $$[I+(C+C_1)P]^{-1}(C+C_1) \in \mathbf{M}(\mathbf{S}).$$

From (15),

17 $$[I+(C+C_1)P]^{-1}C_1 = (I+CP)^{-1}(I+C_1P_1)^{-1}C_1 \in \mathbf{M}(\mathbf{S}),$$

so it only remains to show that

18 $$[I+(C+C_1)P]^{-1}C \in \mathbf{M}(\mathbf{S}).$$

Now, from (5.1.33), $P_1 = N_p\tilde{D}_c$ for an appropriate r.c.f. (N_p,D_p) of P and l.c.f. $(\tilde{D}_c,\tilde{N}_c)$ of C. Hence $P_1C = N_p\tilde{N}_c \in \mathbf{M}(\mathbf{S})$. Further, from (5.2.23), $(I+C_1P_1)^{-1}$ equals $I-RP_1$ for some $R \in \mathbf{M}(\mathbf{S})$. Therefore

19 $$[I+(C+C_1)P]^{-1}C = (I+CP)^{-1}(I+C_1P_1)^{-1}C$$
$$= (I+CP)^{-1}(I-RP_1)C$$
$$= (I+CP)^{-1}C \,-\, (I+CP)^{-1}RP_1C$$
$$\in \mathbf{M}(\mathbf{S})$$

since $(I+CP)^{-1}C$, $(I+CP)^{-1}$, R, $P_1C \in \mathbf{M}(\mathbf{S})$. This completes the proof of (11).

In the second part of the proof, it is shown that if P is strongly stabilizable and $C \in \mathbf{M}(\mathbf{S})$ stabilizes P, then (11) holds with equality. From Lemma (7), $(I-P_1C, P_1)$, $(P_1, I-CP_1)$ are respectively an r.c.f. and an l.c.f. of P. Now apply Theorem (5.2.1) with

20 $$N_p = \tilde{N}_p = P_1,\; D_p = I-CP_1,\; \tilde{D}_p = I-P_1C,\; X = C,\; Y = I.$$

This gives

21 $$S(P) = \{(I-RP_1)^{-1}[C+R(I-P_1C)]:$$
$$R \in \mathbf{M}(\mathbf{S}) \text{ and } |I-RP_1| \neq 0\}$$

On the other hand, since $P_1 \in \mathbf{M}(\mathbf{S})$, a slight modification of Corollary (5.2.21) gives

22 $$S(P_1) = \{(I-RP_1)^{-1}R\colon R \in \mathbf{M}(\mathbf{S}) \text{ and } |I-RP_1| \neq 0\}.$$

Finally, note that

23 $$(I-RP_1)^{-1}R \,+\, C = (I-RP_1)^{-1}[R+(I-RP_1)C]$$
$$= (I-RP_1)^{-1}[C+R(I-P_1C)]$$

This last equation shows that

24 $$S(P) = C \,+\, S(P_1).$$

The last part of the proof consists of showing that if $C \notin \mathbf{M}(\mathbf{S})$, then the inclusion in (11) is strict. This is shown by constructing a $\bar{C} \in S(P)$ such that $\bar{C} - C \notin S(P_1)$. Since $C \in S(P)$, by Corollary (5.1.30) there exist an r.c.f. (N_p, D_p) of P and an l.c.f. $(\tilde{D}_c, \tilde{N}_c)$ of C such that $\tilde{D}_c D_p + \tilde{N}_c N_p = I$. Moreover $\tilde{D}_c$ is *not* unimodular since C is unstable. Now let $\bar{C} = (1+\alpha)C$, where α is a positive number less than $1/\| \tilde{N}_c N_p \|$. Then $(\tilde{D}_c, (1+\alpha)\tilde{N}_c)$ is an l.c.f. of $\bar{C}$, and $(\tilde{D}_c, \alpha\tilde{N}_c)$ is an l.c.f. of $\bar{C} - C = \alpha C$. Now $\bar{C} \in S(P)$ since

25 $$\tilde{D}_c D_p + (1+\alpha)\tilde{N}_c N_p = I + \alpha \tilde{N}_c N_p$$

is unimodular by virtue of the condition $\alpha < 1/\| \tilde{N}_c N_p \|$. On the other hand, it is claimed that $\bar{C} - C = \alpha C$ does *not* stabilize P. By (5.1.33), $P_1 = N_p \tilde{D}_c$, so that $(N_p \tilde{D}_c, I)$ is an r.c.f. of P. Since $(\tilde{D}_c, \alpha\tilde{N}_c)$ is an l.c.f. of αC, applying the stability criterion of Theorem (5.1.25) gives the "return difference" matrix

26 $$\Delta(P_1, \alpha C) = \tilde{D}_c + \alpha \tilde{N}_c N_p \tilde{D}_c = (I + \alpha \tilde{N}_c N_p)\tilde{D}_c .$$

However, this matrix Δ is not unimodular since it is a multiple of the nonunimodular matrix $\tilde{D}_c$. Hence $\alpha C \notin S(P_1)$. □

5.4 SIMULTANEOUS STABILIZATION

In this section, we consider the problem of designing a compensator that stabilizes each of a given family of plants. Specifically, suppose $P_0, \cdots, P_l$ are given plants; we would like to know whether or not there exists a common stabilizing controller for this set of plants. This is referred to as the *simultaneous stabilization problem.* It turns out that the problem of stabilizing l plants with a common compensator is equivalent to that of stabilizing $l-1$ plants using a common *stable* compensator. In particular, the problem of simultaneously stabilizing two plants is equivalent to that of stabilizing a single plant using a stable compensator, which is studied in the previous section.

The simultaneous stabilization problem arises naturally in the synthesis of control systems with "integrity." In this set-up, P_0 corresponds to the nominal description of the plant to be controlled, while $P_1, \cdots, P_l$ correspond to the plant description after some sort of structural change, such as a loss of sensors, actuators, transducers, etc. The problem of synthesizing a compensator that continues to stabilize the plant even after a contingency of this sort is precisely that of finding a compensator that stabilizes each of the plants $P_0, \cdots, P_l$. Another application for this problem arises when one tries to use a compensator for a nonlinear system. As the operating point for the nonlinear system changes, so does its linearized model around its operating point. If each of these linearized models can be stabilized using a common compensator, then a fixed (i.e. independent of operating point) compensator can be used for the nonlinear system, with a saving in complexity.

To bring out the basic ideas clearly, we begin with the problem of simultaneously stabilizing two plants, which can be stated as follows: Given two plants P_0 and P_1, when does there exist a compensator C such that (P_0,C) and (P_1,C) are both stable? If such a C exists, we say that P_0,P_1 are *simultaneously stabilizable.* To facilitate the presentation of the results, we recall a few definitions.

Suppose $P \in \mathbf{M}(\mathbb{R}(s))$. Then a doubly coprime factorization of P is an r.c.f. (N,D) and an l.c.f. $(\tilde{D},\tilde{N})$ of P, together with matrices $X,Y,\tilde{X},\tilde{Y} \in \mathbf{M}(\mathbf{S})$ such that

1 $$\begin{bmatrix} Y & X \\ -\tilde{N} & \tilde{D} \end{bmatrix} \begin{bmatrix} D & -\tilde{X} \\ N & \tilde{Y} \end{bmatrix} = I.$$

Such matrices can always be found (see Theorem (4.1.60)). Throughout the rest of this section, it is assumed that doubly coprime factorizations of P_0 and P_1 are available, and symbols such as Y_0,N_1 etc. are used without further definition.

If $C \in \mathbf{M}(\mathbf{S})$, then (C,I), (I,C) are respectively an r.c.f. and an l.c.f. of C. Hence, for a plant $P \in \mathbf{M}(\mathbb{R}(s))$, the following are equivalent (see Theorem (5.1.25)):

(i) There exists $C \in \mathbf{M}(\mathbf{S})$ such that (P,C) is stable.

(ii) There exists $R \in \mathbf{M}(\mathbf{S})$ such that $D+RN$ is unimodular, where (N,D) is any r.c.f. of P.

(iii) There exists $S \in \mathbf{M}(\mathbf{S})$ such that $\tilde{D}+\tilde{N}S$ is unimodular, where $(\tilde{D},\tilde{N})$ is any l.c.f. of P.

The main result of this section is presented next.

2 *Theorem* Given two plants P_0 and P_1, define

3 $$A = Y_0 D_1 + X_0 N_1, \; B = -\tilde{N}_0 D_1 + \tilde{D}_0 N_1.$$

Then P_0 and P_1 can be simultaneously stabilized if and only if there exists an $M \in \mathbf{M}(\mathbf{S})$ such that $A+MB$ is unimodular.

Remarks It can be shown (see Problem 1) that A and B are right-coprime. Hence the theorem states that P_0 and P_1 can be simultaneously stabilized if and only if the associated system BA^{-1} is strongly stabilizable. Once A and B have been computed, one can use Theorem (5.3.2) to determine whether or not there exists an $M \in \mathbf{M}(\mathbf{S})$ such that $A+MB \in \mathbf{U}(\mathbf{S})$. This leads to the parity condition.

Proof As a first step, assume that P_0 and P_1 are both strictly proper. This assumption is removed later on. The sets of compensators that stabilize P_0 and P_1 are given by (see Lemma (5.2.18))

4 $$S(P_0) = \{(Y_0 - R_0\tilde{N}_0)^{-1}(X_0 + R_0\tilde{D}_0): R_0 \in \mathbf{M}(\mathbf{S})\},$$

5 $$S(P_1) = \{(Y_1 - R_1\tilde{N}_1)^{-1}(X_1 + R_1\tilde{D}_1): R_1 \in \mathbf{M}(\mathbf{S})\}.$$

Hence P_0 and P_1 can be simultaneously stabilized if and only if there exist R_0 and R_1 such that

6 $$(Y_0 - R_0\tilde{N}_0)^{-1}(X_0 + R_0\tilde{D}_0) = (Y_1 - R_1\tilde{N}_1)^{-1}(X_1 + R_1\tilde{D}_1)$$

Let C denote the compensator defined in the above equation. Since both sides of (6) give *l.c.f.'s* of C, (6) holds if and only if there is a unimodular matrix U such that

7 $$Y_0 - R_0\tilde{N}_0 = U(Y_1 - R_1\tilde{N}_1),\ X_0 + R_0\tilde{D}_0 = U(X_1 + R_1\tilde{D}_1).$$

Hence P_0 and P_1 can be simultaneously stabilized if and only if there exist $R_0, R_1 \in \mathbf{M}(\mathbf{S})$, $U \in \mathbf{U}(\mathbf{S})$ such that (7) holds. Now (7) can be rewritten as

8 $$[I \;\; R_0] \begin{bmatrix} Y_0 & X_0 \\ -\tilde{N}_0 & \tilde{D}_0 \end{bmatrix} = U[I \;\; R_1] \begin{bmatrix} Y_1 & X_1 \\ -\tilde{N}_1 & \tilde{D}_1 \end{bmatrix},$$

which is equivalent to

9 $$[I \;\; R_0] \begin{bmatrix} Y_0 & X_0 \\ -\tilde{N}_0 & \tilde{D}_0 \end{bmatrix} \begin{bmatrix} D_1 & -\tilde{X}_1 \\ N_1 & \tilde{Y}_1 \end{bmatrix} = U[I \;\; R_1],$$

or

10 $$[I \;\; R_0] \begin{bmatrix} A & S \\ B & T \end{bmatrix} = U[I \;\; R_1],$$

where

11 $$\begin{bmatrix} A & S \\ B & T \end{bmatrix} = \begin{bmatrix} Y_0 & X_0 \\ -\tilde{N}_0 & \tilde{D}_0 \end{bmatrix} \begin{bmatrix} D_1 & -\tilde{X}_1 \\ N_1 & \tilde{Y}_1 \end{bmatrix}.$$

Thus P_0 and P_1 can be simultaneously stabilized if and only if there exist $R_0, R_1 \in \mathbf{M}(\mathbf{S})$, $U \in \mathbf{U}(\mathbf{S})$ such that (10) holds. Therefore the theorem is proved if we can establish that (10) holds if and only if $A + MB$ is unimodular for some $M \in \mathbf{M}(\mathbf{S})$.

Accordingly, suppose first that (10) holds for suitable R_0, R_1, U. Then $A + R_0 B = U$ is unimodular. Conversely, suppose $A + MB$ is unimodular for some M. Then (10) holds with $U = A + MB$, $R_0 = M$, $R_1 = U^{-1}(S + R_0 T)$.

To complete the proof, consider the case where P_0 and P_1 are not necessarily strictly proper. In this case, P_0 and P_1 are simultaneously stabilizable if and only if there exist $R_0, R_1 \in \mathbf{M}(\mathbf{S})$ such that

$|Y_0 - R_0\tilde{N}_0| \neq 0$, $|Y_1 - R_1\tilde{N}_1| \neq 0$, and (6) holds. The preceding discussion shows that *if* (6) holds for some R_0, R_1, then $A + R_0 B$ is unimodular. Hence, if P_0, P_1 can be simultaneously stabilized, then there exists an M such that $A + MB$ is unimodular. To prove the converse, suppose $A + M_0 B \in \mathbf{U}(\mathbf{S})$ for some $M_0 \in \mathbf{M}(\mathbf{S})$. If we define $R_0 = M_0$, $R_1 = (A + M_0 B)^{-1}(S + M_0 T)$, then it is conceivable that $|Y_0 - R_0\tilde{N}_0|$ or $|Y_1 - R_1\tilde{N}_1|$ is zero. If so, the idea is to perturb these elements slightly without destroying the unimodularity of $A + MB$. Proceed as follows: Since the map $M \rightarrow A + MB$ is continuous and since $\mathbf{U}(\mathbf{S})$ is an open subset of $\mathbf{M}(\mathbf{S})$, it folows that there is a ball $\mathbf{B}(M_0, \varepsilon)$ centered at M_0 such that $A + MB \in \mathbf{U}(\mathbf{S})$ $\forall M \in \mathbf{B}(M_0, \varepsilon)$. Now by Lemma (5.2.11), the sets $\{R_0 \colon |Y_0 - R_0\tilde{N}_0| \neq 0\}$, $\{R_1 \colon |Y_1 - R_1\tilde{N}_1| \neq 0\}$ are both open dense subsets of $\mathbf{M}(\mathbf{S})$. Hence, for some $M \in \mathbf{B}(M_0, \varepsilon)$, $|Y_0 - M\tilde{N}_0|$ must be nonzero, and $|Y_1 - R_1\tilde{N}_1|$ must also be nonzero with $R_1 = (A + MB)^{-1}(S + MT)$. □

The next result gives a more intuitive interpretation of simultaneous stabilization.

12 *Corollary* Suppose P_0 is stable and P_1 is arbitrary. Then P_0 and P_1 can be simultaneously stabilized if and only if $P_1 - P_0$ is strongly stabilizable.

Proof It is left to the reader to give a proof based on Theorem (2) (see Problem 2); an independent proof is given below.

"if" Suppose (N_1, D_1) is an r.c.f. of P_1. Then $(N_1 - P_0 D_1, D_1)$ is an r.c.f. of $P_1 - P_0$, since

13 $$X_1(N_1 - P_0 D_1) + (Y_1 + X_1 P_0) D_1 = I.$$

Suppose $R \in \mathbf{M}(\mathbf{S})$ stabilizes $P_1 - P_0$. Then from Theorem (5.1.25) it follows that

14 $$D_1 + R(N_1 - P_0 D_1) \in \mathbf{U}(\mathbf{S}).$$

Now rearrange (14) as

15 $(I - RP_0)D_1 + RN_1 \in \mathbf{U}(\mathbf{S})$,

and define $C = (I - RP_0)^{-1}R$. Then (15) shows that C stabilizes P_1. By Corollary (5.2.21), C also stabilizes P_0. Hence P_1 and P_0 can be simultaneously stabilized.

"only if" This part consists essentially of reversing the above reasoning. Suppose P_0 and P_1 are simultaneously stabilizable, and let C be a compensator that stabilizes both. By Corollary (5.2.21), since C stabilizes P_0, C has an l.c.f. of the form $(I - RP_0, R)$ for some $R \in \mathbf{M}(\mathbf{S})$. Since C also stabilizes P_1, Theorem (5.1.25) implies (15), which in turn is the same as (14). Finally, (14) shows that R stabilizes $P_1 - P_0$, so that $P_1 - P_0$ is strongly stabilizable. □

Theorem (2) shows that the problem of simultaneously stabilizing two plants can be reduced to that of stabilizing a single plant using a stable compensator. The converse is also true: Observe that $C \in \mathbf{M}(\mathbb{R}(s))$ is stable if and only if it stabilizes the plant $P = 0$ (see (5.1.6)). Hence the problem of stabilizing a plant P using a stable compensator is equivalent to simultaneously stabilizing 0 and P using the same compensator.

The discussion of the simultaneous stabilization of two plants is concluded with a derivation of alternate conditions that are equivalent to the one in Theorem (2).

16 *Lemma* Suppose P_0, P_1 have no common C_{+e}-poles. Then the C_{+e}-blocking zeros of $B = -\tilde{N}_0 D_1 + \tilde{D}_0 N_1$ are precisely the C_{+e}-blocking zeros of $P_1 - P_0$.

Proof It is first shown that if $B(s) = 0$ for some $s \in C_{+e}$, then $|D_1(s) \neq 0$ and $|\tilde{D}_0(s)| \neq 0$. Accordingly suppose $s \in C_{+e}$ and that $B(s) = 0$. By the hypothesis, $|D_1|$ and $|\tilde{D}_0|$ are coprime, which implies that either $|D_1(s)| \neq 0$ or $|\tilde{D}_0(s)| \neq 0$. Suppose to be specific that $|D_1(s)| \neq 0$; the other case is entirely similar. Then

17 $$B(s) = 0 \Longleftrightarrow (\tilde{N}_0 D_1)(s) = (\tilde{D}_0 N_1)(s)$$
$$\Longleftrightarrow \tilde{N}_0(s) = \tilde{D}_0(s) P_1(s).$$

Hence

18 $$[\tilde{N}_0(s) \;\; \tilde{D}_0(s)] = \tilde{D}_0(s)[P_1(s) \;\; I].$$

Now, if $|\tilde{D}_0(s)| = 0$, then $[\tilde{N}_0(s) \;\; \tilde{D}_0(s)]$ does not have full row rank. But this contradicts the left-coprimeness of the pair $(\tilde{D}_0, \tilde{N}_0)$. Hence $|\tilde{D}_0(s)| \neq 0$. It can be similarly established that if $|\tilde{D}_0(s)| \neq 0$, then $|D_1(s)| \neq 0$.

Next, suppose $s \in C_{+e}$ and that $B(s) = 0$. It has already been established that $|\tilde{D}_0(s)| \neq 0$ and $|D_1(s)| \neq 0$. Hence, from (17), we get

19 $$B(s) = 0 \Longleftrightarrow P_0(s) = P_1(s) = 0 \Longleftrightarrow P_1(s) - P_0(s) = 0.$$

This concludes the proof. □

20 *Lemma* Suppose P_0, P_1 have no common C_{+e}-poles. Then P_0, P_1 are simultaneously stabilizable if and only if the following interlacing property holds: Let $\sigma_1, \cdots, \sigma_l$ denote the real C_{+e}-blocking zeros of the plant $P_1 - P_0$; then the number of poles of P_0 and of P_1 in every interval (σ_i, σ_{i+1}), counted according to their McMillan degrees, is even.

Proof By Theorem (2), P_0 and P_1 can be simultaneously stabilized if and only if the auxiliary plant BA^{-1} is strongly stabilizable. By Corollary (5.2.6), this is the case if and only if $|A|$ has the same sign at all real C_{+e}-blocking zeros of B. Next, at such a blocking zero,

21 $$A(s) = [Y_0 D_1 + X_0 N_1](s)$$
$$= [Y_0 + X_0 N_1 D_1^{-1}](s) D_1(s)$$
$$= [Y_0 + X_0 N_0 D_0^{-1}](s) D_1(s)$$

$$= [D_0(s)]^{-1}D_1(s)$$

where the last two equations follow respectively from the facts that: (i) $P_0(s) = P_1(s)$, and (ii) $Y_0+X_0N_0D_0^{-1} = [Y_0D_0+X_0N_0]D_0^{-1} = D_0^{-1}$. Thus, if $\sigma_1, \cdots, \sigma_l$ are the real C_{+e}-blocking zeros of P_1-P_0, then P_0, P_1 can be simultaneously stabilized if and only if $|D_0(\sigma_1)|^{-1}|D_1(\sigma_1)|, \cdots, |D_0(\sigma_l)|^{-1}|D_1(\sigma_l)|$ are all of the same sign. Since $|D_0(\sigma)|^2 \geq 0 \ \forall \sigma$, this condition holds if and only if $|D_0(\sigma_1)| \cdot |D_1(\sigma_1)|, \cdots, |D_0(\sigma_l)| \cdot |D_1(\sigma_l)|$ all have the same sign. But this is true if and only if the function $|D_0(\cdot)| \cdot |D_1(\cdot)|$ has an even number of zeros in the interval (σ_i, σ_{i+1}) for all i. Since there is a one-to-one correspondence between the zeros of $|D_i(\cdot)|$ and the poles of P_i (see Theorem (4.3.22)), the result follows. □

Now we turn to the simultaneous stabilization problem with more than two plants. Given plants $P_0, \cdots, P_l$, we would like to know whether or not there exists a compensator C that stabilizes all of them. This problem can be reduced to one of simultaneously stabilizing l plants using a *stable* compensator.

22 *Theorem* Suppose $P_0, \cdots, P_l$ are given plants. Define

23 $$A_i = Y_0D_i + X_0N_i,\ B_i = -\tilde{N}_0D_i + \tilde{D}_0N_i,\ i = 1, \cdots, l.$$

Then $P_0, \cdots, P_l$ can be simultaneously stabilized if and only if there exists an $M \in \mathbf{M}(\mathbf{S})$ such that A_i+MB_i is unimodular for $i = 1, \cdots, l$.

Proof Since the proof closely parallels that of Theorem (2), it is only sketched here. There exists a C such that (P_i, C) is stable for $i = 0, \cdots, l$ if and only if there exist matrices $R_i \in \mathbf{M}(\mathbf{S})$ for $i = 1, \cdots, l$ and $U_i \in \mathbf{U}(\mathbf{S})$ for $i = 0, \cdots, l$ such that

24 $$[I \ \ R_0] \begin{bmatrix} Y_0 & X_0 \\ -\tilde{N}_0 & \tilde{D}_0 \end{bmatrix} = U_i[I \ \ R_i] \begin{bmatrix} Y_i & X_i \\ -\tilde{N}_i & \tilde{D}_i \end{bmatrix}, i = 1, \cdots, l,$$

(see (8)), or equivalently,

25 $$[I \;\; R_0] \begin{bmatrix} Y_0 & X_0 \\ -\tilde{N}_0 & \tilde{D}_0 \end{bmatrix} \begin{bmatrix} D_i & -\tilde{X}_i \\ N_i & \tilde{Y}_i \end{bmatrix} = U_i[I \;\; R_i], i = 1, \cdots ,l.$$

The first of these equations gives

26 $$A_i + R_0 B_i = U_i, i = 1, \cdots ,l,$$

while the second can be written as

27 $$R_i = U_i^{-1}[I \;\; R_0] \begin{bmatrix} Y_0 & X_0 \\ -\tilde{N}_0 & \tilde{D}_0 \end{bmatrix} \begin{bmatrix} -\tilde{X}_i \\ \tilde{Y}_i \end{bmatrix}, i = 1, \cdots ,l.$$

It is left to the reader to complete the proof. □

See Section 7.6 for some results concerning the genericity of simultaneous stabilizability.

Finally, the problem of simultaneously stabilizing l plants $P_1, \cdots ,P_l$ using a stable compensator is equivalent to the problem of simultaneously stabilizing $l+1$ plants $P_0 = 0, P_1, \cdots ,P_l$.

Problems

1. Show that A and B in (3) are right-coprime.

2. Prove Corollary (12) from Theorem (2). (Hint: If P_0 is strictly proper and stable, then one may take

$$\begin{bmatrix} Y_0 & X_0 \\ -\tilde{N}_0 & \tilde{D}_0 \end{bmatrix} = \begin{bmatrix} I & 0 \\ -P_0 & I \end{bmatrix}.$$

Now apply the theorem.)

5.5 MULTI-COMPENSATOR CONFIGURATION

The problem studied in this section is in a sense the dual of the simultaneous stabilization problem studied in the previous section. In the latter, the emphasis is on determining whether there exists a single compensator that stabilizes each of two (or more) plants. In contrast, the objective of this section is to determine conditions under which a single plant is stabilized by two compensators, each acting alone or both acting together.

To state the problem precisely, consider the system shown in Figure 5.3, where $P, C_1, C_2 \in \mathbf{M}(\mathbb{R}(s))$ and have compatible dimensions. The relationship between $e = (e_1, e_2, e_3)$ and $u = (u_1, u_2, u_3)$ is given by

1 $$\begin{bmatrix} e_1 \\ e_2 \\ e_3 \end{bmatrix} = H(P,C_1,C_2) \begin{bmatrix} u_1 \\ u_2 \\ u_3 \end{bmatrix}$$

where, letting C denote $C_1 + C_2$,

2 $$H(P,C_1,C_2) = \begin{bmatrix} (I+PC)^{-1} & -PC_2(I+PC)^{-1} & -P(I+CP)^{-1} \\ -PC_1(I+PC)^{-1} & (I+PC)^{-1} & -P(I+CP)^{-1} \\ C_1(I+PC)^{-1} & C_2(I+PC)^{-1} & (I+CP)^{-1} \end{bmatrix}.$$

The *Reliable Stabilization Problem* (RSP) investigated in this section can now be stated precisely: Given P, find C_1 and C_2 such that

(i) $H(P,C_1,C_2) \in \mathbf{M}(\mathbf{S})$,

(ii) $H(P,C_1,0) \in \mathbf{M}(\mathbf{S})$,

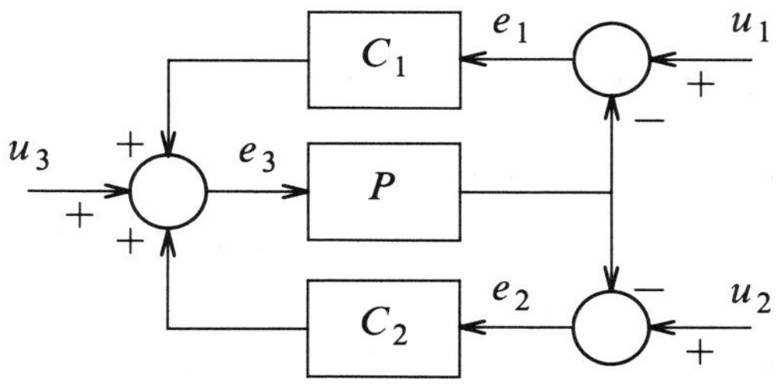

Figure 5.3. Multi-Compensator Configuration.

(iii) $H(P,0,C_2) \in \mathbf{M}(\mathbf{S})$.

If such C_1, C_2 can be found, then the system of Figure 5.3 is externally and internally stable whenever both controllers C_1 and C_2 are in operation, as well as when only one (either one) of the controllers is operational and the other is set to zero. Such a system becomes unstable only when *both* controllers fail simultaneously, and is therefore reliable against a single controller failure.

Comparing the present problem with that in Section 5.4, we see that in the latter case the objective is to maintain stability in the face of possible structural changes in the plant, whereas here the objective is to guard against possible compensator failure. In a practical problem, the nature and location of the expected failures would dictate which mathematical formalism is more appropriate.

It is easy to see from (2) that $H(P,C_1,0) \in \mathbf{M}(\mathbf{S})$ if and only if $C_1 \in S(P)$; similarly, $H(P,0,C_2) \in \mathbf{M}(\mathbf{S})$ if and only if $C_2 \in S(P)$. However, $H(P,C_1,C_2) \in \mathbf{M}(\mathbf{S})$ implies, but is not necessarily implied by $C_1 + C_2 \in S(P)$. The reason is that if $H(P,C_1,C_2) \in \mathbf{M}(\mathbf{S})$, then

3 $$(I+PC)^{-1} = H_{11} \in \mathbf{M}(\mathbf{S}),$$

$$(I+CP)^{-1} = H_{33} \in \mathbf{M}(\mathbf{S}),$$

$$P(I+CP)^{-1} = -H_{13} \in \mathbf{M}(\mathbf{S}),$$

$$C(I+PC)^{-1} = H_{31}+H_{32} \in \mathbf{M}(\mathbf{S}).$$

As a result, $C \in S(P)$. However, $H_{31}+H_{32} \in \mathbf{M}(\mathbf{S})$ does not necessarily imply that H_{31} and H_{32} individually belong to $\mathbf{M}(\mathbf{S})$. Therefore $C_1+C_2 \in S(P)$ does not necessarily imply that $H(P,C_1,C_2) \in \mathbf{M}(\mathbf{S})$. Hence the RSP is not quite the same thing as the modified RSP, which can be stated as follows: Given a plant P, find C_1, $C_2 \in S(P)$ such that $C_1+C_2 \in S(P)$. Nevertheless, we shall find it profitable to approach the RSP via the modified RSP.

We present at once the first main result of this section.

4 *Theorem* Suppose $P \in \mathbf{M}(\mathbb{R}(s))$ and $C_1 \in S(P)$. Then there exists a $C_2 \in S(P)$ such that C_1 and C_2 together solve the Reliable Stabilization Problem.

In other words, not only does the RSP have a solution for *every* plant P, but *one* of the two stabilizing controllers can be arbitrarily specified.

5 *Theorem* Suppose $P \in \mathbf{M}(\mathbb{R}(s))$. Then there exists a $C \in S(P)$ such that $C_1 = C_2 = C$ solve the RSP.

This theorem shows that it is always possible to find a solution to the RSP in the form of a "duplicate" controller, in the sense that $C_1 = C_2$. It should be noted however that the multi-controller configuration of Figure 5.3 is *not* the same as having a back-up controller; rather, in the present set-up, *both* controllers are connected at all times.

The proof of Theorem 1 requires the following lemma.

6 *Lemma* Suppose A, $B \in \mathbf{M}(\mathbb{R}(s))$, and that the products AB, BA are both defined (and hence square). Then there exists a $Q \in \mathbf{M}(\mathbf{S})$such that $I-AB+QBAB$ is unimodular.

This lemma means that, whatever be the matrices A,B, the plant $BAB(I-AB)^{-1}$ is strongly stabilizable.

Proof Recall that the norm on $\mathbf{M}(\mathbf{S})$ is defined by

7 $$\| F \| = \sup_{\omega} \bar{\sigma}(F(j\omega)), \forall F \in \mathbf{M}(\mathbf{S}),$$

where $\bar{\sigma}(\cdot)$ denotes the largest singular value of a matrix. Then $I + F$ is unimodular whenever $\| F \| < 1$. In particular, $I - rAB$ is unimodular whenever $| r | < \| AB \|^{-1}$. Let k be an integer larger than $\| AB \|$. Then $(I - k^{-1}AB)$ is unimodular, and so is $(I - k^{-1}AB)^k$. By the binomial expansion,

8 $$(I - k^{-1}AB)^k = I - AB + \sum_{i=2}^{k} f_i (AB)^i,$$

where the f_i are appropriate real numbers. Now define

9 $$Q = \sum_{i=0}^{k-2} f_{i+2}(AB)^i A \in \mathbf{M}(\mathbf{S}).$$

Then clearly

10 $$I - AB + QBAB = (I - k^{-1}AB)^k$$

is unimodular. □

Proof of Theorem (4) Suppose $P \in \mathbf{M}(\mathbb{R}(s))$ and $C_1 \in S(P)$ are specified. Let (N, D), $(\tilde{D}, \tilde{N})$ be any right-coprime factorization and left-coprime factorization, respectively, of P over $\mathbf{M}(\mathbf{S})$. The fact that $C_1 \in S(P)$ implies, by Theorem 5.2.1, that $C_1 = Y^{-1}X = \tilde{X}\tilde{Y}^{-1}$, where $X, Y, \tilde{X}, \tilde{Y} \in \mathbf{M}(\mathbf{S})$ satisfy $XN + YD = I$, $\tilde{N}\tilde{X} + \tilde{D}\tilde{Y} = I$. Moreover, $Y^{-1}X = \tilde{X}\tilde{Y}^{-1}$ implies that $Y\tilde{X} = X\tilde{Y}$.

Using Lemma (6), select a matrix $Q \in \mathbf{M}(\mathbf{S})$ such that $I - XN + QNXN$ is unimodular. From Theorem (5.2.1), the controller C defined by

11 $$C = (Y - Q\tilde{Y}\tilde{N})^{-1}(X + Q\tilde{Y}\tilde{D})$$

is in $S(P)$. Let $C_2 = C - C_1$. We now show that C_2 is also in $S(P)$, which shows that C_1 and C_2 together solve the modified RSP. Now

12 $$C_2 = C - C_1 = (Y - Q\tilde{Y}\tilde{N})^{-1}(X + Q\tilde{Y}\tilde{D}) - \tilde{X}\tilde{Y}^{-1}$$
$$= (Y - Q\tilde{Y}\tilde{N})^{-1}[(X + Q\tilde{Y}\tilde{D})\tilde{Y} - (Y - Q\tilde{Y}\tilde{N})\tilde{X}]\tilde{Y}^{-1}$$
$$= (Y - Q\tilde{Y}\tilde{N})^{-1}[X\tilde{Y} - Y\tilde{X} + Q\tilde{Y}(\tilde{D}\tilde{Y} + \tilde{N}\tilde{X})]\tilde{Y}^{-1}$$
$$= (Y - Q\tilde{Y}\tilde{N})^{-1}Q\tilde{Y}\tilde{Y}^{-1}, \text{ by (7) and (8)}$$
$$= (Y - Q\tilde{Y}\tilde{N})^{-1}Q$$
$$= \tilde{D}_{c_2}^{-1}\tilde{N}_{c_2},$$

where $\tilde{D}_{c_2} = Y - Q\tilde{Y}\tilde{N}$, $\tilde{N}_{c_2} = Q$. At this stage, it has not been shown that $\tilde{D}_{c_2}$, $\tilde{N}_{c_2}$ are left-coprime. But let us anyway compute the "return difference" matrix $\tilde{D}_{c_2}D + \tilde{N}_{c_2}N$. This gives

13 $$\tilde{D}_{c_2}D + \tilde{N}_{c_2}N = (Y - Q\tilde{Y}\tilde{N})D + QN$$
$$= YD - Q\tilde{Y}\tilde{N}D + QN$$
$$= YD - Q\tilde{Y}\tilde{D}N + QN, \text{ since } \tilde{N}D = \tilde{D}N.$$

Next, note that

14 $$\begin{bmatrix} Y & X \\ -\tilde{N} & \tilde{D} \end{bmatrix} \begin{bmatrix} D & -\tilde{X} \\ N & \tilde{Y} \end{bmatrix} = I.$$

Thus the two matrices in (14) are the inverses of each other. Hence their product in the opposite order is also equal to the identity matrix; that is,

15 $$\begin{bmatrix} D & -\tilde{X} \\ N & \tilde{Y} \end{bmatrix} \begin{bmatrix} Y & X \\ -\tilde{N} & \tilde{D} \end{bmatrix} = I.$$

In particular, $NX + \tilde{Y}\tilde{D} = I$, so that $\tilde{Y}\tilde{D} = I - NX$; similarly, $YD = I - XN$. Substituting these identities in (13) gives

16 $$\tilde{D}_{c_2}D + \tilde{N}_{c_2}N = I - XN - QN - QNXN + QN$$

$$= I - XN + QNXN,$$

which is unimodular by construction. Hence $C_2 \in S(P)$. This also shows, *a fortiori*, the left-coprimeness of $(\tilde{D}_{c2}, \tilde{N}_{c_2}) = (Y - Q\tilde{Y}\tilde{N}, Q)$.

To complete the proof, it is necessary to show that the transfer matrix $H(P,C_1,C_2) \in \mathbf{M}(\mathbf{S})$. For this purpose, it is more convenient to work with the quantity $K(P,C_1,C_2)$, which is the transfer matrix between $(u_1, u_3, u_1 - u_2)$ and $(e_1, e_3, e_1 - e_2)$. Routine calculations show that

17 $$\begin{bmatrix} e_1 \\ e_3 \\ e_1 - e_2 \end{bmatrix} = \begin{bmatrix} u_1 \\ u_3 \\ u_1 - u_3 \end{bmatrix} - \begin{bmatrix} 0 & P & 0 \\ -(C_1 + C_2) & 0 & C_2 \\ 0 & 0 & 0 \end{bmatrix} \begin{bmatrix} e_1 \\ e_3 \\ e_1 - e_2 \end{bmatrix}$$

Since $K(P,C_1,C_2)$ is obtained by pre- and post-multiplying $H(P,C_1,C_2)$ by unimodular (and in fact nonsingular constant) matrices, it follows that $H(P,C_1,C_2) \in \mathbf{M}(\mathbf{S})$ if and only if $K(P,C_1,C_2) \in \mathbf{M}(\mathbf{S})$. Now from (17),

18 $$K(P,C_1,C_2) = \begin{bmatrix} I & P & 0 \\ -(C_1 + C_2) & I & C_2 \\ 0 & 0 & I \end{bmatrix}^{-1}.$$

The matrix in (18) is block-lower triangular, which makes it easy to invert. Also, by construction,

19 $$H(P,C_1 + C_2) = H(P,C)$$

$$= \begin{bmatrix} I & -P \\ C_1 + C_2 & I \end{bmatrix}^{-1}$$

$$= \begin{bmatrix} I - N(X + Q\tilde{Y}\tilde{D}) & -N(Y - Q\tilde{Y}\tilde{N}) \\ D(X + Q\tilde{Y}\tilde{D}) & D(Y - Q\tilde{Y}\tilde{N}) \end{bmatrix}$$

It now follows from (18) and (19) that

20 $$K(P,C_1,C_2) = \begin{bmatrix} I-N(X+Q\tilde{Y}\tilde{D}) & -N(Y-Q\tilde{Y}\tilde{N}) & -NQ \\ D(X+Q\tilde{Y}\tilde{D}) & D(Y-Q\tilde{Y}\tilde{N}) & DQ \\ 0 & 0 & I \end{bmatrix}$$

which belongs to $\mathbf{M}(\mathbf{S})$. Therefore C_1 and C_2 solve the RSP. □

The above method of proof makes it clear that, whenever $\tilde{D}_{c_2}$ divides $\tilde{D}_c$, we have that $H(P,C) \in \mathbf{M}(\mathbf{S})$ implies that $H(P,C_1,C_2) \in \mathbf{M}(\mathbf{S})$. Thus, in order to prove Theorem (5), it is enough to prove the existence of a controller C such that $C, 2C \in S(P)$. The proof of Theorem (5) depends on the following lemma.

21 *Lemma 2* Given a plant $P \in \mathbf{M}(\mathbb{R}(s))$, let (N,D), $(\tilde{D},\tilde{N})$ be any r.c.f. and l.c.f. of P, and let $X,Y \in \mathbf{M}(\mathbf{S})$ be solutions of $XN + YD = I$. Then there exists an $R \in \mathbf{M}(\mathbf{S})$ such that $I+XN+R\tilde{D}N$ is unimodular.

Proof It is first shown that the matrices $I+XN$, $\tilde{D}N$ are right-coprime. From Theorem (4.1.60), one can select matrices $\tilde{X},\tilde{Y} \in \mathbf{M}(\mathbf{S})$ such that (14) and (15) hold. Suppose now that M is a right divisor of both $I+XN$ and $\tilde{D}N$, denoted by $M \mid (I+XN)$, $M \mid \tilde{D}N$. This implies, successively, that

22 $M \mid \tilde{Y}\tilde{D}N$, $M \mid (I-NX)N$, since $NX+\tilde{Y}\tilde{D} = I$,

$M \mid N(I-XN)$, $M \mid (I+NX)N$,

$M \mid N$ since $N = [(I+NX)N+(I-NX)N]/2$,

$M \mid XN$,

$M \mid I$ since $M \mid (I+XN)$, $M \mid XN$.

This last step shows that M is unimodular.

Now let C_{+e} denote the extended closed right half-plane, i.e. $C_{+e} = \{s: Re\ s \geq 0\} \bigcup \{\infty\}$. The next step is to show that $|I+X(s)N(s)| > 0$ whenever $s \in C_{+e}$ is real and $\tilde{D}(s)N(s) = 0$. It would then follow from Theorem (5.3.1) that $I+XN+R\tilde{D}N$ is unimodular for some $R \in \mathbf{M}(\mathbf{S})$. Suppose $(\tilde{D}N)(s) = 0$. Then

23 $$(\tilde{Y}\tilde{D}N)(s) = 0 \Longrightarrow [(I-NX)N](s) = 0$$
$$\Longrightarrow N(s) = NXN(s)$$
$$\Longrightarrow XN(s) = XNXN(s) = [XN(s)]^2.$$

Let $\alpha = \sqrt{2}-1 \approx .414$. Then it is easy to verify that $1-2\alpha = \alpha^2$. Thus

24 $$I+XN(s) = I+2\alpha XN(s) + \alpha^2(XN)(s)$$
$$= I+2\alpha(XN)(s) + \alpha^2[(XN)(s)]^2, \text{ by (22)}$$
$$= [I+\alpha(XN)(s)]^2.$$

25 $$|I+(XN)(s)| = |I+\alpha(XN)(s)|^2 \geq 0.$$

However, since $I+XN$ and $\tilde{D}N$ are right-coprime, the smallest invariant factor of $\tilde{D}N$ and $|I+XN|$ are coprime. Hence $|I+(XN)(s)| \neq 0$, which implies in conjunction with (25) that $|I+(XN)(s)| > 0$. □.

Proof of Theorem (5) Let $C = (Y-R\tilde{N})^{-1}(X+R\tilde{D})$. Then $2C = (Y-R\tilde{N})^{-1} \cdot 2(X+R\tilde{D})$. Clearly $C \in S(P)$. The return difference matrix corresponding to P and $2C$ is

26 $$(Y-R\tilde{N})D + 2\cdot(X+R\tilde{D})N = I+XN+R\tilde{D}N,$$

which is unimodular by construction. Thus $2C \in S(P)$. By earlier remarks, this is enough to show that $C_1 = C_2 = C$ solve the RSP. □

We summarize the results presented up to now by providing two algorithms for reliable stabilization.

Algorithm 1 Given a plant P and a controller $C_1 \in S(P)$, to find another controller C_2 which, together with C_1, reliably stabilizes P.

Step 1 Find r.c.f.'s (N,D) of P and $(\tilde{X},\tilde{Y})$ of C_1 and l.c.f.'s $(\tilde{D},\tilde{N})$ of P and (Y,X) of C_1 such that $XN + YD = I$ and $\tilde{N}\tilde{X} + \tilde{D}\tilde{Y} = I$ are satisfied.

Step 2 Find a Q such that $I-XN+QNXN$ is unimodular, either by using the procedure described in the proof of Lemma (6), or by any other means.

Step 3 Let $C_2 = (Y-Q\tilde{Y}\tilde{N})^{-1}Q$.

Algorithm 2 Given a plant P, to find a "duplicate" controller C such that $C_1=C_2=C$ solves the reliable stabilization problem.

Step 1 Find any r.c.f. (N,D) and any l.c.f. $(\tilde{D},\tilde{N})$ of P, together with a particular solution (X,Y) of $XN + YD = I$.

Step 2 Find an $R \in \mathbf{M}(\mathbf{S})$ such that $I+XN+R\tilde{D}N$ is unimodular.

Step 3 Let $C = (Y-R\tilde{N})^{-1}(X+R\tilde{D})$.

27 *Example 1* Suppose it is desired to reliably stabilize the plant

$$p(s) = \frac{s^2-1}{s^2-4}.$$

Since this plant is nonminimum phase, robustness recovery techniques cannot be used to solve the problem. To use Algorithm 1, express p in the form $p = n/d$, where

$$n(s) = \frac{s-1}{s+2},\quad d(s) = \frac{s-2}{s+1}.$$

A particular solution to the Bezout identity $xn+yd = 1$ is given by

$$x(s) = \frac{3(s+2)}{s+1},\quad y(s) = -2.$$

Hence the controller

$$c_1 = \frac{x}{y} = -1.5\frac{s+2}{s+1}$$

stabilizes p. We shall now find another controller c_2 which, together with c_1, reliably stabilizes p. The first step is to find a q such that $1-xn+qnxn$ is a unit. This can be done using the method of Lemma (6). Note that

$$(xn)(s) = 3\frac{s-1}{s+1},$$

so that $\| xn \| = 3$. Hence $(1-0.25xn)^4$ is a unit, and is of the form $1+xn+qnxn$. In fact a routine computation yields that

$$q = x \{\frac{3}{8} - \frac{xn}{16} + \frac{xn^2}{256}\}.$$

The expression for $c_2 = q/(y-yqn) = q/y(1-qn)$ is easily obtained by using the variable $z = (s-1)/(s+1)$, since in this case xn is just $3z$. This leads to

$$c_2 = \frac{3(z-2)(9z^2-48z+96)}{4(-27z^3+144z^2-288z+256)},$$

which is a third-order controller.

5.6 TWO-PARAMETER COMPENSATORS

In this section, the feedback system shown in Figure 5.1 is replaced by a more general type of system, which employs what is generally referred to as a two-degrees-of-freedom compensator. For reasons that will become clear shortly, it is referred to here as a *two-parameter compensator.*

Suppose a plant P is specified, and it is desired to stabilize P using a feedback compensator. If e and y denote the plant input and output, respectively, and u denotes the external input, then the *most general* feedback compensation scheme is given by

1 $$e = C(u,y),$$

where C is a (possibly nonlinear) operator. The most general linear time-invariant dynamical scheme is

2 $$e = C_1u - C_2y,$$

where $C_1,C_2 \in \mathbf{M}(\mathbb{R}(s))$ and have the same number of rows.[2] If

[2] The minus sign before C_2 is to represent negative feedback from y to e.

$C_1 = C_2 = C$, then (2) represents the feedback configuration of Figure 5.1, but otherwise the compensation law (2) is more general. For this reason, the compensator (2), which can also be written as

3 $$e = [C_1 \;\; -C_2] \begin{bmatrix} u \\ y \end{bmatrix},$$

is referred to as a two-parameter compensator. It is of course understood that the two inputs u and y can themselves be vectors. It is worth emphasizing again that (3) represents the most general feedback compensation scheme involving the external input and the plant output. Thus there is no "three-parameter" compensator, for example, since *all* external inputs are included in u.

One way in which the general feedback law (2) can be implemented is shown in Figure 5.4. As shown, this implementation makes no sense unless C_1 is stable: If C_1 is unstable, then certain bounded inputs u will produce unbounded outputs $C_1 u$, and as a result the system can never be internally stable. Thus we search for an alternate implementation of (2) that avoids this problem.

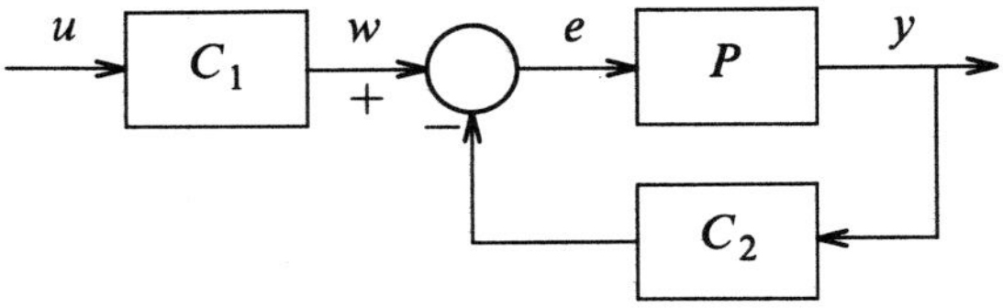

Figure 5.4. Infeasible Implementation of
a Two-Parameter Compensator.

To arrive at a feasible implementation of the two-parameter compensator, define $C = [C_1 \;\; C_2]$, and let $(\tilde{D}_c, [\tilde{N}_{c_1} \;\; \tilde{N}_{c_2}])$ be an l.c.f. of C. Thus $C_1 = \tilde{D}_c^{-1}\tilde{N}_{c_1}$ and $C_2 = \tilde{D}_c^{-1}\tilde{N}_{c_2}$. However, the left-coprimeness of $\tilde{D}_c$ and $[\tilde{N}_{c_1} \;\; \tilde{N}_{c_2}]$ does *not* imply that $\tilde{D}_c$ and $\tilde{N}_{c_1}$ are left-coprime; i.e. $(\tilde{D}_c, \tilde{N}_{c_1})$ need not be an l.c.f. of C_1. Similar remarks apply to $\tilde{D}_c, \tilde{N}_{c_2}$. In fact, suppose

$(\tilde{D}_1,\tilde{N}_1)$ (resp. $(\tilde{D}_2,\tilde{N}_2)$) is an l.c.f. of C_1 (resp. C_2). Then $\tilde{D}_c$ is a least common left multiple of $\tilde{D}_1$ and $\tilde{D}_2$; moreover, if $\tilde{D}_c = \tilde{A}\tilde{D}_1 = \tilde{B}\tilde{D}_2$, then $[\tilde{N}_{c_1} \ \tilde{N}_{c_2}] = [\tilde{A}\tilde{N}_1 \ \tilde{B}\tilde{N}_2]$ (see Problem 1).

Once $C = [C_1 \ C_2]$ has been factorized as $\tilde{D}_c^{-1}[\tilde{N}_{c_1} \ \tilde{N}_{c_2}]$, the feedback law (2) can be implemented as shown in Figure 5.5. In this figure, it is assumed that the plant has been factored as $N_p D_p^{-1}$ where $N_p, D_p \in \mathbf{M}(\mathbf{S})$ are right-coprime. Further, exogenous signals u_2 and u_3 are included to model the presence of errors in measuring the outputs of C and P, respectively.

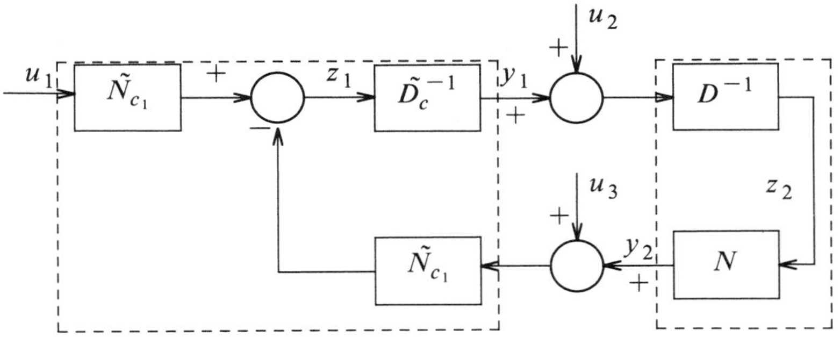

Figure 5.5. Feasible Implementation of a Two-Parameter Compensator.

The system of Figure 5.5 is *stable* if the matrix W mapping (u_1,u_2,u_3) into (y_1,y_2) belongs to $\mathbf{M}(\mathbf{S})$. It can be shown that this is equivalent to requiring the matrix mapping (u_1,u_2,u_3) into (z_1,z_2,y_1,y_2) to belong to $\mathbf{M}(\mathbf{S})$ (see Problem 2). Thus the present notion of stability corresponds to requiring all "internal" quantities z_1,z_2,y_1,y_2 to belong to $\mathbf{M}(\mathbf{S})$ whenever $u_1,u_2,u_3 \in \mathbf{M}(\mathbf{S})$. Lemma (4) below gives necessary and sufficient conditions for the system of Figure 5.5 to be stable, in terms of the factorizations of C and P.

4 *Lemma* Let (N_p, D_p) be a r.c.f. of P and let $(\tilde{D}_c, [\tilde{N}_{c_1} \; \tilde{N}_{c_2}])$ be an l.c.f. of $C = [C_1 \; C_2]$. Then the system of Figure 5.5 is stable if and only if

5 $$\Delta := \tilde{D}_c D_p + \tilde{N}_{c_2} N_p \in \mathbf{U}(\mathbf{S}).$$

Remarks 1) Clearly a *necessary* condition for Δ to be unimodular is that $\tilde{D}_c$ and $\tilde{N}_{c_2}$ be left-coprime. In view of earlier discussion, this can be interpreted as follows: As before, let $(\tilde{D}_1, \tilde{N}_1)$, $(\tilde{D}_2, \tilde{N}_2)$ be l.c.f.'s of C_1 and C_2 respectively, and suppose $\tilde{D}_c = \tilde{A}\tilde{D}_1 = \tilde{B}\tilde{D}_2$ is a least common left multiple of $\tilde{D}_1$ and $\tilde{D}_2$. Then, since $\tilde{N}_{c_2} = \tilde{B}\tilde{N}_2$, we see that $\tilde{D}_c = \tilde{B}\tilde{D}_2$, $\tilde{N}_{c_2} = \tilde{B}\tilde{N}_2$ are left-coprime (if and) only if $\tilde{B}$ is unimodular. Therefore, $\tilde{D}_c = \tilde{D}_2$ must be a least common left multiple of $\tilde{D}_1$, $\tilde{D}_2$; in other words, $\tilde{D}_2$ is a left multiple of $\tilde{D}_1$. To summarize, the two-parameter compensator (3) cannot stabilize P unless the left denominator of C_1 divides the left denominator of C_2. This means that every C_{+e}-pole of C_1 must also be a pole of C_2 of at least the same McMillan degree.

Proof The equations describing the system of Figure 5.5 can be written in terms of the quantities y_1 and z_2, which are the outputs of the "inversion" elements $\tilde{D}_c^{-1}$ and D_p^{-1}, respectively. Thus

6 $$e_2 = D_p z_2 = y_1 + u_2 \Rightarrow y_1 - D_p z_2 = -u_2,$$

7 $$z_1 = \tilde{D}_c y_1 = \tilde{N}_{c_1} u_1 - \tilde{N}_{c_2}(u_3 + y_2) = \tilde{N}_{c_1} u_1 - \tilde{N}_{c_2} u_3 - \tilde{N}_{c_2} N_p z_2$$

$$\Rightarrow \tilde{D}_c y_1 + \tilde{N}_{c_2} N_p z_2 = \tilde{N}_{c_1} u_1 - \tilde{N}_{c_2} u_3.$$

Equations (6) and (7) can be combined in the form

8 $$\begin{bmatrix} I & -D_p \\ \tilde{D}_c & \tilde{N}_{c_2} N_p \end{bmatrix} \begin{bmatrix} y_1 \\ z_2 \end{bmatrix} = \begin{bmatrix} 0 & -I & 0 \\ \tilde{N}_{c_1} & 0 & -\tilde{N}_{c_2} \end{bmatrix} \begin{bmatrix} u_1 \\ u_2 \\ u_3 \end{bmatrix}.$$

Now the outputs y_1 and y_2 are given by

$$9 \qquad \begin{bmatrix} y_1 \\ y_2 \end{bmatrix} = \begin{bmatrix} I & 0 \\ 0 & N_p \end{bmatrix} \begin{bmatrix} y_1 \\ z_2 \end{bmatrix}.$$

It is immediate from (8) and (9) that

$$10 \qquad \begin{bmatrix} y_1 \\ y_2 \end{bmatrix} = N_r D^{-1} \tilde{N}_l \begin{bmatrix} u_1 \\ u_2 \\ u_3 \end{bmatrix},$$

where

$$11 \qquad N_r = \begin{bmatrix} I & 0 \\ 0 & N_p \end{bmatrix}, D = \begin{bmatrix} I & -D_p \\ \tilde{D}_c & \tilde{N}_{c_2} N_p \end{bmatrix}, \tilde{N}_l = \begin{bmatrix} 0 & -I & 0 \\ \tilde{N}_{c_1} & 0 & \tilde{N}_{c_2} \end{bmatrix}.$$

The next step is to show that N_r, D are right-coprime, and that $\tilde{N}_l, D$ are left-coprime. For this purpose, select $X_p, Y_p, \tilde{X}_{c_1}, \tilde{X}_{c_2}, \tilde{Y}_c \in \mathbf{M}(\mathbf{S})$ such that

$$12 \qquad X_p N_p + Y_p D_p = I, \, [\tilde{N}_{c_1} \;\; \tilde{N}_{c_2}] \begin{bmatrix} \tilde{X}_{c_1} \\ \tilde{X}_{c_2} \end{bmatrix} + \tilde{D}_c \tilde{Y}_c = I.$$

Then, recalling that $\Delta = \tilde{D}_c D_p + \tilde{N}_{c_2} N_p$, it is routine to verify that

$$13 \qquad \begin{bmatrix} \Delta Y_p + I & \Delta X_p \\ Y_p & X_p \end{bmatrix} \begin{bmatrix} I & 0 \\ 0 & N_p \end{bmatrix} + \begin{bmatrix} \tilde{D}_c - \Delta Y_p & -I \\ -Y_p & 0 \end{bmatrix} \begin{bmatrix} I & -D_p \\ \tilde{D}_c & \tilde{N}_{c_2} N_p \end{bmatrix}$$

$$= I,$$

$$14 \qquad \begin{bmatrix} 0 & -I & 0 \\ \tilde{N}_{c_1} & 0 & -\tilde{N}_{c_2} \end{bmatrix} \begin{bmatrix} \tilde{X}_{c_1} \Delta & \tilde{X}_{c_1} \\ \tilde{Y}_c \Delta - I & \tilde{Y}_c \\ -\tilde{X}_{c_2} \Delta & -\tilde{X}_{c_2} \end{bmatrix}$$

$$+ \begin{bmatrix} I & -D_p \\ \tilde{D}_c & \tilde{N}_c N_p \end{bmatrix} \begin{bmatrix} \tilde{Y}_c \Delta - D_p & \tilde{Y}_c \\ -I & 0 \end{bmatrix} = I.$$

By definition, the system under study is stable if and only if $N_r D^{-1} \tilde{N}_l \in \mathbf{M}(\mathbf{S})$. By Corollary (4.3.19), this is so if and only if $D \in \mathbf{U}(\mathbf{S})$. The proof is completed by showing that $D \in \mathbf{U}(\mathbf{S})$ if and only if $\Delta \in \mathbf{U}(\mathbf{S})$. From (11), we see that Δ is just the Schur complement of D with respect to the identity matrix in the upper left-hand corner. Hence, from Fact (B.1.32), we get $|D| = |I| \cdot |\Delta| = |\Delta|$. Therefore $D \in \mathbf{U}(\mathbf{S})$ if and only if $\Delta \in \mathbf{U}(\mathbf{S})$ (if and only if $|D| = |\Delta|$ is a unit of $\mathbf{S}$). □

Now we come to the main result of this section.

15 *Theorem* Let P be a given plant. Suppose (N_p, D_p), $(\tilde{D}_p, \tilde{N}_p)$ are any r.c.f. and any l.c.f. of P, and that $X, Y \in \mathbf{M}(\mathbf{S})$ satisfy $XN_p + YD_p = I$. Then the set of all two-parameter compensators that stabilize P is given by

16 $$S_2(P) = \{(Y - R\tilde{N}_p)^{-1} [Q \;\; X + R\tilde{D}_p]\},$$

where $Q \in \mathbf{M}(\mathbf{S})$ is arbitrary, and $R \in \mathbf{M}(\mathbf{S})$ is arbitrary but for the constraint that $|Y - R\tilde{N}_p| \neq 0$. The set of all possible stable transfer matrices from (u_1, u_2, u_3) to (y_1, y_2) is the set of all matrices of the form

17 $$\begin{bmatrix} D_p Q & D_p(Y - R\tilde{N}_p) - I & -D_p(X + R\tilde{D}_p) \\ N_p Q & N_p(Y - R\tilde{N}_p) & -N_p(X + R\tilde{D}_p) \end{bmatrix}$$

$$= \begin{bmatrix} D_p \\ N_p \end{bmatrix} [Q \;\; R] \begin{bmatrix} I & 0 & 0 \\ 0 & -\tilde{N}_p & -\tilde{D}_p \end{bmatrix} + \begin{bmatrix} 0 & D_p Y - I & -D_p X \\ 0 & N_p Y & -N_p X \end{bmatrix},$$

where $Q, R \in \mathbf{M}(\mathbf{S})$ are arbitrary but for the constraint that $|Y - R\tilde{N}_p| \neq 0$.

Proof Based on the results of Section 5.2, specifically Theorem (5.2.1), it follows that the return difference matrix Δ of (5) is unimodular if and only if $\tilde{D}_c^{-1}\tilde{N}_{c_2}$ equals $(Y-R\tilde{N}_p)^{-1}(X+R\tilde{D}_p)$ for some $R \in \mathbf{M}(\mathbf{S})$ such that $|Y-R\tilde{N}_p| \neq 0$. Moreover, Δ is independent of $\tilde{N}_{c_1}$. This leads at once to (16). Now (17) follows by routine computation. □

Several remarks are in order concerning this theorem.

1) Equation (17) shows that the transfer matrix from (u_1,u_2,u_3) to (y_1,y_2) involves *two independent* parameters Q and R. This is the most important feature of the two-parameter compensator (and it incidentally explains the name of the scheme). For comparison, recall that from (5.2.10) that the set of all stable transfer matrices from (u_1,u_2) to (y_1,y_2) in a one-parameter scheme, such as in Figure 5.1, is

18 $$\begin{bmatrix} D_p(X+R\tilde{D}_p) & D_p(Y-R\tilde{N}_p)-I \\ N_p(X+R\tilde{D}_p) & N_p(Y-R\tilde{N}_p) \end{bmatrix}.$$

Thus the two-parameter scheme offers greater flexibility in that the transfer matrix from u_1 to the outputs can be adjusted "independently" of that between u_2 and the outputs. This is not so in the case of the one-parameter scheme. This flexibility is further illustrated in subsequent remarks.

2) Suppose P is a *stable* plant; then (P,I), (I,P) are respectively an r.c.f. and an l.c.f. of P, and one can take $X = 0$, $Y = I$. Then the set of all possible stable transfer matrices between u_1 and (y_1,y_2), subject to internal stability, is

19 $$\begin{bmatrix} Q \\ PQ \end{bmatrix}, \; Q \in \mathbf{M}(\mathbf{S})$$

in *both* schemes. Thus, in the case of stable plants, any stable transfer matrix between u_1 and the outputs that can be achieved using a two-parameter scheme can also be achieved using a one-parameter scheme. This is not true if P is unstable: If $\tilde{D}_p$ is not unimodular, then the set $\{D_p(X+R\tilde{D}_p)\colon R \in \mathbf{M}(\mathbf{S})\}$ is a *proper* subset of $\{D_pQ\colon Q \in \mathbf{M}(\mathbf{S})\}$. To see this, note that $D_p A = D_p B$ if and only if $A = B$, since $|D_p| \neq 0$. Thus

20 $$\{D_p(X+R\tilde{D}_p): R \in \mathbf{M}(\mathbf{S})\} = \{D_p Q: Q \in \mathbf{M}(\mathbf{S})\}$$

$$\Longleftrightarrow \{X+R\tilde{D}_p: R \in \mathbf{M}(\mathbf{S})\} = \mathbf{M}(\mathbf{S})$$

$$\Longleftrightarrow \{R\tilde{D}_p: R \in \mathbf{M}(\mathbf{S})\} = \mathbf{M}(\mathbf{S}).$$

But this last equality holds if and only if $\tilde{D}_p$ is unimodular. Hence, if P is unstable, there exist stable transfer matrices between u_1 and the outputs that can be achieved with a two-parameter scheme but not a one-parameter scheme. In fact, (17) shows that by using a two-parameter scheme, the set of achievable transfer matrices between the external input u_1 and the plant output y_2 is just $\{N_p Q: Q \in \mathbf{M}(\mathbf{S})\}$, which is the same as if the plant were stable and equal to N_p. *Thus, in a two-parameter compensation scheme, the achievable performance between the external input and the plant output is not limited by the plant poles.* This point is discussed further in Section 6.7.

3) Let us think of u_2 and u_3 as sensor noises at the outputs of the compensator and plant, respectively. Then (17) and (18) show that, in terms of achievable performance between the sensor noises and the plant outputs, there is no difference between the two-parameter and the one-parameter schemes.

4) Suppose $M \in \mathbf{M}(\mathbf{S})$ is a right multiple of N_p. Then (17) shows that M can be realized as the transfer matrix between u_1 and y_2 by an appropriate two-parameter scheme. If N_p has full column rank, then there is a *unique* Q such that $M = N_p Q$. Nevertheless, there are *infinitely many* two-parameter schemes that realize the transfer matrix M, namely,

21 $$[C_1 \;\; C_2] = (Y - R\tilde{N}_p)^{-1}[Q \;\; X + R\tilde{D}_p],\; R \in \mathbf{M}(\mathbf{S}).$$

In other words, specifying M fixes Q but not R. This "free" choice of R can be used to advantage in one of several ways: (i) It is possible *simultaneously* to achieve "good" transfer matrices $H_{y_2 u_1}$ and $H_{y_2 u_2}$, or $H_{y_2 u_1}$ and $H_{y_2 u_3}$. This is made quantitatively precise in Section 6.6. (ii) If P is strongly stabilizable, then *any* right multiple of N_p can be realized as $H_{y_2 u_1}$ using only *stable* C_1

and C_2.

5) Problems such as strong stabilization, simultaneous stabilization, etc. are no different in the two-parameter case from the one-parameter case. This is because the "feedback" compensator C_2 has to stabilize P in the sense of Section 5.1 in order to achieve stability, irrespective of which scheme is used.

An important special case of the two-parameter compensator is the observer-controller configuration of Figure 5.6. In it, P is a given plant with an r.c.f. (N_p, D_p), and the matrices $X, Y \in \mathbf{M}(\mathbf{S})$ are chosen to satisfy $XN_p + YD_p = I$. The observer reconstructs the "internal state" z (sometimes referred to also as the "partial state"), and the controller feeds back z after multiplying it by M.

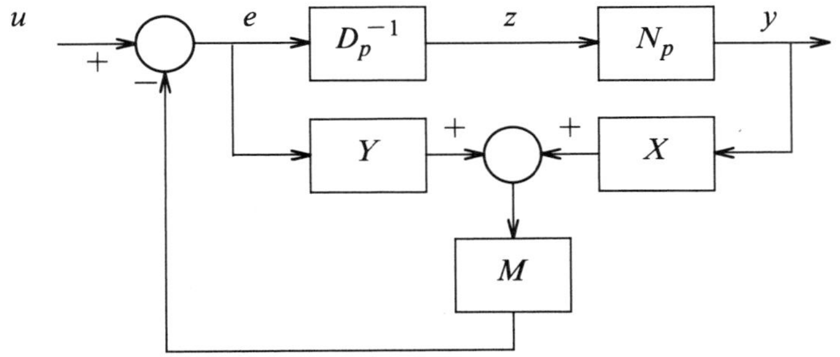

Figure 5.6. Observer-Controller Configuration.

To compare the observer-controller configuration of Figure 5.6 with the two-parameter compensator of Figure 5.5, let us redraw the former as shown in Figure 5.7. This results in a two-parameter compensator with

22 $\tilde{N}_{c_1} = I, \tilde{N}_{c_2} = MX, \tilde{D}_c = I + MY.$

Clearly $[\tilde{N}_{c_1} \ \tilde{N}_{c_2}]$ and $\tilde{D}_c$ are left-coprime, since

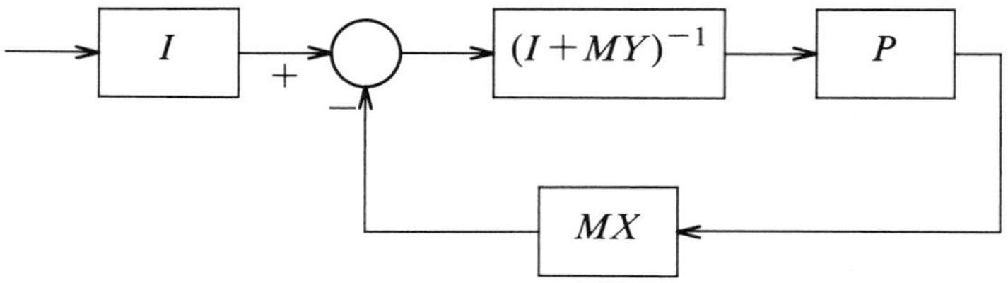

Figure 5.7. Rearrangement of the Observer-Controller Configuration.

23 $$[I \;\; MX]\begin{bmatrix} I \\ 0 \end{bmatrix} + (I+MY)\cdot 0 = I.$$

Hence, from Lemma (4), this system is stable if and only if

24 $$\Delta = (I+MY)D_p + MXN_p = D_p + M \in \mathbf{U}(\mathbf{S}).$$

Problems

1. Suppose $(\tilde{D}_1, \tilde{N}_1)$, $(\tilde{D}_2, \tilde{N}_2)$ are l.c.f.'s of C_1 and C_2 respectively. Suppose $\tilde{D}_c = \tilde{A}\tilde{D}_1 = \tilde{B}\tilde{D}_2$ is a least common left multiple of $\tilde{D}_1$ and $\tilde{D}_2$. Show that $(\tilde{D}_c, [\tilde{A}\tilde{N}_1 \;\; \tilde{B}\tilde{N}_2])$ is an l.c.f. of $[C_1 \;\; C_2]$. (Hint: First show that $\tilde{A}, \tilde{B}$ are left-coprime.)

2. For the system of Figure 5.5, define the transfer matrices W and V by

$$\begin{bmatrix} z_1 \\ z_2 \\ y_1 \\ y_2 \end{bmatrix} = V \begin{bmatrix} u_1 \\ u_2 \\ u_3 \end{bmatrix}, \quad \begin{bmatrix} y_1 \\ y_2 \end{bmatrix} = W \begin{bmatrix} u_1 \\ u_2 \\ u_3 \end{bmatrix}.$$

Show that $W \in \mathbf{M}(\mathbf{S})$ if and only if $V \in \mathbf{M}(\mathbf{S})$.

5.7 REGULATION AND DECOUPLING

In this section, the focus is on control problems that can be reduced to the solvability of a set of linear equations over $\mathbf{S}$. Examples of such problems include regulation (tracking and disturbance rejection) and decoupling.

Consider the linear equation $AR = B$, where A and B are given matrices of compatible dimensions and R is unknown. If A, B, R are matrices over a field, then the following result is well-known: There exists an R such that $AR = B$ if and only if rank A = rank $[A \;\; B]$. However, if A, B, R are matrices over a ring, the rank condition alone is not enough.

1 *Lemma* Suppose $\mathbf{R}$ is a commutative ring with identity, and that $A, B \in \mathbf{M}(\mathbf{R})$ have the same number of rows. Then there exists an $R \in \mathbf{M}(\mathbf{R})$ such that $AR = B$ if and only if $[A \;\; 0]$ and $[A \;\; B]$ are right associates.

Proof "if" Suppose $U \in \mathbf{U}(\mathbf{R})$ is such that $[A \;\; B] = [A \;\; 0]U$. Partitioning U as a block 2×2 matrix in the obvious way leads to

2 $$[A \;\; B] = [A \;\; 0] \begin{bmatrix} U_{11} & U_{12} \\ U_{21} & U_{22} \end{bmatrix},$$

or $B = AU_{12}$.

"only if" Suppose $B = AR$ for some $R \in \mathbf{M}(\mathbf{R})$. Then

3 $$[A \;\; B] = A[I \;\; R] =_R A[I \;\; R] \begin{bmatrix} I & -R \\ 0 & I \end{bmatrix}$$

$$= A[I \;\; 0] = [A \;\; 0],$$

where $=_R$ denotes right equivalence (see Section B.1). Hence $[A \;\; B]$ and $[A \;\; 0]$ are right associates. □

The next lemma presents another set of conditions for the equation $AR = B$ to have a solution. It is a bit more explicit than Lemma (1). However, it requires that the ring $\mathbf{R}$ be a principal ideal domain.

4 *Lemma* Suppose $\mathbf{R}$ is a p.i.d., and that $A,B \in \mathbf{M}(\mathbf{R})$ have the same number of rows. Let r denote the rank of A, and let $a_1, \cdots, a_r$ denote the invariant factors of A. Suppose $U,V \in \mathbf{U}(\mathbf{R})$ are unimodular matrices such that $UAV = S$, where

5 $$S = \begin{bmatrix} a_1 & & & & \\ & \cdot & & & 0 \\ & & \cdot & & \\ & & & \cdot & \\ & & & a_r & \\ & 0 & & & 0 \end{bmatrix}$$

is the Smith form of A. Then there exists an $R \in \mathbf{M}(\mathbf{R})$ such that $AR = B$ if and only if

(i) For $i = 1, \cdots, r$, a_i divides each element in the i-th row of UB, and

(ii) All other rows of UB (if any) are zero.

Proof Note that $A = U^{-1}SV^{-1}$, so that the equation $AR = B$ is equivalent to $SV^{-1}R = UB$. Since V is unimodular, we can replace $V^{-1}R$ by another unknown matrix T, so that the equation now becomes $ST = UB$. From this (i) and (ii) follow readily. □

To set up the regulation and decoupling problems, consider the configuration shown in Figure 5.8, where P is a given plant, and T and R are specified reference signal generators. To achieve the greatest possible generality, a two-parameter compensation scheme is shown in this figure. The objective is to find a $C \in S(P)$ that achieves one of the following additional features:

(i) *Tracking:* $(I - H_{yt})T \in \mathbf{M}(\mathbf{S})$.

(ii) *Disturbance Rejection:* $H_{yd}R \in \mathbf{M}(\mathbf{S})$.

(iii) *Decoupling:* H_{yt} is block-diagonal (and nonzero).

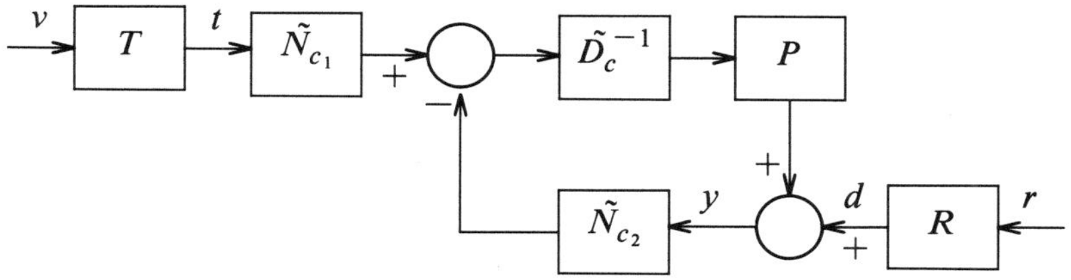

Figure 5.8. Configuration for Tracking and Disturbance Rejection.

Thus tracking corresponds to the transfer matrix from v to the tracking error $t-y$ being stable; disturbance rejection corresponds to the transfer matrix from d to the plant output y being stable. It is clear that, if the compensation scheme stabilizes the plant, then tracking and disturbance rejection pose additional restrictions only when T and R are unstable. Decoupling corresponds to the transfer matrix from the external input (also taken to be t) to the plant output being diagonal, or more generally block-diagonal.

The following result summarizes a simplification that is used repeatedly.

6 *Lemma* Suppose $P, Q, R \in \mathbf{M}(\mathbf{S})$ with $|Q| \neq 0$ and Q, R left-coprime. Then $PQ^{-1}R \in \mathbf{M}(\mathbf{S})$ if and only if $PQ^{-1} \in \mathbf{M}(\mathbf{S})$.

Proof "if" Obvious.

"only if" Select $A, B \in \mathbf{M}(\mathbf{S})$ such that $QA + RB = I$. Then

7 $$Q^{-1} = A + Q^{-1}RB,$$

8 $$PQ^{-1} = PA + PQ^{-1}RB.$$

Hence, if $PQ^{-1}R \in \mathbf{M}(\mathbf{S})$, then $PQ^{-1} \in \mathbf{M}(\mathbf{S})$. □

We are now in a position to state the main results of this section.

9 *Theorem* Consider the system of Figure 5.8, and let $(\tilde{D}_t, \tilde{N}_t)$, $(\tilde{D}_r, \tilde{N}_r)$ be any l.c.f.'s of T and R, respectively. Let (N, D), $(\tilde{D}, \tilde{N})$ be any r.c.f. and any l.c.f. of P, and let $X, Y \in \mathbf{M}(\mathbf{S})$ be any particular solutions of the identity $XN + YD = I$. Then:

(i) The tracking problem has a solution if and only if there exist $Q, W \in \mathbf{M}(\mathbf{S})$ such that

10 $$NQ + W\tilde{D}_t = I.$$

(ii) The disturbance rejection problem has a solution if and only if there exist $S, V \in \mathbf{M}(\mathbf{S})$ such that

11 $$NS\tilde{D} + W\tilde{D}_r = NX.$$

Proof By Theorem (5.6.15), the set of stable transfer matrices from (t, r) to y is given by

12 $$[NQ \quad -N(X+S\tilde{D})],\ Q, S \in \mathbf{M}(\mathbf{S}).$$

Thus the tracking problem has a solution if and only if there exists a matrix $Q \in \mathbf{M}(\mathbf{S})$ such that

13 $$(I-NQ)\tilde{D}_t^{-1}\tilde{N}_t \in \mathbf{M}(\mathbf{S}),$$

But by Lemma (6), (13) is equivalent to

14 $$(I-NQ)\tilde{D}_t^{-1} \in \mathbf{M}(\mathbf{S})$$

for some $Q \in \mathbf{M}(\mathbf{S})$. Now (14) is the same as

15 $$I-NQ = W\tilde{D}_t$$

for some $Q, W \in \mathbf{M}(\mathbf{S})$. It is clear that (15) and (10) are equivalent. One can derive (11) in the same manner. □

The criteria given in Theorem (9) are testable, since (10) and (11) are just linear equations in the components of the unknown matrices Q, W, S, V, so that their solvability can be ascertained using Lemma (1) and (4).

It is left to the reader to derive the condition analogous to (10) in the case where a one-parameter compensator is used.

Finally, consider the problem of decoupling. It is easy to see that if N is square, then NQ can always be made into a diagonal matrix by an appropriate choice of Q (e.g. $Q = N^{adj}$). More generally, if N has more columns than rows, then once again it is possible to make NQ into a diagonal matrix.

Notes and References

The formulation of the feedback stabilization problem is from [19], which also contains several examples that show that any three of the components of $H(P,C)$ could be stable while the fourth is unstable.

The parametrization of all compensators that stabilize a given plant is given in [107]. This characterization is somewhat different from the one given here, in that it involves a combination of both stable rational functions as well as polynomials. The present treatment is due to [25], with some technical improvements from [97].

The notion of strong stabilization is defined, and necessary and sufficient conditions for it are derived, in [104]. The current approach based on matrix Euclidean division is from [98]. See [2] for an alternate set of conditions for strong stabilizability based on Cauchy indices; these conditions do not require one to compute explicitly the poles and zeros of a plant.

The present results on simultaneous stabilization are from [98]; see [79] for a geometric interpretation of this problem. The interlacing test for simultaneous stabilization is from [62, 80, 81] The idea of reliable stabilization using a multi-compensator configuration is introduced in [83, 84], while the present results are taken from [99]. The treatment of two-parameter compensators follows [22].

CHAPTER 6. FILTERING AND SENSITIVITY MINIMIZATION

6.1 PROBLEM STATEMENT

The problem studied in this chapter is the following: Suppose P is a given plant, not necessarily stable, and W_1, W_2 are given proper stable rational matrices. The objective is to find, among all stabilizing compensators of P, a C that minimizes the *weighted performance measure*

1 $$J = \| W_1(I+PC)^{-1}W_2 \|$$

The significance of the problem can be explained in terms of Figure 6.1.

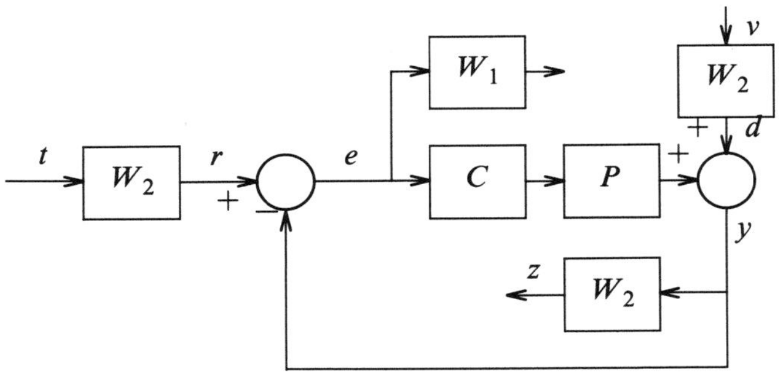

Figure 6.1. Pictorial Representation of the Weighted Performance Measure.

Here W_2 can be interpreted as the generator of the disturbance input d, and $(I+PC)^{-1}W_2$ is the transfer matrix from j to the measured output y. In this case W_1y is the *weighted* measured output, and $W_1(I+PC)^{-1}W_2$ is the transfer matrix from v to the weighted measured output. Alternately, one can think of W_2 as generating a (bounded) reference input r, and $(I+PC)^{-1}W_2$ is the transfer matrix from t to the tracking error e. In this case W_1e is the weighted error. By suitable definition of the norm $\|\cdot\|$, the weighted performance measure J can be made to represent several quantities of practical significance.

Case I. Filtering Suppose r is a random process with power spectral density $\| W_2(j\omega) \|^2$, and that it is desired to minimize the weighted energy of the resulting process e by selecting C appropriately. In this case, t represents white noise and W_2 is a whitening filter; $e(s) = [(I+PC)^{-1}W_2](s)$, and the power spectral density of e is given by

2 $$\| e(j\omega) \|^2 = \| [(I+PC)^{-1}W_2](j\omega) \|^2.$$

In assessing the situation, one may be more concerned about certain frequencies more than others. The role of the weighting matrix W_1 is to reflect the relative importance attached to energy at different frequencies. Thus the weighted error z equals $W_1 e$ and its power spectral density is

3 $$\| z(j\omega) \|^2 = \| [W_1(I+PC)^{-1}W_2](j\omega) \|^2.$$

The objective is to minimize the energy of the process z, i.e.

4 $$J = \frac{1}{2\pi} \int_{-\infty}^{\infty} \| z(j\omega) \|^2 d\omega$$
$$= \frac{1}{2\pi} \int_{-\infty}^{\infty} \| [W_1(I+PC)^{-1}W_2](j\omega) \|^2 d\omega.$$

Case II. Sensitivity Minimization Suppose u is a square-integrable input with an $\mathbf{L}_2$- norm no larger than one, and suppose it is desired to minimize the largest $\mathbf{L}_2$-norm that z may have in response to such an input. Here again the objective function is to minimize the weighted energy of the error e, but in this case the input u is assumed to be any $\mathbf{L}_2$-input of unit norm or less, rather than white noise as in Case I. In this case the objective function to be minimized is

5 $$J = \sup_{\| u \|_2 \leq 1} \| z \|_2$$
$$= \sup_{\| u \|_2 \leq 1} \| W_1(I+PC)^{-1} W_2 u \|_2$$
$$= \sup_{\omega \in \mathbb{R}} \| [W_1(I+PC)^{-1} W_2](j\omega) \| ,$$

where $\| \cdot \|_2$ denotes the $\mathbf{L}_2$-norm, and the last equality is a standard result from [26, 90] as pointed out in the next section. Since in this case the quantity measures the maximum error in response to a unit-norm input u, this problem is sometimes referred to as the *minimax* sensitivity problem.

In both the above cases, as well as in others that the reader may construct, the minimization of J in (1) is required to be carried out *subject to the constraint that* C stabilizes P. As things stand, the compensator C is thus constrained, and also enters the objective function J in a highly nonlinear fashion. However, using the results of Chapter 5, it is possible to transform this into an *unconstrained* optimization problem involving a much simpler objective function.

The set of all compensators C that stabilize a given plant P is neatly parametrized in Theorem (5.2.1). Let (N,D), $(\tilde{D},\tilde{N})$ be any r.c.f. and any l.c.f. of P, and select $\tilde{X},\tilde{Y} \in \mathbf{M}(\mathbf{S})$ such that $\tilde{N}\tilde{X} + \tilde{D}\tilde{Y} = I$. Then, from this theorem,

6 $$S(P) = \{(\tilde{X}+DR)(\tilde{Y}-NR)^{-1} : R \in \mathbf{M}(\mathbf{S}) \text{ and } | \tilde{Y}-NR | \neq 0\}.$$

Moreover, if $C \in S(P)$ equals $(\tilde{X}+DR)(\tilde{Y}-NR)^{-1}$ for a certain $R \in \mathbf{M}(\mathbf{S})$, then (*cf.* (5.2.9))

7 $$(I+PC)^{-1} = (\tilde{Y}-NR)\tilde{D}.$$

Thus the problem of minimizing $J(R) = \| W_1(I+PC)^{-1} W_2 \|$ subject to the constraint that $C \in S(P)$ is equivalent to minimizing

8 $$J(R) = \| W_1(\tilde{Y}-NR)\tilde{D}W_2 \|$$

subject to the constraint that $| \tilde{Y}-NR | \neq 0$. Now, from Lemma (5.2.11),

$|\tilde{Y}-NR| \neq 0$ for all R belonging to an open dense subset of $\mathbf{M}(\mathbf{S})$, and if P is strictly proper, then from Lemma (5.2.18), $|\tilde{Y}-NR| \neq 0$ for *all* $R \in \mathbf{M}(\mathbf{S})$. Hence the original problem of minimizing the weighted performance measure J of (1) is equivalent, in the case where P is strictly proper, to an *unconstrained* minimization of the objective function $J(\cdot)$ of (8) with respect to R. Moreover, the original expression $\| W_1(I+PC)^{-1}W_2 \|$ has been replaced by $\| W_1(\tilde{Y}-NR)\tilde{D}W_2 \|$, where the noteworthy feature is that the quantity inside the norm is *affine* in R. If the plant is not strictly proper, the quantity R is constrained to the extent that it has to satisfy the "nonsingularity" constraint. However, since this constraint is only operational on a nowhere dense set, it can be ignored during the minimization and can be tested once an unconstrained minimum has been found. If by chance the optimal choice of R does not satisfy the nonsingularity constraint, then arbitrarily small perturbations in R will satisfy the constraint and will result in only an arbitrarily small increase in the performance measure (since the latter is a continuous function).

Up to now we have only talked about minimizing the weighted energy of the *error* signal in response to various types of inputs. Depending on the application, one might prefer instead to minimize the weighted energy of some other signal within the loop. Since the transfer matrix between any two points can be expressed as an affine function of R using the results of Chapter 5, such a minimization is qualitatively no different from the problems discussed earlier. For instance, if it is desired to minimize the weighted energy of the plant output in response to an input u, then

9 $$J = \| W_1 PC(I+PC)^{-1}W_2 \| .$$

However, if $C \in S(P)$, then by (5.2.9),

10 $$PC(I+PC)^{-1} = (\tilde{X}+DR)\tilde{N}$$

for some $R \in \mathbf{M}(\mathbf{S})$, so that

11 $$J = W_1(\tilde{X}+DR)\tilde{N}W_2 \| .$$

In fact, it is easy to see that the general problem becomes one of minimizing

12 $$J = \| F-GRH \| ,$$

where $F,G,H \in \mathbf{M}(\mathbf{S})$ are given matrices.

Finally, the above discussion applies even when a two-parameter control scheme is used. In this case, by (5.6.17), all transfer matrices are affine functions of the matrix $S = [Q \quad R]$, and the general cost function assumes the form

13 $$J = \| F-GSH \| ,$$

where $F,G,H \in \mathbf{M}(\mathbf{S})$ are given matrices. The discrete-time analogs of the above problems are simply obtained by replacing $\mathbf{S}$ by $\mathbf{R}_\infty$.

6.2 SOME FACTS ABOUT HARDY SPACES

In the previous section, it was shown that several problems of filtering and sensitivity minimization can be formulated as follows: Minimize

1 $$J(R) = \| F-GRH \| ,$$

where $F,G,H \in \mathbf{M}(\mathbf{S})$ are given and $R \in \mathbf{M}(\mathbf{S})$ is to be chosen. The solution of this class of problems is facilitated by the use of some concepts and results from the theory of $\mathbf{H}_p$- (or Hardy) spaces. The purpose of this section is to present a listing of the basic concepts and results from this theory, mostly without proof. The reader is referred to [32] for an excellent treatment of this theory, and to [55] for some specialized results.

Throughout this section, $\mathbf{D}$ denotes the open unit disc in the complex plane, and $\bar{\mathbf{D}}$ the closed unit disc. For each $p \in [1,\infty)$, the set $\mathbf{H}_p$ consists of all analytic functions f on the open unit disc $\mathbf{D}$ with the property that

2 $$\sup_{r\in(0,1)} \left[\frac{1}{2\pi} \int_0^{2\pi} |f(re^{j\theta})|^p \, d\theta \right]^{1/p} =: \|f\|_p < \infty.$$

The set $\mathbf{H}_\infty$ consists of all analytic functions f on $\mathbf{D}$ such that

3 $$\sup_{r\in(0,1)} \max_{\theta\in[0,2\pi]} |f(re^{j\theta})| =: \|f\|_\infty < \infty.$$

If $f\in\mathbf{H}_p$ for some $p\in[1,\infty]$, then strictly speaking $f(z)$ is only defined when $|z| < 1$. However, for almost all $\theta\in[0,2\pi]$, it is possible to define $f(e^{j\theta})$ as a nontangential limit of $f(z)$ as $z\rightarrow e^{j\theta}$. If $f\in\mathbf{H}_p$ then its boundary value function $f(e^{j\theta})$ belongs to the Lebesgue space $\mathbf{L}_p[0,2\pi]$. Moreover,

4 $$\|f\|_p = \left[\frac{1}{2\pi} \int_0^{2\pi} |f(e^{j\theta})|^p \, d\theta \right]^{1/p} \text{ if } 1\le p<\infty,$$

$$= \text{ess.} \sup_{\theta\in[0,2\pi]} |f(e^{j\theta})| \text{ if } p = \infty.$$

In other words, the norm $\|f\|_p$ of a function $f\in\mathbf{H}_p$ can be computed from its boundary value function.

For each $p\in[1,\infty]$, the space $\mathbf{H}_p$ is a Banach space. In addition, $\mathbf{H}_2$ is a (complex) Hilbert space with inner product defined by

5 $$<f,g> = \frac{1}{2\pi} \int_0^{2\pi} \bar{f}(e^{j\theta}) \, g(e^{j\theta}) \, d\theta,$$

where the bar denotes complex conjugation. It turns out that the set of functions $\{z^i\}_{i=0}^{\infty}$ forms an orthonormal basis for $\mathbf{H}_2$. Hence a function $f(\cdot)$ analytic in $\mathbf{D}$ belongs to $\mathbf{H}_2$ if and only if it can be expanded in the form

6 $$f(z) = \sum_{i=0}^{\infty} f_i z^i,$$

where the sequence $\{f_i\}$ is square-summable and the infinite sum converges in the sense of the $\mathbf{H}_2$-norm. Thus one can think of $\mathbf{H}_2$ as the set of z-transforms of square-summable sequences. Similarly, if $\{f_i\}$ is an *absolutely* summable sequence, then

7 $$f(z) = \sum_{i=0}^{\infty} f_i z^i$$

belongs to $\mathbf{H}_\infty$, and in fact

8 $$\| f \|_\infty \leq \sum_{i=0}^{\infty} | f_i | .$$

In general, the converse of the above statement is false: Not every function in $\mathbf{H}_\infty$ is of the form (7) where the coefficient sequence is absolutely summable. However, it is easy to verify that a *rational* function belongs to $\mathbf{H}_\infty$ if and only if it is the z-transform of an absolutely summable sequence. If the product of two $\mathbf{H}_\infty$-functions is defined to be their pointwise product, then $\mathbf{H}_\infty$ becomes a Banach algebra.

Suppose $f \in \mathbf{H}_\infty$. Then one can associate with f a continuous linear operator T_f mapping $\mathbf{H}_2$ into itself by

9 $$(T_f g)(z) = f(z) g(z)$$

In other words, an application of T_f consists simply of pointwise multiplication by $f(\cdot)$. It can be shown that this association is norm-preserving. That is, the operator norm of T_f equals $\| f \|_\infty$; or in other words,

10 $$\| T_f \| := \sup_{\| g \|_2 \leq 1} \| fg \|_2 = \| f \|_\infty .$$

Suppose m is a nonnegative integer, and that $\{a_i\}$ is a sequence of nonzero complex numbers such that $|a_i| < 1$ for all i, and the sum $\sum_{i=1}^{\infty}(1-|a_i|)$ converges. Then the function

$$b(z) = z^m \prod_{i=1}^{\infty} \frac{|a_i|}{a_i} \frac{z-a_i}{1-\bar{a}_i z} \tag{11}$$

is called the *Blaschke product* corresponding to m and the sequence $\{a_i\}$. The function $b(\cdot)$ is an element of $\mathbf{H}_\infty$ and satisfies $|b(z)| < 1$ for all z in $\mathbf{D}$, $|b(e^{j\theta})| = 1$ for almost all $\theta \in [0,2\pi]$. Moreover, the set of zeros of $b(\cdot)$ in $\mathbf{D}$ consists precisely of the set $\{a_i\}$, where repeated occurrence of a particular complex number corresponds to multiple zeros of b. If m is a nonnegative integer and $\{a_1, \cdots, a_n\}$ is a *finite* set of points (not necessarily distinct) such that $0 < |a_i| < 1$ for all i, then

$$b(z) = z^m \prod_{i=1}^{n} \frac{|a_i|}{a_i} \frac{z-a_i}{1-\bar{a}_i z} \tag{12}$$

is called the associated (finite) Blaschke product. It too belongs to $\mathbf{H}_\infty$ and satisfies $|b(z)| < 1$ for all $z \in \mathbf{D}$ and $|b(e^{j\theta})| = 1 \ \forall \theta \in [0,2\pi]$.

Suppose f is any *rational* function in $\mathbf{H}_\infty$. Let m denote the multiplicity of $z = 0$ as a zero of f (if $f(0) \neq 0$ set $m = 0$), and let $a_1, \cdots, a_n$ denote the other zeros of f in $\mathbf{D}$. If f has a multiple zero at some point in $\mathbf{D}$, repeat this point in the sequence $\{a_1, \cdots, a_n\}$ as many times as the multiplicity. If $b(\cdot)$ is now defined as in (12), then it is clear that f can be factored in the form $f(z) = a(z)b(z)$, where both $a, b \in \mathbf{H}_\infty$, b is a Blaschke product, and a has no zeros in the open unit disc. Further, if f has no zeros *on* the unit circle, then a is actually a *unit* of $\mathbf{H}_\infty$, i.e. it has an inverse in $\mathbf{H}_\infty$.

The factorization described above is a special case of what is known as an *inner-outer factorization*. A function $g \in \mathbf{H}_\infty$ is *inner* if it satisfies $|g(e^{j\theta})| = 1$ for almost all $\theta \in [0,2\pi]$. The only *rational* inner functions are finite Blaschke products multiplied by a complex number of unit modulus. In general, an inner function can be expressed as a product of a Blaschke product and a so-called singular inner function; we will not require the latter concept in this book. The definition of an outer function is more technical: A function $f \in \mathbf{H}_\infty$ is *outer* if the range of the map T_f defined in (9) is dense in $\mathbf{H}_2$. However, a *rational* function $f \in \mathbf{H}_\infty$ is outer if and only if it has no zeros in the *open* unit disc (but it could have zeros on the unit circle). With these definitions, we have the follwing result:

13 *Fact* Every $f \in \mathbf{H}_\infty$ can be factored as ab where a is outer and b is inner. If f is rational and has no zeros on the unit circle, then f can be factored in the form ab where a is a unit of $\mathbf{H}_\infty$ and b is a finite Blaschke product.

In the case of *matrices* over $\mathbf{H}_p$-spaces, the essential results for our purposes are as follows: Let $n > 1$ be an integer. Then $\mathbf{H}_2^n$ is a Hilbert space with the inner product

14 $$\langle f, g \rangle = \frac{1}{2\pi} \int_0^{2\pi} f^*(e^{j\theta})\, g(e^{j\theta})\, d\theta,$$

where * denotes the conjugate transpose. Corresponding to every $F \in \mathbf{H}_\infty^{n \times m}$ one can define a continuous linear operator $T_F : \mathbf{H}_2^m \to \mathbf{H}_2^n$ by

15 $$(T_F g)(z) = F(z) g(z) \ \forall z \in \mathbf{D}.$$

Moreover, the norm of the operator T_F equals the norm of F; that is,

16 $$\| T_F \| = \sup_{\| g \|_2 \le 1} \| T_F g \|_2 = \| F \|_\infty = \sup_{\theta \in [0, 2\pi]} \bar{\sigma}(F(e^{j\theta})),$$

where $\bar{\sigma}(\cdot)$ denotes the largest singular value.

The matrix analog of Fact (13) is stated next. A matrix $G \in \mathbf{H}_\infty^{n \times m}$ is *inner* if $G^*(e^{j\theta}) G(e^{j\theta}) = I$ for almost all $\theta \in [0, 2\pi]$ (in this definition it is important to note that n need not equal m). A matrix $F \in \mathbf{H}_\infty^{m \times n}$ is *outer* if the range of the map T_F defined in (15) is dense in $\mathbf{H}_2^n$. In particular, a rational matrix $F \in \mathbf{H}_\infty^{n \times m}$ is outer if and only if rank $F(z) = n$ for all $z \in \mathbf{D}$. Note that if G inner (resp. outer) then $n \ge m$ (resp. $n \le m$).

To state the next result precisely, we recall the set $\mathbf{R}_\infty$ introduced in Section 2.1, which consists of all rational functions that have no poles in the closed unit disc. It is easy to see that $\mathbf{R}_\infty$ is just the set of rational functions in $\mathbf{H}_\infty$ (this explains the symbolism), and that the norm on $\mathbf{R}_\infty$ defined in (2.2.24) is consistent with (4).

17 *Fact* Suppose $G \in \mathbf{H}_\infty^{n \times m}$ and has rank $\min\{n,m\}$.[1] Then G can be factored as $G_i G_o$ where G_i is inner and G_o is outer. If $n \geq m$, then G_o is square, whereas if $n \leq m$, then G_i is square.

Fact (17) has several interesting consequences. First, suppose $G \in \mathbf{R}_\infty^{n \times m}$ has rank n and that $n \leq m$. Then, by applying Fact (17) to $F = G'$, we see that G can be factored as AB where $A \in \mathbf{R}_\infty^{n \times n}$ is outer and $B \in \mathbf{R}_\infty^{n \times m}$ is the transpose of an inner matrix; This is quite distinct from the factorization in Fact (17). Second, noting that the transpose of a *square* inner matrix is also inner, we see that a square matrix $G \in \mathbf{M}(\mathbf{R}_\infty)$ of full rank can be factored as $G_i G_o$ and also as $F_o F_i$ where G_i, F_i are inner and G_o, F_o are outer.

Finally, observe that if G is a square inner matrix, then $|G|$ is an inner function, and the adjoint matrix of G is also a square inner matrix. Also, note that left multiplication by an inner matrix preserves norms in both $\mathbf{H}_2$ and $\mathbf{H}_\infty$. That is, if $F \in \mathbf{M}(\mathbf{H}_\infty)$ is inner, $G \in \mathbf{M}(\mathbf{H}_\infty)$ and $H \in \mathbf{M}(\mathbf{H}_2)$, then $\| FG \|_\infty = \| G \|_\infty$ and $\| FH \|_2 = \| H \|_2$.

One can also define the class $\mathbf{H}_p$ over the open right half-plane, rather than the open unit disc as is the case above. Suppose $f: C_+ \rightarrow C$ is analytic; then f belongs to the class $\mathbf{H}_p(C_+)$, where $1 \leq p < \infty$, if

18 $$\sup_{\sigma > 0} \left[\frac{1}{2\pi} \int_{-\infty}^{\infty} |f(\sigma + j\omega)|^p \, d\omega \right]^{1/p} =: \| f \|_{p+} < \infty.$$

The function f belongs to $\mathbf{H}_\infty(C_+)$ if

[1] This means that at least one full-size minor of G is nonzero.

19 $$\sup_{\sigma>0} \sup_{\omega} |f(\sigma+j\omega)| =: \|f\|_{\infty+} < \infty.$$

If $f \in \mathbf{H}_p(C_+)$, then its domain of definition can be extended to the $j\omega$-axis, and the function $\omega \to f(j\omega)$ belongs to the Lebesgue space $\mathbf{L}_p(\mathbb{R})$. Moreover,

20 $$\|f\|_{p+} = \left[\frac{1}{2\pi} \int_{-\infty}^{\infty} |f(j\omega)|^p \, d\omega \right]^{1/p} \quad \text{if } p \in [1,\infty),$$

$$\|f\|_{\infty+} = \text{ess.} \sup_{\omega} |f(j\omega)| \text{ if } p = \infty.$$

$\mathbf{H}_2(C_+)$ is a Hilbert space, with the inner product of two functions f and g defined by

21 $$<f,g>_+ = \frac{1}{2\pi} \int_{-\infty}^{\infty} \bar{f}(j\omega) g(j\omega) d\omega.$$

The notions of Blaschke product, inner and outer functions all have their counterparts in this setting. In particular, a function $f \in \mathbf{H}_\infty(C_+)$ is *inner* if $|f(j\omega)| = 1$ for almost all ω; a rational $f \in \mathbf{H}_\infty(C_+)$ with $f(j\omega) \neq 0$ for all ω is *outer* if it is a unit. Facts (13) and (17) have their obvious analogs for the space $\mathbf{H}_\infty(C_+)$, obtained by replacing "unit circle" by "imaginary axis" and "$e^{j\theta}$" by "$j\omega$" throughout.

Suppose $f \in \mathbf{H}_\infty$ (of the unit disc). Then the function $g: C_+ \to C$ defined by

22 $$g(s) = f[(s-1)/(s+1)]$$

belongs to $\mathbf{H}_\infty(C_+)$, and satisfies $\|g\|_{\infty+} = \|f\|_\infty$. However, not all functions in $\mathbf{H}_\infty(C_+)$ can be generated in this manner. For example, $g(s) = e^{-s}$ belongs to $\mathbf{H}_\infty(C_+)$, but is not of the form (22) for any $f \in \mathbf{H}_\infty$. The situation is worse when $p < \infty$: Even if $f \in H_p$, the corresponding g might not belong to $H_p(C_+)$. An easy way to see this is to note that every $g \in H_p(C_+)$ must be strictly proper. Thus, if $f \in H_p$ is rational and $f(1) \neq 0$, then from (22) $g(\infty) = f(1) \neq 0$, so that $g \notin H_p(C_+)$. Over the unit disc, $\mathbf{H}_\infty$ is a subset of H_p for all $p<\infty$, and indeed H_p is a subset of H_q whenever $p>q$; this follows from the compactness of the interval $[0,2\pi]$. However, $H_p(C_+)$ and $H_q(C_+)$ are distinct sets whenever $p \neq q$, with neither being a subset of the other.

There is however an interesting relationship, which is given next.

23 *Fact* Suppose $f \in \mathbf{H}_p$ for some $p \in [1,\infty)$, and define g by (22). Let

24 $$h(s) = \frac{2^{1/p} g(s)}{(1+s)^{2/p}}.$$

Then $h \in \mathbf{H}_p(C_+)$ and $\| h \|_{p+} = \| f \|_p$.

Proof By definition

25 $$\| f \|_p^p = \frac{1}{2\pi} \int_0^{\infty} | f(e^{j\theta}) |^p d\theta,$$

26 $$\| h \|_{p+}^p = \frac{1}{2\pi} \int_{-\infty}^{\infty} \left| \frac{g(j\omega)}{(1+\omega^2)^{1/p}} \right|^p 2\, d\omega$$

$$= \frac{1}{2\pi} \int_{-\infty}^{\infty} | f([(j\omega-1)/(j\omega+1)] |^p \frac{2d\omega}{1+\omega^2}.$$

To relate (25) and (26), let $e^{j\theta} = (j\omega-1)/(j\omega+1)$. Then

27 $$j\omega = \frac{1+e^{j\theta}}{1-e^{j\theta}} = \frac{e^{-j\theta/2} + e^{j\theta/2}}{e^{-j\theta/2} - e^{j\theta/2}}$$

$$= j \cot \theta/2.$$

Hence

28 $$\theta = 2 \cot^{-1}\omega;\ d\theta = \frac{2d\omega}{1+\omega^2}.$$

Combining (28) and (26) completes the proof. □

Finally, if $f \in \mathbf{H}_2$, then its boundary function $\theta \rightarrow f(e^{j\theta})$ belongs to $\mathbf{L}_2[0,2\pi]$. Let $\mathbf{H}_2^+[0,2\pi]$ denote the set of boundary values of functions in $\mathbf{H}_2$. Then $\mathbf{H}_2^+[0,2\pi]$ is a closed subspace of $\mathbf{L}_2[0,2\pi]$. Let $\mathbf{H}_2^-[0,2\pi]$ denote the orthogonal complement of $\mathbf{H}_2^+[0,2\pi]$ in $\mathbf{L}_2[0,2\pi]$. Then every function $f \in \mathbf{L}_2$ can be expressed uniquely in the form $f = f_+ + f_-$ where $f_+ \in \mathbf{H}_2^+$ and $f_- \in \mathbf{H}_2^-$ (we drop the $[0,2\pi]$ for brevity). This splitting up has a very intuitive interpretation. The collection of functions $\{e^{jn\theta}, n \in \mathbf{Z}\}$ forms an orthonormal basis for $\mathbf{L}_2$, so that every $f \in \mathbf{L}_2$ has the Fourier series

29 $$f(e^{j\theta}) = \sum_{n=-\infty}^{\infty} f_n \, e^{jn\theta}.$$

Now $\mathbf{H}_2^+$ is the span of the functions $\{e^{jn\theta}, n \in \mathbf{Z}_+\}$, while $\mathbf{H}_2^-$ is the span of the set $\{e^{jn\theta}, n < 0\}$. Hence

30 $$f_+(e^{j\theta}) = \sum_{n=0}^{\infty} f_n \, e^{jn\theta}, f_-(e^{j\theta}) = \sum_{n=-\infty}^{-1} f_n \, e^{jn\theta}.$$

Note that the function $z \rightarrow z^{-1} f_-(z^{-1})$ belongs to $\mathbf{H}_2$. Such a function is called *coanalytic*. In case $f(\cdot)$ is rational and has no poles on the unit circle, the above decomposition is even more familiar: Expand f into partial fractions; then f_+ (resp. f_-) is the sum of the terms corresponding to poles outside (resp. inside) the unit disc.

The above discussion applies, *mutatis mutandis,* to the space $\mathbf{L}_2(\mathbb{R})$.

6.3 FILTERING

The problem studied in this section is the following: Suppose a plant $P \in \mathbf{M}(\mathbb{R}(s))$ is specified, together with weighting matrices W_1, W_2. The objective is to minimize the weighted sensitivity measure

1 $$J = \| W_1 (I + PC)^{-1} W_2 \|_{2+}$$

over all $C \in S(P)$. At this stage, several assumptions are made to facilitate a solution. First, note that J might not even be defined for all $C \in S(P)$, because $W_1(I+PC)^{-1}W_2$ might not be strictly proper. To get around this problem, it is assumed that either W_1 or W_2 is strictly proper. Next, it is assumed that W_1, W_2 are square, and that $W_1(s) \neq 0$, $W_2(s) \neq 0$ for all $s \in C_+$. Finally, for technical reasons, it is assumed that the plant P has at least as many inputs as outputs. This is the so-called "square" or "fat" plant case; the case of "tall" plants is discussed in Section 6.6.

As pointed out in Section 6.1, this problem can be reformulated in a more manageable form. Let (N,D), $(\tilde{D},\tilde{N})$ be an r.c.f. and an l.c.f. of P, and let $\tilde{X},\tilde{Y} \in \mathbf{M}(\mathbf{S})$ be any solutions of $\tilde{N}\tilde{X} + \tilde{D}\tilde{Y} = I$. Then, by Theorem (5.2.1), every $C \in S(P)$ is of the form $(\tilde{X}+DR)(\tilde{Y}-NR)^{-1}$ for some $R \in \mathbf{M}(\mathbf{S})$ such that $|\tilde{Y}-NR| \neq 0$. Moreover, the corresponding weighted sensitivity measure equals

2
$$J(R) = \| W_1(\tilde{Y}-NR)\tilde{D}W_2 \|_2 = \| W_1\tilde{Y}\tilde{D}W_2 - W_1NR\tilde{D}W_2 \|_2.$$

Hence the problem is to minimize $J(R)$ by a suitable choice of $R \in \mathbf{M}(\mathbf{S})$.[2] This problem has a very nice geometrical interpretation. As R varies over $\mathbf{M}(\mathbf{S})$, the product $W_1NR\tilde{D}W_2$ generates a subspace of $M(\mathbf{H}_2(C_+))$. To minimize $J(\cdot)$, one has to find the closest point in this subspace to $W_1\tilde{Y}\tilde{D}W_2$. Now, since at least one of the matrices W_1 or W_2 is strictly proper, it turns out that the subspace $\{W_1NR\tilde{D}W_2 : R \in \mathbf{M}(\mathbf{S})\}$ is not closed. As a result, there may not exist a closest point in this subspace to $W_1\tilde{Y}\tilde{D}W_2$. In other words, the infimum of $J(\cdot)$ is not attained by any R, except in trivial cases. It is nevertheless possible to compute the infimum of $J(\cdot)$, and to construct a family (R_ε) of functions in $\mathbf{M}(\mathbf{S})$ such that $J(R_\varepsilon)$ approaches this infimum as $\varepsilon \rightarrow 0^+$. This procedure is based on the following lemma.

3 *Lemma* Suppose $F,G,H \in \mathbf{M}(\mathbf{S})$, and suppose in addition that (i) F is strictly proper, (ii) G has full row rank at all points on the $j\omega$-axis and at infinity,[3] and (iii) H is square and nonsingular at all points on the $j\omega$-axis and at infinity. Define

---

[2] The constraint that $|\tilde{Y}-NR| \neq 0$ is ignored since it is satisfied by almost all R (see Lemma (5.2.11)).

4 $$J_m(R) = \| F - GRH \|_2,$$

and let $J_m(R) = \infty$ if $F - GRH \notin \mathbf{M}(\mathbf{H}_2(C_+))$. Factorize G, H as $G = G_i G_o$, $H = H_o H_i$, where G_i, H_i are inner and G_o, H_o are outer. For any $A \in \mathbf{M}(\mathbf{S})$, let A^* denote the rational matrix $A^*(s) = A'(-s)$.[4] Then

5 $$\min_R J_m(R) = \| (G_i{}^* F H_i)_- \|_2,$$

and this minimum is attained by any R satisfying

6 $$G_o R H_o = (G_i{}^* F H_i{}^*)_+.$$

Proof First of all, observe that the equation $G_o R H_o = A$ has a solution for R corresponding to *every* $A \in \mathbf{M}(\mathbf{S})$. The reason is that G_o has a right inverse and H_o is a unit.

To prove the lemma, note that $G_i{}^*(s) G_i(s) = I$, $H_i{}^*(s) H_i(s) = H_i(s) H_i{}^*(s) = I \ \forall s$, since both G_i and H_i are inner matrices. In particular, $G_i{}^*(j\omega)$ is a unitary matrix for all ω, and as a result for any $A \in \mathbf{M}(\mathbf{H}_2(C_+))$ we have

7 $$\| A \|_2^2 = \| G_i{}^* A \|_2^2,$$

where the norm on the left is the $\mathbf{H}_2$-norm or the $\mathbf{L}_2$-norm, while the norm on the right side is strictly the $\mathbf{L}_2$-norm (since $G_i{}^* A$ need not belong to $\mathbf{M}(\mathbf{H}_2(C_+))$). In the same way, multiplication by $H_i{}^*$ is also norm-preserving.

Using this fact leads to

8 $$(J_m(R))^2 = \| F - GRH \|_2^2 = \| F - G_i G_o R H_o H_i \|_2^2$$
$$= \| G_i{}^* F H_i{}^* - G_o R H_o \|_2^2.$$

If $G_i{}^* F H_i$ is decomposed as $(G_i{}^* F H_i{}^*)_+ + (G_i{}^* F H_i{}^*)_-$, then by orthogonality,

[3] This means that G has at least as many columns as rows.

[4] In particular, if $s = j\omega$, then $A^*(j\omega)$ is the conjugate transpose of $A(j\omega)$, so that this nota-

9 $$(J_m(R))^2 = \| (G_i{}^*FH_i{}^*)_- \|_2^2 + \| (G_i{}^*FH_i{}^*)_+ - G_oRH_o \|_2^2$$
$$\geq \| (G_i{}^*FH_i{}^*)_- \|_2^2.$$

On the other hand, equality can be achieved in (9) by choosing R to satisfy (6). □

If we attempt to apply Lemma (3) to minimize the function $J(\cdot)$ of (2), we run into the difficulty that the functions $G = W_1N$ and $H = \tilde{D}W_2$ may not satisfy the hypotheses of Lemma (3). In particular, if $P(j\omega)$ has less than full row rank for some ω, then so does $N(j\omega)$ (i.e. $G(j\omega)$). Similarly, any $j\omega$-axis poles of P correspond to zeros of $|\tilde{D}|$ and of $|H|$. Finally, since W_1 or W_2 is assumed to be strictly proper, either $|G(\infty)| = 0$ or $|H(\infty)| = 0$. To get around these difficulties, the following modification of Lemma (3) is required.

10 *Lemma* Suppose $F,G,H \in \mathbf{M}(\mathbf{S})$, and suppose in addition that

(i) G has the form $G = G_1D_g$ where $G_1 \in \mathbf{M}(\mathbf{S})$ has full row rank at all points on the $j\omega$-axis and at infinity, while $D_g \in \mathbf{M}(\mathbf{S})$ is a diagonal marix $D_g = \text{Diag}\ \{d_{g_1}, \cdots, d_{g_n}\}$, where the only C_{+e}-zeros of d_{g_i} (if any) are at infinity or on the $j\omega$-axis.

(ii) H has the form $H = D_hH_1$, where $H_1 \in \mathbf{M}(\mathbf{S})$ is square and nonsingular at all points on the $j\omega$-axis and at infinity, while $D_h \in \mathbf{M}(\mathbf{S})$ is a diagonal matrix of the form $D_h = \text{Diag}\ \{d_{h_1}, \cdots, d_{h_m}\}$, where the only C_{+e}-zeros of d_{h_i} (if any) are at infinity or on the $j\omega$-axis.

Under these conditions,

11 $$\inf_{R \in \mathbf{M}(\mathbf{S})} \| F - GRH \|_2 = \inf_{R \in \mathbf{M}(\mathbf{S})} \| F - G_1RH_1 \|_2.$$

Proof As shown by Lemma (3), the infimum on the right side of (11) is actually attained. Let $R_0 \in \mathbf{M}(\mathbf{S})$ be an element that achieves this minimum. Now every matrix of the form GRH is also of the form $G_1R_1H_1$ (just let $R_1 = D_gRD_h$). Thus the infimum on the left side of (11) can be no smaller than that on the right side. Hence, in order to prove the lemma, it is enough to

---

tion is consistent with earlier usage.

construct a family of functions $(R_\varepsilon)_{\varepsilon>0}$ in $\mathbf{M}(\mathbf{S})$ such that $\| F-GR_\varepsilon H \|_2 \to \| F-G_1R_0H_1 \|_2$ as $\varepsilon \to 0^+$.

Suppose $a \in \mathbf{S}$ has the property that $a(s) \neq 0$ whenever $\text{Re } s > 0$, so that the only possible C_{+e}-zeros of a are at infinity or on the $j\omega$-axis. (Note that all the d_{g_i}, d_{h_i} are of this type.) To be specific, let $j\omega_1, \cdots, j\omega_l$ (where ω_l may be infinite) denote the distinct $j\omega$-axis zeros of a, with multiplicities $m_1, \cdots, m_l$, respectively. It is claimed that there exists a family of functions $(b_\varepsilon)_{\varepsilon>0}$ in $\mathbf{S}$ such that

(i) b_ε/a is a unit of $\mathbf{S}$ for each $\varepsilon > 0$.

(ii) $b_\varepsilon(j\omega) \to 1$ uniformly on all compact subsets of $(-\infty,\infty)$ that do not contain any of the ω_i's, as $\varepsilon \to 0^+$.

To see this, note that $a(s)$ is of the form

12 $$a(s) = u(s) \prod_{i=1}^{l} \left[\frac{s-j\omega_i}{s+1} \right]^{m_i} \cdot \left[\frac{1}{s+1} \right]^n$$

for some unit $u \in \mathbf{S}$. Hence

13 $$b_\varepsilon(s) = \prod_{i=1}^{l} \left[\frac{s-j\omega_i}{s+\varepsilon-j\omega_i} \right]^{m_i} \cdot \left[\frac{1}{\varepsilon s+1} \right]^n$$

satisfies the requirements (i) and (ii).

In view of this, it follows that there exists families $B_{\varepsilon g}$, $B_{\varepsilon h}$ of diagonal matrices in $\mathbf{M}(\mathbf{S})$ such that

(i) $D_g^{-1}B_{\varepsilon g}$ is unimodular for all $\varepsilon > 0$.

(ii) $B_{\varepsilon h}D_h^{-1}$ is unimodular for all $\varepsilon > 0$.

(iii) $B_{\varepsilon g}(j\omega) \to I$, $B_{\varepsilon h}(j\omega) \to I$ as $\varepsilon \to 0^+$ uniformly on all compact subsets of $(-\infty,\infty)$ that do not contain the zeros of $|D_g|$, $|D_h|$.

Now define $R_\varepsilon = D_g^{-1}B_{\varepsilon g}R_0B_{\varepsilon h}D_h^{-1}$, where $R_0 \in \mathbf{M}(\mathbf{S})$ attains the infimum on the right side of (11). Then

14 $$\| F-GR_{\varepsilon}H \|_2 = \| F - G_1 D_g \cdot D_g^{-1} B_{\varepsilon g} R_0 B_{\varepsilon h} D_h^{-1} \cdot D_h H_1 \|_2$$
$$= \| F - G_1 B_{\varepsilon g} R_0 B_{\varepsilon h} H_1 \|_2.$$

The proof of the lemma is complete if it can be shown that the last quantity in (14) approaches $\| F-G_1 R_0 H_1 \|_2$ as $\varepsilon \to 0^+$. But this follows from the limiting behavior of $B_{\varepsilon g}$, $B_{\varepsilon h}$ as described in (iii) above. □

We are now ready to tackle the general problem of minimizing $J(\cdot)$ as defined in (2). Let G_s denote the Smith form of $W_1 N$ and H_s the Smith form of $\tilde{D} W_2$, and suppose

15 $$U_g G_s V_g = W_1 N,\ U_h H_s V_h = \tilde{D} W_2,$$

where $U_g, V_g, U_h, V_h \in \mathbf{U}(\mathbf{S})$. Then

16 $$J(R) = \| W_1 \tilde{Y} \tilde{D} W_2 - W_1 N R \tilde{D} W_2 \|_2$$
$$= \| W_1 \tilde{Y} \tilde{D} W_2 - U_g G_s V_g R U_h H_s V_h \|_2.$$

Since V_g, U_h are unimodular, one can replace $V_g R U_h$ by another "free" parameter, say R_1. Moreover, G_s (and H_s) can be expressed as a product of two diagonal matrices E_g and D_g (resp. E_h and D_h) such that $| E_g |$ is nonzero at all points on the $j\omega$-axis and at infinity, while $| D_g(s) | \neq 0$ whenever $s \in C_+$. Now, defining

17 $$G_1 = U_g E_g,\ H_1 = E_h V_h$$

shows that

18 $$J(R) = \| W_1 \tilde{Y} \tilde{D} W_2 - G_1 D_g R_1 D_h H_1 \|_2$$

where $R_1 = V_g R U_h$. Hence by Lemma (10),

19 $$\inf_R J(R) = \inf_R \| W_1 \tilde{Y} \tilde{D} W_2 - G_1 R H_1 \|_2,$$

which can be computed using Lemma (2).

The situation is much cleaner in the case of discrete-time systems. First, since the interval $[0,2\pi]$ is compact, every $\mathbf{H}_\infty$-function is also in $\mathbf{H}_2$. Further, a *strictly causal* plant, which is the discrete analog of a strictly proper plant, is one whose transfer matrix is zero at $z = 0$. As a result, in the formulation of the discrete-time filtering problem, the weighting matrices W_1 and W_2 can be any rational $\mathbf{H}_\infty$-matrices and need not be strictly causal. Moreover, an optimal controller always exists and the approximation procedure of Lemma (10) is not needed.

6.4 SENSITIVITY MINIMIZATION: SCALAR CASE

Because the sensitivity minimization problem is much more involved technically than the filtering problem, the discussion of the former is divided into three parts, namely the scalar case, the fat plant case, and the general case. Accordingly, the problem studied in this section is the following: Suppose p is a given scalar plant, and w is a given element of $\mathbf{S}$. The objective is to find a $c \in S(p)$ that minimizes the weighted sensitivity measure[5]

1 $$J = \sup_{\omega \in \mathbb{R}} \left| \frac{w(j\omega)}{1+p(j\omega)c(j\omega)} \right|.$$

As in the previous section, one can transform this problem into an affine minimization problem. Let (n,d) be a coprime factorization of p, and let $x,y \in \mathbf{S}$ be such that $xn + yd = 1$. Then the problem becomes: Minimize

2 $$J(r) = \sup_{\omega \in \mathbb{R}} \left| [(y-rn)dw](j\omega) \right|$$
$$= \| (y-rn)dw \|_\infty$$

by suitable choice of $r \in \mathbf{S}$. Moreover, it can be assumed without loss of generality that the weighting function w is outer. To see this, factor w in the form $w = w_i w_o$ where w_i and w_o are inner and outer, respectively. Then,

[5] Note that the two weights w_1 and w_2 are combined into one, since we are dealing with the scalar case.

since multiplication by the inner function w_i is norm-preserving, we have $J(r) = \| (y-rn)dw_o \|_\infty$.

In order to take full advantage of the available results on $\mathbf{H}_p$-spaces, a bilinear transformation is now used to map the set $\mathbf{S}$ into $\mathbf{H}_\infty$. Suppose $f \in \mathbf{S}$, and define the function $\tilde{f}$ by

3 $$\tilde{f}(z) = f[(1+z)/(1-z)]$$

Since the bilinear transformation $s = (1+z)/(1-z)$ maps the open unit disc into the open right half-plane, it follows that $\tilde{f}$ is a rational element of $\mathbf{H}_\infty$, i.e. $f \in \mathbf{R}_\infty$. Moreover,

4 $$\| \tilde{f} \|_\infty = \sup_{\theta \in [0,2\pi]} | \tilde{f}(e^{j\theta}) |$$

$$= \sup_{\omega \in \mathbb{R}} | f(j\omega) | = \| f \|_{\mathbf{S}},$$

since the unit circle maps onto the imaginary axis plus the point at infinity. Conversely, suppose $\tilde{g} \in \mathbf{H}_\infty$ is rational, and define g by

5 $$g(s) = \tilde{g}[(s-1)/(s+1)].$$

Then $g \in \mathbf{S}$ and $\| \tilde{g} \|_\infty = \| g \|_{\mathbf{S}}$. In this way one can set up a one-to-one norm-preserving correspondence between functions in $\mathbf{S}$ and functions in $\mathbf{R}_\infty$. If this procedure is adopted, the problem under study becomes the following: Find a function $\tilde{r} \in \mathbf{R}_\infty$ that minimizes

6 $$J = \| (\tilde{y} - \tilde{r}\tilde{n})\tilde{d}\tilde{w} \|_\infty.$$

At this stage we drop the tilde and state the problem in $\mathbf{H}_\infty$ rather than in $\mathbf{R}_\infty$; the reason for the latter is that $\mathbf{H}_\infty$ is a Banach space while $\mathbf{R}_\infty$ is merely a normed space. The problem in essence is as follows: Given two rational functions $f, g \in \mathbf{H}_\infty$, find a rational function $r \in \mathbf{H}_\infty$ that minimizes

7 $$J(r) = \| f - gr \|_\infty.$$

The definitions of f and g when J is as in (6) are obvious. The problem of minimizing J has a very simple geometric interpretation: Let $g\mathbf{H}_\infty$ denote the subspace $\{gr: r \in \mathbf{H}_\infty\}$. Then the problem is to find the closest point to f in the subspace $g\mathbf{H}_\infty$, where distances are measured using the norm $\|\cdot\|_\infty$.

It turns out that the subspace $g\mathbf{H}_\infty$ is *closed* if and only if g has no zeros on the unit circle. If $g\mathbf{H}_\infty$ is a closed subspace, then it can be shown that, corresponding to each rational $f \in \mathbf{H}_\infty$ there exists a *unique* closest point in $g\mathbf{H}_\infty$ (more on this shortly). Unfortunately, if the plant p has zeros on the $j\omega$-axis or at infinity, then so does n. As a result, $g = \tilde{n}\tilde{d}\tilde{w}$ has one or more zeros on the unit disc, and the subspace $g\mathbf{H}_\infty$ is *not* closed. As a consequence, if one defines

$$8 \qquad J_0 = \inf_{r \in \mathbf{H}_\infty} \|f - gr\|_\infty,$$

then the infimum in (8) may or may not be attained, depending on f. In the original context, this means that in general there does *not* exist a single compensator that minimizes the weighted sensitivity measure (1), in the practically important case where the plant is strictly proper (or has $j\omega$-axis zeros).

In view of the above discussion, one would nevertheless like to know the *value* of J_0, as well as a *sequence* $\{r_i\}$ of functions in $\mathbf{H}_\infty$ such that $J(r_i) \rightarrow J_0$. This information can then be translated back into a greatest lower bound on the weighted sensitivity measure of (1), as well as an *optimizing sequence* $\{c_i\}$ of compensators.

In order to do this, first carry out an inner-outer factorization of g. Thus $g = bg_0$ where b is a finite Blaschke product and g_0 is outer. Now factor g_0 as vu where $v \in \mathbf{H}_\infty$ has all of its zeros *on* the unit circle and u is a unit of $\mathbf{H}_\infty$. Then

$$9 \qquad J_0 = \inf_{r \in \mathbf{H}_\infty} \|f - bdur\|_\infty = \inf_{r \in \mathbf{H}_\infty} \|f - bvr\|_\infty,$$

since the unit factor u can be absorbed into the "free" parameter r. Now consider the modified problem: Minimize $\|f - rb\|_\infty$ by a suitable choice of $r \in \mathbf{H}_\infty$. Define

10 $$J_m = \inf_{r \in \mathbf{H}_\infty} \| f - rb \|_\infty$$

as the value of the modified problem. In view of earlier discussion, the infimum in (10) is actually attained, since b has no zeros *on* the unit circle. Now, since every function of the form bvr is also a multiple of b, it is clear that $J_m \leq J_0$. The next lemma gives conditions under which equality holds.

11 *Lemma* Let $e^{j\theta_1}, \cdots, e^{j\theta_l}$ denote the distinct zeros of v on the unit circle. Then $J_0 = J_m$ if and only if $| f(e^{j\theta_i}) | \leq J_m \ \forall i$.

Proof "only if" Since $v(e^{j\theta_i}) = 0 \ \forall i$, we have $(bvr)(e^{j\theta_i}) = 0 \ \forall i$, $\forall r \in \mathbf{H}_\infty$.[6] Hence $(f - bvr)(e^{j\theta_i}) = f(e^{j\theta_i}) \ \forall i$, and as a result

12 $$\begin{aligned} \| f - bvr \|_\infty &= \sup_{\theta \in [0, 2\pi]} | [f - bvr](e^{j\theta}) | \\ &\geq \max_i | [f - bvr](e^{j\theta_i}) | \\ &= \max_i | f(e^{j\theta_i}) |, \quad \forall r \in \mathbf{H}_\infty. \end{aligned}$$

Combining (12) with the hypothesis that $J_0 = J_m$ leads to

13 $$J_m = J_0 = \inf_{r \in \mathbf{H}_\infty} \| f - bvr \|_\infty \geq \max_i | f(e^{j\theta_i}) | .$$

"if" Select an $r_0 \in \mathbf{H}_\infty$ such that $J_m = \| f - r_0 b \|_\infty$; such an r_0 exists, by earlier remarks. Let $f_0 = f - r_0 b$. Next, introduce the function

14 $$h_{\theta,\varepsilon}(z) = (z - e^{j\theta}) / [z - (1+\varepsilon) e^{j\theta}],$$

and define

15 $$h_\varepsilon(z) = \prod_{i=1}^{l} [h_{\theta_i,\varepsilon}(z)]^{m_i},$$

where m_i is the multiplicity of $e^{j\theta_i}$ as a zero of $v(\cdot)$. Now observe that (i)

[6] Strictly speaking, $r(e^{j\theta_i})$ is undefined for a general $\mathbf{H}_\infty$-function, and this statement is only valid under more restrictive conditions, e.g. r is rational. These technicalities can be addressed, but spelling out the details would obscure the main ideas of the proof.

h_ε/v is a unit of $\mathbf{H}_\infty$ for each $\varepsilon > 0$, and (ii) $h_\varepsilon(e^{j\theta}) \to 1$ uniformly on all compact subsets of $[0,2\pi]$ that do not contain any of the θ_i's, as $\varepsilon \to 0^+$. Consider the family $(r_\varepsilon)_{\varepsilon>0}$ in $\mathbf{H}_\infty$ defined by $r_\varepsilon = r_0 h_\varepsilon / v$. Then by (9),

16 $\quad \| f - bvr_\varepsilon \|_\infty \geq J_0$, for each $\varepsilon > 0$.

On the other hand, $f - bvr_\varepsilon = f - br_0 h_\varepsilon$, so that $|(f - bvr_\varepsilon)(e^{j\theta})| \to |f_0(e^{j\theta})| \leq J_m$ uniformly on all compact subsets of $[0,2\pi]$ that do not contain any of the θ_i's, and $|(f - bvr_\varepsilon)(e^{j\theta_i})| = |f(e^{j\theta_i})| \leq J_m$ by hypothesis. This means that

17 $\quad \limsup_{\varepsilon \to 0^+} \| f - bvr_\varepsilon \|_\infty \leq J_m.$

Combining (16), (17) and the fact that $J_0 \geq J_m$ leads to the conclusion that $J_0 = J_m$ and that

18 $\quad \| f - bvr_\varepsilon \|_\infty \to J_m$ as $\varepsilon \to 0^+$.

This completes the proof. □

In the above proof, it is easy to see that there is nothing special about the function h_ε of (15), and that *any* function h_ε can be used so long as (i) $h_\varepsilon \in \mathbf{H}_\infty \ \forall \varepsilon > 0$, (ii) $h_\varepsilon(e^{j\theta_i}) = 0 \ \forall i$, and (iii) $h_\varepsilon(e^{j\theta}) \to 1$ uniformly on all compact subsets of $[0,2\pi]$ that do not contain any of the θ_i's.

Note that if $f(e^{j\theta_i}) = 0 \ \forall i$, then the condition of Lemma (11) is automatically satisfied and $J_0 = J_m$. Going back to (6), we have $f = \tilde{y}\tilde{d}\tilde{w}$ and $g = \tilde{n}\tilde{d}\tilde{w}$. Hence if the weighting function w is chosen such that every unit circle zero of $\tilde{n}$ is also a zero of $\tilde{w}$, then $J_0 = J_m$. (Note that unit circle zeros of $\tilde{d}$ are already common to both f and g.) Translating this condition back to the s-domain yields the result that $J_0 = J_m$ provided w is strictly proper whenever p is, and vanishes at the $j\omega$-axis zeros of the plant. Once such a weighting function is chosen, then any unit circle zeros of $\tilde{n}\tilde{d}\tilde{w}$ can be "deleted" without affecting the infimum of J.

19 *Example* Let p be a strictly proper, nonminimum phase plant with the transfer function

$$p(s) = \frac{s-1}{(s-3)(s^2+1)}.$$

Let the weighting function w equal

$$w(s) = \frac{1}{s+1}.$$

Then $w(\infty) = p(\infty) = 0$, so that $J_0 = J_m$ in this case. Now, substituting $s = (1+z)/(1-z)$ gives

$$\tilde{p}(z) = \frac{z(z-1)^2}{2(2z-1)(z^2+1)} = \frac{\tilde{n}(z)}{\tilde{d}(z)},$$

where $\tilde{n}$, $\tilde{d}$ are the numerator and denominator polynomials of $\tilde{p}$, respectively. A particular solution of $\tilde{x}\tilde{n} + \tilde{y}\tilde{d} = 1$ can be obtained by Euclidean division and back substitution, namely

$$\tilde{x}(z) = 6z^2 + \frac{13}{2}, \tilde{y}(z) = -\frac{1}{4}(6z^2-9z+2).$$

Finally,

$$\tilde{w}(z) = \frac{1-z}{2}.$$

Hence

$$J(r) = \| f - rg \|_\infty,$$

where

$$f(z) = (\tilde{w}\tilde{y}\tilde{d})(z) = \frac{1}{4}(z-1)(2z-1)(z^2+1)(6z^2-9z+2),$$

$$g(z) = (\tilde{w}\tilde{n}\tilde{d})(z) = -z(z-1)(2z-1)(z^2+1).$$

If we factor g in the form $g = bvu$ where b is a Blaschke product, v has all of its zeros on the unit circle, and u is a unit, then

$$b(z) = \frac{z(z-0.5)}{1-0.5z},$$

$$v(z) = (z-1)^3(z^2+1),$$

$$u(z) = -2(1-0.5z).$$

Now Lemma (11) allows us to ignore v and u, in that

$$\inf_r J(r) = J_m = \inf_r \| f - rb \|_\infty .$$

Lemma (11) provides conditions under which, by solving the minimization problem in (10) (which does have a solution), it is possible to determine the value of the infimum in (9). Actually, it turns out that it is possible to compute J_m (the value of the problem in (10)) *without* actually computing the r_0 that attains the minimum. If the conditions of Lemma (11) hold, this J_m is also the infimum in (9). In this way, it is possible to obtain rapidly a figure of merit against which suboptimal choices of r in (9) can be compared. If the optimization problem in (10) is solved completely, i.e. if r_0 is also determined, then a family r_ε such that $\| f - gr_\varepsilon \|_\infty \to J_m$ can be computed as in the proof of Lemma (11).

The next step then is to give a procedure for computing J_m. We begin with a result that is now obvious, but has some practical significance.

20 *Lemma* Suppose the plant p is strictly proper and stable, and has no zeros in the open right half-plane; suppose also that the weighting function w is stable, strictly proper and vanishes at all $j\omega$-axis zeros (if any) of p. Then

21 $$\inf_{c \in S(p)} \sup_\omega \left| \left[\frac{w}{1+pc} \right](j\omega) \right| = 0.$$

Remarks In other words, Lemma (20) states that if the plant is strictly proper, stable, and "minimum phase,"[7] then the weighted sensitivity measure

[7] Normally, the term "minimum phase" implies that the plant has no zeros in the *closed* right half-plane; here the usage is slightly different.

can be made arbitarily small provided the weighting function is also stable and vanishes at all RHP zeros of the plant.

Proof In this case $\tilde{d}$ is a unit, and all zeros of $\tilde{n}$ in the closed unit disc are in fact on the unit circle. As a result, the finite Blaschke product b equals 1, and the condition of Lemma (11) is satisfied. Finally, it is clear that

22 $$\inf_r \| f - rb \|_\infty = \inf_r \| f - r \|_\infty = 0,$$

corresponding to the choice $r = f$. □

In the general case the infimum J_m in (10) can be computed by associating, with each $\mathbf{H}_\infty$-function $f - rb$, the corresponding continuous linear operator T_{f-rb} mapping $\mathbf{H}_2$ into itself, as defined in (6.2.9). Since $\| T_h \| = \| h \|_\infty \; \forall h \in \mathbf{H}_\infty$, it follows that

23 $$J_m = \inf_{r \in \mathbf{H}_\infty} \| T_{f-rb} \| .$$

Let $b\mathbf{H}_2$ denote the range of the operator T_b; i.e.

24 $$b\mathbf{H}_2 = \{bg: g \in \mathbf{H}_2\},$$

and let $\mathbf{N}$ denote the orthogonal complement of $b\mathbf{H}_2$. Thus

25 $$\mathbf{N} = \{h \in \mathbf{H}_2: \langle h, bg \rangle = 0 \; \forall g \in \mathbf{H}_2\}.$$

Finally, let Π denote the orthogonal projection mapping $\mathbf{H}_2$ onto $\mathbf{N}$. Now, as r varies over $\mathbf{H}_\infty$, the operator T_{f-rb} ranges over various operators. However, the operator ΠT_{f-rb} is the same for all r, and in fact equals ΠT_f. To see this, let $g \in \mathbf{H}_2$ be arbitrary; then

26 $$\Pi T_{f-rb} g = \Pi[(f - rb)g] = \Pi(fg) - \Pi(rbg)$$
$$= \Pi(fg) - \Pi(brg) = \Pi(fg) = \Pi T_f g,$$

since $brg \in b\mathbf{H}_2$ and hence its projection onto $\mathbf{N}$ is zero. Using this idea, one can prove the following result (see Sarason [82] for the proof):

27 *Fact* Define a linear map $A: \mathbf{N} \rightarrow \mathbf{N}$ to be the restriction of ΠT_f to $\mathbf{N}$. Then

28 $$\inf_{r \in \mathbf{H}_\infty} \| f - rb \|_\infty = \| A \| .$$

Moreover, if $f(e^{j\theta})$ is a continuous function of θ, then there is a unique function $r_0 \in \mathbf{H}_\infty$ that attains the infimum in (30). Let $f_0 = f - r_0 b$, and let $g \in \mathbf{N}$ be a function such that $\| Ag \|_2 = \| g \|_2$. Then $f_0 / \| A \|$ is inner, and

29 $$f_0(z) = \frac{(Ag)(z)}{g(z)} .$$

Fact (27) is very useful because it is possible to compute all the quantities involved *using only linear algebra.* First of all, the subspace $\mathbf{N}$ is finite-dimensional, and in fact its dimension equals the number of zeros of the Blaschke product b. Next, one can readily find a basis for $\mathbf{N}$, which can be converted to an orthonormal basis using the Gram-Schmidt procedure. Let $\{v_1, \cdots, v_n\}$ denote this orthonormal basis. Then one can compute a *matrix representation* of the linear operator A with respect to this basis; call it A_0. Now (28) states that the infimum of $\| f - rb \|_\infty$ is precisely the largest singular value of A_0. Further, if $p = [p_1 \cdots p_n]'$ denotes the (column) eigenvector of $A_0{}^* A_0$ corresponding to the eigenvalue $(\bar{\sigma}(A_0))^2$, then

30 $$f_0(z) = \frac{\sum_{i=1}^{n} (A_0 p)_i v_i(z)}{\sum_{i=1}^{n} p_i v_i(z)} .$$

Let us elaborate each of these points in turn.

31 *Lemma* Let b be a finite Blaschke product, and define $\mathbf{N}$ as in (25). Let $\{\lambda_1, \cdots, \lambda_n\}$ denote the distinct zeros of b, with multiplicities $m_1, \cdots, m_n$, respectively. Then a (nonorthonormal) basis for $\mathbf{N}$ is given by the functions

32 $$\left\{ \frac{z^{k-1}}{(1-\bar{\lambda}_i z)^k} \right\}, k = 1, \cdots, m_i; i = 1, \cdots, n.$$

The proof follows readily from the following lemma.

33 *Lemma* Suppose λ is a complex number with $|\lambda| < 1$, k is an integer greater than or equal to 1, and define

34 $$\phi(z) = \frac{z^{k-1}}{(1-\bar{\lambda}z)^k}.$$

Then $\phi \in \mathbf{H}_2$; moreover, for any $g \in \mathbf{H}_2$,

35 $$<\phi, g> = \frac{g^{(k-1)}(\lambda)}{(k-1)!}.$$

Proof Clearly $\phi \in \mathbf{H}_2$, since its only singularity is at $1/\bar{\lambda}$. By definition (6.2.5), we have

36 $$\begin{aligned} <\phi, g> &= \frac{1}{2\pi} \int_0^{2\pi} \bar{\phi}(e^{j\theta}) g(e^{j\theta}) d\theta \\ &= \frac{1}{2\pi} \int_0^{2\pi} \frac{e^{-j(k-1)\theta}}{(1-\lambda e^{-j\theta})^k} g(e^{j\theta}) d\theta \\ &= \frac{1}{2\pi} \int_0^{2\pi} \frac{e^{j\theta}}{(e^{j\theta}-\lambda)^k} g(e^{j\theta}) d\theta \\ &= \frac{1}{2\pi j} \int_C \frac{g(z)}{(z-\lambda)^k} dz, \end{aligned}$$

where C denotes the unit circle oriented counterclockwise (note that $z = e^{j\theta}$ on C, so that $dz = je^{j\theta} d\theta$). Finally, by the Cauchy integral formula, the last term in (36) equals $g^{(k-1)}(\lambda)/(k-1)!$. □

Proof of Lemma (31) The subspace $\mathbf{N}$ is defined by (25). Let $g \in \mathbf{H}_2$ be arbitrary. Then $bg \in \mathbf{H}_2$ has a zero of multiplicity m_i at λ_i, for $i = 1, \cdots, n$. Hence, if ϕ_{i_k} denotes the function

37 $$\phi_{i_k}(z) = \frac{z^{k-1}}{(1-\bar{\lambda}_i z)^k},$$

then, by Lemma (33),

38 $$<\phi_{i_k}, bg> = \frac{(bg)^{(k-1)}(\lambda_i)}{(k-1)!} = 0$$

whenever $1 \le k \le m_i$. Thus each of the functions ϕ_{i_k} belongs to $\mathbf{N}$. It is routine to verify that they are linearly independent. Finally, in order to show that they span $\mathbf{N}$, it is enough to show that the orthogonal complement of the span of the ϕ_{i_k}'s is contained in $b\mathbf{H}_2$. Accordingly, suppose $h \in \mathbf{H}_2$ is orthogonal to each ϕ_{i_k}. Then, by Lemma (33), h has a zero of multiplicity m_i at λ_i for all i, and is hence of the form bg for some $g \in \mathbf{H}_2$. □

Once a basis has been found for $\mathbf{N}$, it is easy to construct an orthonormal basis using the Gram-Schmidt procedure. Let us drop the double subscript notation, let n denote the dimension of the null space $\mathbf{N}$, and let $\{\phi_1, \cdots, \phi_n\}$ denote the basis for $\mathbf{N}$ given by Lemma (31). Let $\Gamma \in C^{n \times n}$ denote the associated Grammian matrix; i.e.

39 $$\gamma_{ij} = <\phi_i, \phi_j>.$$

Note that the constants γ_{ij} can be readily computed using Lemma (33). For instance, if b has only simple poles at $\lambda_1, \cdots, \lambda_n$, then Lemma (31) shows that the set

40 $$\phi_i(z) = \frac{1}{1-\bar{\lambda}_i z}$$

is a basis for $\mathbf{N}$. Next, by Lemma (33),

41 $$\gamma_{ij} = <\phi_i, \phi_j> = \phi_j(\lambda_i) = \frac{1}{1-\bar{\lambda}_j \lambda_i}.$$

Since Γ is the Grammian of a linearly independent set, it is Hermitian and positive definite. Hence it is possible to factor Γ in the form B^*B where B is nonsingular. Let $X = B^{-1}$, and define

42 $$v_j(z) = \sum_{i=1}^{n} \phi_i(z)\, x_{ij},\ j = 1, \cdots, n.$$

Then $\{v_1, \cdots, v_n\}$ is an orthonormal basis for $\mathbf{N}$. If the factor B is selected to be upper triangular, then this is just the Gram-Schmidt procedure.

The next step is to obtain a matrix representation A_0 for the operator $A = \Pi T_f$ with respect to the basis $\{v_i\}$. This is given by

43 $$a_{ij} = \langle v_i, A v_j \rangle = \langle v_i, T_f v_j \rangle = \langle v_i, f v_j \rangle .$$

Now, by Lemma (33), a quantity of the form $\langle \phi_i, f \phi_j \rangle$ is very easy to compute. Hence it is desirable to expand a_{ij} as a summation of terms of the form $\langle \phi_i, f \phi_j \rangle$ using (42). Let Φ denote the matrix in $C^{n \times n}$ whose ij-th element is $\langle \phi_i, f \phi_j \rangle$. Then routine computations show that

44 $$A_0 = X^* \Phi X.$$

Now Fact (27) asserts that the infimum J_m in (10) equals $\| A_0 \|$. Further, if $p \in C^n$ is a column vector such that $\| A_0 p \|_2 = \| p \|_2$, then the optimal function f_0 is given by (30).

Finally, as is the case with the filtering problem, the solution of the sensitivity minimization problem is a lot cleaner in the discrete-time case than in the continuous-time case. If $p(z)$ is the transfer function of a strictly-causal plant, then $p(0) = 0$. If, in addition, $p(\cdot)$ has no poles nor zeros on the unit circle, and the weighting function $w(\cdot)$ has no zeros on the unit circle, then an optimal controller always exists.

This section is concluded with an illustrative example.

45 *Example* Consider once again the plant of Example (19), namely

$$p(s) = \frac{s-1}{(s-3)(s^2+1)},$$

together with the weighting function

$$w(s) = \frac{1}{s+1}.$$

By previous discussion, the sensitivity minimization problem reduces to finding an $r \in \mathbf{H}_\infty$ that minimizes $\| f - rb \|_\infty$, where f and b are computed in Example (19). The first objective of this example is to show that it is unnecessary to compute f and b explicitly. Now b is the Blaschke product part of $\tilde{n}\tilde{d}\tilde{w}$. Hence, to form b, it is only necessary to know the locations of the zeros of $\tilde{n},\tilde{d},\tilde{w}$ inside the *open* unit disc. In the s-domain, this translates to knowing the open RHP zeros and poles of the plant, as well as the open RHP zeros of the weighting function. Moreover, as stated earlier, it can be assumed without loss of generality that the weighting function is outer, so that it has no open RHP zeros. In the problem at hand, p has a zero at $s = 1$, and a pole at $s = 3$. Under the bilinear transformation $z = (s-1)/(s+1)$, these points correspond to $z = 0, 0.5$, respectively. Hence

$$b(z) = \frac{z(z-0.5)}{1-0.5z}.$$

Next, in order to apply Fact (27) to compute the optimal function f_0, it is only necessary to know the values of f (and its derivatives in case b has repeated zeros) at the zeros of b. Since b is the inner part of $\tilde{n}\tilde{d}\tilde{w}$ and $f = \tilde{y}\tilde{d}\tilde{w}$, it follows that if z is a zero of b, then $f(z) = \tilde{w}(z)$ or 0, depending on whether z is a zero of $\tilde{n}$ (i.e. a plant zero) or a zero of $\tilde{d}$ (i.e. a plant pole). In the present instance, $f(0) = \tilde{w}(0) = w(1) = 0.5$, $f(0.5) = 0$. All of this discussion is made more precise in the next section, where it is shown that the sensitivity minimization problem is one of interpolation by a function of minimum norm.

To compute the optimal function f_0, we first find a basis for the null space $\mathbf{N}$ of (25). By Lemma (31), such a basis is given by

$$\phi_1(z) = 1, \; \phi_2(z) = \frac{1}{1-0.5z}.$$

From (41), the Grammian matrix is

$$\Gamma = \begin{bmatrix} 1 & 1 \\ 1 & 4/3 \end{bmatrix},$$

which can be factored as

$$\Gamma = \begin{bmatrix} 1 & 0 \\ 1 & 1/\sqrt{3} \end{bmatrix} \begin{bmatrix} 1 & 1 \\ 0 & 1/\sqrt{3} \end{bmatrix} = B'B.$$

Hence

$$X = B^{-1} = \begin{bmatrix} 1 & -\sqrt{3} \\ 0 & \sqrt{3} \end{bmatrix},$$

and an orthonormal basis for **N** is given by

$$v_1(z) = \phi_1(z) = 1,\ v_2(z) = \sqrt{3}(\phi_2(z) - \phi_1(z)) = \sqrt{3}\frac{z}{2-z}$$

Next we need to compute $A_0 = X'\Phi X$ where $\phi_{ij} = <\phi_i, f\phi_j>$. From Lemma (33),

$$<\phi_1, f\phi_1> = (f\phi_1)(0) = \tilde{w}(0) = 0.5,$$

$$<\phi_1, f\phi_2> = (f\phi_2)(0) = 0.5,$$

$$<\phi_2, f\phi_1> = (f\phi_1)(0.5) = 0,$$

$$<\phi_2, f\phi_2> = (f\phi_2)(0.5) = 0.$$

Therefore

$$A_0 = X'\Phi X = \begin{bmatrix} 0.5 & 0 \\ -0.5\sqrt{3} & 0 \end{bmatrix}.$$

It is routine to compute that $\overline{\sigma}(A_0) = 1$, and that the associated vector p is $[10]'$, which leads to $A_0 p = [0.5 \ \ -0.5\sqrt{3}]'$. Hence by (30) the optimal function f_0 is given by

$$f_0(z) = \frac{0.5v_1(z) - 0.5\sqrt{3}v_2(z)}{v_1(z)}$$

$$= \frac{1-2z}{2-z} .$$

6.5 SENSITIVITY MINIMIZATION: FAT PLANT CASE

The sensitivity minimization problem for multivariable systems is much deeper than the corresponding problem for scalar systems. At present, a clean solution to the problem is available only in the case where the plant has at least as many inputs as outputs. Not all of the details of this solution are given in this section, as they require some background concepts from functional analysis not covered previously.

Suppose P is a given strictly proper plant with at least as many inputs as outputs, and that $W_1, W_2 \in \mathbf{M}(\mathbf{S})$ are weighting matrices such that W_1 has full row rank (as a rational matrix) and W_2 is square and nonsingular. The objective is to minimize the weighted sensitivity measure

1 $$J = \| W_1(I+PC)^{-1}W_2 \|_\infty .$$

Let (N,D), $(\tilde{D},\tilde{N})$ be any r.c.f. and any l.c.f. of P, and let $\tilde{X},\tilde{Y} \in \mathbf{M}(\mathbf{S})$ be any solutions of $\tilde{N}\tilde{X} + \tilde{D}\tilde{Y} = I$. Then, as shown in Section 6.2, one can replace (1) by the equivalent problem of minimizing

2 $$J(R) = \| W_1\tilde{Y}\tilde{D}W_2 - W_1NR\tilde{D}W_2 \|_\infty$$

as the "free" parameter R varies over $\mathbf{M}(\mathbf{S})$. Now define rational $\mathbf{H}_\infty$-matrices F, G, H by

3 $$F(z) = (W_1\tilde{Y}\tilde{D}W_2)[(1+z)/(1-z)],$$

4 $$G(z) = (W_1N)[(1+z)/(1-z)],$$

5 $$H(z) = (\tilde{D}W_2)[(1+z)/(1-z)].$$

Then the problem becomes one of minimizing

6 $$J(R) = \| F - GRH \|_{\infty}$$

by suitably choosing R from the set of rational $\mathbf{H}_{\infty}$-matrices.

Since the plant is strictly proper, we have that $G(1) = 0$. As a result, the infimum of J is not attained in general. To deal with this situation, it is necessary to extend Lemma (6.4.11) to the matrix case. For this purpose, define

7 $$J_0 = \inf_{R \in \mathbf{M}(\mathbf{H}_{\infty})} \| F - GRH \|_{\infty}.$$

Suppose G and H have the following properties: G is of the form $G_1 g$ where $G_1 \in \mathbf{M}(\mathbf{H}_{\infty})$, $g \in \mathbf{H}_{\infty}$, and in addition $G_1(e^{j\theta})$ has full row rank at all $\theta \in [0, 2\pi]$, and all zeros of $g(\cdot)$ are on the unit circle. Similarly, H is of the form $H_1 h$ where $H_1 \in \mathbf{M}(\mathbf{H}_{\infty})$, $h \in \mathbf{H}_{\infty}$, and in addition $H_1(e^{j\theta})$ is square and nonsingular at all $\theta \in [0, 2\pi]$, and all zeros of $h(\cdot)$ are on the unit circle. This means in particular that if $G(e^{j\theta})$ has less than full rank for some θ, then $G(e^{j\theta})$ must in fact equal the zero matrix; similar remarks apply to H. Note that the assumptions on G and H are trivially satisfied in the scalar case. Now define

8 $$J_m = \inf_{R \in \mathbf{M}(\mathbf{H}_{\infty})} \| F - G_1 R H_1 \|_{\infty}.$$

It can be shown that the infimum in (8) *is* achieved, owing to the rank conditions on G_1 and H_1.

The next result is a generalization of Lemma (6.4.11).

9 *Lemma* Let $e^{j\theta_1}, \cdots, e^{j\theta_l}$ denote the distinct zeros of the product gh on the unit circle. Then $J_0 = J_m$ if and only if $\| F(e^{j\theta_i}) \| \leq J_m \ \forall i$.

Proof Since every matrix of the form GRH is also of the form $G_1 R H_1$, it is easy to see that $J_m \leq J_0$.

"only if" Since $gh(e^{j\theta_i}) = 0 \ \forall i$, it follows that $GRH(e^{j\theta_i}) = 0 \ \forall i$, $\forall R \in \mathbf{M}(\mathbf{H}_{\infty})$. Hence $(F - GRH)(e^{j\theta_i}) = F(e^{j\theta_i}) \ \forall i$, $\forall R \in \mathbf{M}(\mathbf{H}_{\infty})$, and as a result

10 $$\| F-GRH \|_\infty = \sup_{\theta\in[0,2\pi]} \| (F-GRH)(e^{j\theta}) \| \geq \max_i \| (F-GRH)(e^{j\theta_i}) \| = \max_i \| F(e^{j\theta_i}) \|, \ \forall R\in\mathbf{M}(\mathbf{H}_\infty).$$

Combining (10) with the hypothesis that $J_0 = J_m$ leads to

11 $$J_m = J_0 = \inf_{R\in\mathbf{M}(\mathbf{H}_\infty)} \| F-GRH \|_\infty \geq \max_i \| F(e^{j\theta_i}) \|.$$

"if" Select an $R_0\in\mathbf{M}(\mathbf{H}_\infty)$ such that $J_m = \| F-G_1R_0H_1 \|_\infty$. Such an R_0 exists, by earlier remarks. Now recall the function (cf. (6.4.14))

12 $$h_{\theta,\varepsilon}(z) = \frac{z-e^{j\theta}}{z-(1+\varepsilon)e^{j\theta}},$$

and define

13 $$h_\varepsilon(z) = \prod_{i=1}^{l} [h_{\theta_i,\varepsilon}(z)]^{m_i},$$

where m_i is the multiplicity of $e^{j\theta_i}$ as a zero of gh. Define the family (R_ε) in $\mathbf{M}(\mathbf{H}_\infty)$ by $R_\varepsilon = R_0 h_\varepsilon / gh$. Then $F-GR_\varepsilon H = F-G_1R_0H_1h_\varepsilon$. Moreover, as in the proof of Lemma (6.4.11), it will follow that $\| F-G_1R_0H_1h_\varepsilon \|_\infty \to \| F-G_1R_0H_1 \|_\infty = J_m$ as $\varepsilon\to 0^+$. Hence $J_0 = J_m$. □

To interpret Lemma (9) in the context of the sensitivity minimization problem, suppose the plant P has blocking zeros[8] at $s = \pm j\omega_1, \cdots, \pm j\omega_n$ and at infinity, but has full row rank at all other points on the $j\omega$-axis. Suppose also that P has no $j\omega$-axis poles. Select the weighting matrices W_1, W_2 such that W_1 has the same blocking zeros as P on the extended $j\omega$-axis, but has full row rank at all other points on the $j\omega$-axis, while $W_2(j\omega)$ is square and nonsingular at all points on the extended $j\omega$-axis. Then $F(e^{j\theta_i}) = 0 \ \forall i$ and the condition $\| F(e^{j\theta_i}) \| \leq J_m$ is automatically

[8] Recall that s is a *blocking zero* of P if $P(s) = 0$.

satisfied; as a result, $J_0 = J_m$. Moreover, if R_0 is a choice of R that attains the infimum in (8), then the family $(R_\varepsilon = R_0 h_\varepsilon / gh)$ is an optimizing family for $J(\cdot)$. This leads to a family of compensators (C_ε) for P such that $\| W_1(I+PC_\varepsilon)^{-1} W_2 \|_\infty \rightarrow J_m$ as $\varepsilon \rightarrow 0^+$.

Let us therefore turn our attention to the problem of computing the infimum in (8) and an R that attains this infimum. As a first step, factorize G_1 and H_1 in the form $G_1 = G_i G_o$, $H_1 = H_o H_i$, where G_i, H_i are inner, $G_o \in \mathbf{M}(\mathbf{H}_\infty)$ has a right inverse, and $H_o \in \mathbf{M}(\mathbf{H}_\infty)$ is a unit. This is possible in view of the rank conditions on G_1 and H_1 (see Fact (6.2.17)). Now, for any $V \in \mathbf{M}(\mathbf{H}_\infty)$, the equation $V = G_o R H_o$ has a solution for R in $\mathbf{M}(\mathbf{H}_\infty)$. Hence

14 $$\inf_{R \in \mathbf{M}(\mathbf{H}_\infty)} \| F - G_1 R H_1 \|_\infty = \inf_{V \in \mathbf{M}(\mathbf{H}_\infty)} \| F - G_i V H_i \|_\infty .$$

Next, let G_i^{adj}, H_i^{adj} denote the adjoint matrices of G_i, H_i, respectively, and note that these adjoint matrices are themselves inner. Let $b = |G_i| \cdot |H_i|$, and note that b is an inner function. Now, since multiplication by inner matrices is norm-preserving, it follows that

15 $$\| F - G_i V H_i \|_\infty = \| G_i^{adj} F H_i^{adj} - G_i^{adj} G_i V H_i H_i^{adj} \|_\infty$$
$$= \| F_0 - bV \|_\infty, \ \forall V \in \mathbf{M}(\mathbf{H}_\infty),$$

where $F_0 = G_i^{adj} F H_i^{adj}$. Hence, from (8), (14) and (15),

16 $$J_m = \inf_{V \in \mathbf{M}(\mathbf{H}_\infty)} \| F_0 - bV \|_\infty .$$

As a final simplification, we may as well assume that b is a (finite) Blaschke product: Since b is inner, it equals a constant times a Blaschke product; the constant can be absorbed into the "free" matrix V.

Let us now switch notation and restate the problem. Given a rational matrix $F \in \mathbf{M}(\mathbf{H}_\infty)$ and a finite Blaschke product b, find an $R \in \mathbf{M}(\mathbf{H}_\infty)$ that minimizes $\| F - bR \|_\infty$. There are two aspects to this problem: (i) computing the minimum, and (ii) finding an R that attains the minimum. Each aspect is discussed separately.

Backtracking through the various manipulations, the reader can verify that the matrices F and R have the same dimensions, and that they have at least as many columns as rows. To be specific, suppose they are of dimension $r \times s$, where $r \leq s$. Recall from Section 6.2 that $\mathbf{H}_2^s$ is a Hilbert space under the inner product

17 $$<f,g> = \frac{1}{2\pi} \int_0^{2\pi} f^*(e^{j\theta}) g(e^{j\theta}) d\theta.$$

As in the scalar case, with each matrix $H \in \mathbf{H}_\infty^{s \times s}$ one can associate a continuous linear operator $T_H : \mathbf{H}_2^s \rightarrow \mathbf{H}_2^s$ in the obvious way, namely:

18 $$(T_H f)(z) = H(z) f(z) \ \forall z \in \mathbf{D}, f \in \mathbf{H}_2^s.$$

Moreover, this association is norm-preserving; that is,

19 $$\| T_H \| = \sup_{f \in \mathbf{H}_2^s - 0} \frac{\| T_H f \|_2}{\| f \|_2}$$

$$= \| H \|_\infty = \text{ess. sup}_{\theta \in [0, 2\pi]} \bar{\sigma}(H(e^{j\theta})).$$

It turns out that Fact (6.4.27) has a complete analog in the multivariable case in so far as computing the infimum of $\| F - bR \|_\infty$. This analog is based on a preliminary result taken from [S2]. Some notation smoothens the way to a presentation of the latter. Let

20 $$b\mathbf{H}_2^s = \{ bf : f \in \mathbf{H}_2^s \},$$

and let $\mathbf{N}$ denote the orthogonal complement of $b\mathbf{H}_2^s$ in $\mathbf{H}_2^s$; thus

21 $$\mathbf{N} = \{ g \in \mathbf{H}_2^s : <g, bf> = 0 \ \forall f \in \mathbf{H}_2^s \}.$$

Finally, let Π denote the orthogonal projection mapping $\mathbf{H}_2^s$ onto $\mathbf{N}$.

22 *Lemma* For each $\Phi \in \mathbf{H}_\infty^{s \times s}$, let Λ_Φ denote the map ΠT_Φ form $\mathbf{H}_2^s$ into $\mathbf{N}$. Let Λ denote the collection $\{ \Lambda_\Phi : \Phi \in \mathbf{H}_\infty^{s \times s} \}$. Then, corresponding to each map M in the collection Λ, there exists a matrix $\Phi \in \mathbf{H}_\infty^{s \times s}$ such that $\| M \| = \| \Phi \|_\infty$ and $\Lambda_\Phi = M$.

Proof See [S2, Theorem 3].

23 *Theorem* Suppose $F \in \mathbf{H}_\infty^{r \times s}$ with $r \leq s$. If $r < s$, define

24 $$F_0 = \begin{bmatrix} F \\ 0_{r-s \times s} \end{bmatrix}$$

so that $F_0 \in \mathbf{H}_\infty^{s \times s}$. Define a linear mapping $A : \mathbf{N} \rightarrow \mathbf{N}$ as the restriction of ΠT_{F_0} to $\mathbf{N}$. Then

25 $$\inf_{R \in \mathbf{H}_\infty^{s \times s}} \| F - bR \|_\infty = \| A \|.$$

Remarks As in the scalar case, the subspace $\mathbf{N}$ is finite-dimensional, so that $\| A \|$ can be computed by purely algebraic means. More on this later.

Proof As a first step, suppose $r = s$ so that F is square. As R varies over $\mathbf{H}_\infty^{s \times s}$, the operatoar T_{F-bR} ranges over various operators. However, the operator ΠT_{F-bR} is the same for all R, and equals ΠT_F. To see this, let $g \in \mathbf{H}_2^s$ be arbitrary. Then

26 $$\Pi T_{F-bR}\, g = \Pi(F-bR)g = \Pi(Fg) - \Pi(bRg) = \Pi T_F g,$$

since $bRg \in \mathbf{H}_2^s$ and hence its projection onto $\mathbf{N}$ is zero. Now ΠT_F is what is called Λ_F in Lemma (22), which states that there exists a matrix $\Phi \in \mathbf{H}_\infty^{s \times s}$ such that $\Lambda_\Phi = \Lambda_F$ and such that $\| \Phi \|_\infty = \| \Lambda_F \|$ $(= \| \Pi T_F \|)$. But $\Lambda_\Phi = \Lambda_F$ is equivalent to $\Pi(F - \Phi) = 0$, i.e.

27 $$(F - \Phi) g \in \mathbf{H}_2^s \ \forall g \in \mathbf{H}_2^s.$$

Let e_i denote the i-th elementary unit vector, which has a "1" in the i-th position and zeros elsewhere, and let

28 $$g(z) = e_i, \ \forall z \in \mathbf{D}.$$

Then $g \in \mathbf{H}_\infty^s$ (which also implies that $g \in \mathbf{H}_2^s$); moreover, (27) now implies that the i-th column of the matrix $F - \Phi$ is a multiple (in $\mathbf{H}_\infty$) of the inner function b. Repeating this reasoning for all indices i shows that $F - \Phi \in b\mathbf{H}_\infty^{s \times s}$, i.e. that $\Phi = F - bR_0$ for some $R_0 \in \mathbf{H}_\infty^{s \times s}$. It is now claimed that

29 $$\| F - bR_0 \|_\infty = \inf_{R \in \mathbf{H}_\infty^{s \times s}} \| F - bR \|_\infty .$$

To prove (29), note that

30 $$\| F - bR \|_\infty = \| T_{F-bR} \| \text{ by (19)},$$

$$\geq \| \Pi T_{F-bR} \| \text{ since } \| \Pi \| \leq 1,$$

$$= \| \Pi T_F \| \text{, by (26)},$$

$$= \| \Phi \|_\infty = \| F - bR_0 \|_\infty .$$

The preceding discussion can be summarized as follows: If F is square, then

31 $$\inf_{R \in \mathbf{H}_\infty^{s \times s}} \| F - bR \|_\infty = \| \Pi T_F \| .$$

To complete the proof of the theorem, it only remains to show that ΠT_F restricted to $\mathbf{N}$ has the same norm as ΠT_F. Suppose $g \in \mathbf{H}_2^s$. Then g can be expressed uniquely in the form $f + bh$ where $f \in \mathbf{N}$ and $h \in \mathbf{H}_2^s$ (since $\mathbf{H}_2^s$ is the direct sum of $\mathbf{N}$ and $b\mathbf{H}_2^s$). Now[9]

32 $$\Pi T_F g = \Pi(Ff + Fbh) = \Pi(Ff + bFh) = \Pi Ff.$$

Moreover, $\| g \|_2 \geq \| f \|_2$, so that

33 $$\frac{\| \Pi F g \|_2}{\| g \|_2} \leq \frac{\| \Pi F f \|_2}{\| f \|_2} .$$

Hence A (which is the restriction of ΠT_F to $\mathbf{N}$) has the same norm as ΠT_F.

Next, the restriction that F be square is removed. Suppose $r < s$ and define F_0 as in (24). Partition any $R_0 \in \mathbf{H}_\infty^{s \times s}$ as $R_0 = [R' \ \ R'_1]'$ where $R \in \mathbf{H}_\infty^{r \times s}$. Then, for any $R_0 \in \mathbf{H}_\infty^{s \times s}$, we have

[9] Note that this argument fails if b is replaced by an inner *matrix*.

34 $$\| F_0 - bR_0 \|_\infty = \| \begin{bmatrix} F-bR \\ -bR_1 \end{bmatrix} \|_\infty$$

$$= \text{ess.} \sup_{\theta \in [0,2\pi]} \bar{\sigma} \begin{bmatrix} (F-bR)(e^{j\theta}) \\ -bR_1(e^{j\theta}) \end{bmatrix}.$$

Now for any two complex matrices A, B of compatible dimensions, it is easy to show that

35 $$\bar{\sigma} \begin{bmatrix} A \\ B \end{bmatrix} \geq \bar{\sigma}(A).$$

Hence, from (34), $\| F_0 - bR_0 \|_\infty \geq \| F - bR \|_\infty$, so that

36 $$\inf_{R_0 \in \mathbf{H}_\infty^{s \times s}} \| F_0 - bR_0 \|_\infty \geq \inf_{R \in \mathbf{H}_\infty^{r \times s}} \| F - bR \|_\infty.$$

To prove the reverse inequality, let R be any matrix in $\mathbf{H}_\infty^{s \times s}$ that attains the infimum on the right side of (36), and define $R_0 = [R' \;\; 0_{s \times r-s}]'$. Then $\| F_0 - bR_0 \|_\infty = \| F - bR \|_\infty$. Hence the inequality in (36) is actually an equality. Moreover, from the previous paragraph, the infimum on the left side of (36) equals the norm of ΠT_{F_0} when restricted to $\mathbf{N}$. □

As in the scalar case, it is easy to compute a basis for $\mathbf{N}$.

37 *Lemma* Let b be a finite Blaschke product with distinct zeros $\lambda_1, \cdots, \lambda_n$, of multiplicities $m_1, \cdots, m_n$, respectively. For $i \in \{1, \cdots, s\}$, let e_i denote the i-th $s \times 1$ elementary unit vector. Define $\mathbf{N}$ as in (21). Then a (nonorthonormal) basis for $\mathbf{N}$ is given by the vector-valued functions

38 $$v_{kl}^i(z) = \frac{z^{k-1}}{1-\bar{\lambda}_l z} \cdot e_i, \; 1 \leq i \leq s; \; 1 \leq k \leq m_l; \; 1 \leq l \leq n.$$

Proof By Lemma (6.4.33) and (17),

39 $$\langle v_{kl}^i, f \rangle = f_i^{(k-1)}(z_l), \; \forall f \in \mathbf{H}_2^s.$$

Since $f \in b\mathbf{H}_2^s$ if and only if the right side of (39) is zero for all i, k, l, the lemma follows. □

Once a basis for $\mathbf{N}$ has been determined, it is possible to convert it into an orthonormal basis, compute a matrix representation of A_0 of A with respect to this basis, and finally to compute $\overline{\sigma}(A_0)$; the details are exactly the same as in the scalar case (see (6.4.39) *et. seq.*). However, if one is only interested in computing $\overline{\sigma}(A_0)$, a short-cut is available, and is described next.

It should be emphasized that the procedure described next is equally applicable to the scalar case, and can be used to compute $\overline{\sigma}(A_0)$ more readily than by using the formula (6.4.44). HOwevr, there would be no advantage gained by this: In the scalar case, Fact (6.4.27) gives a method for computing the infimum of $\| f - br \|_\infty$ *as well as for finding an* r that attains the infimum, all in one shot. In contrast, the procedure given below only determines the infimum, and a separate interpolation procedure has to be used to find an optimizing r. This is no disadvantage in the multivariable case, however, since such a two-stage procedure is all that is available at present.

Now for the short-cut: Theorem (23) states that the infimum of $\| F - bR \|_\infty$ equals $\| A \|$, where $A: \mathbf{N} \rightarrow \mathbf{N}$ is the restriction of ΠT_F to $\mathbf{N}$. Lemma (37) provides a (nonorthonormal) basis for $\mathbf{N}$. To compute $\| A \|$, the first step is to observe that $\| A \| = \| A^* \|$, where $A^*: \mathbf{N} \rightarrow \mathbf{N}$ is the adjoint map of A. Now compute a representation of A^* with respect to the basis given in Lemma (37), and denote it by $\mathbf{\Phi}$. Also, let Γ denote the Grammian matrix of the basis in Lemma (37). Then it is easy to show that

40 $$\| A \| = \inf \{ \lambda : \lambda \Gamma - \mathbf{\Phi}^* \mathbf{\Phi} \text{ is positive definite} \}.$$

Alternately, $\| A \|$ is the largest solution to the generalized eigenvalue problem $| \lambda \Gamma - \mathbf{\Phi}^* \mathbf{\Phi} | = 0$.

The effectiveness of the above short-cut derives from the fact that the matrix representation $\mathbf{\Phi}$ is block-diagonal, as shown in Lemma (41) below. For the sake of clarity of notation, Lemma (41) is stated only for the case where the Blaschke product b has simple zeros; the general case is conceptually nor more difficult and is left to the reader.

41 *Lemma* Suppose b is a finite Blaschke product with distinct simple zeros $\lambda_1, \cdots, \lambda_m$, and that $F \in \mathbf{H}_\infty^{r \times s}$ is rational. Define $\Phi \in C^{ms \times mr}$, $\Gamma \in C^{mr \times mr}$ by

42 $$\Phi = \text{Block Diag } \{F^*(\lambda_1), \cdots, F^*(\lambda_m)\},$$

43 $$\Gamma = (\Gamma_{ij}),\ i,j = 1, \cdots, m,$$

where

44 $$\Gamma_{ij} = \frac{1}{1-\bar{\lambda}_j \lambda_i} \cdot I_r.$$

Then

45 $$\inf_{R \in \mathbf{H}_\infty^{s \times s}} \| F - bR \|_\infty = \inf \{\lambda : \lambda\Gamma - \Phi^*\Phi \text{ is positive definite}\}.$$

Proof Define $\mathbf{N}$ as in (21). Then $\mathbf{N}$ hs dimension ms, and a basis for $\mathbf{N}$ is given by Lemma (37) as

46 $$v_{ij}(z) = \frac{1}{1-\bar{\lambda}_i z} e_j,\ j = 1, \cdots, s;\ i = 1, \cdots, m.$$

Thus any $h \in \mathbf{N}$ can be expanded as

47 $$h(z) = \sum_{i=1}^{m} \sum_{j=1}^{s} h_{ij} v_{ij}(z).$$

Arrange the coordinates h_{ij} in an $ms \times 1$ column vector as follows:

48 $$\begin{bmatrix} h_{11} \\ \vdots \\ h_{1s} \\ \vdots \\ h_{m1} \\ \vdots \\ h_{ms} \end{bmatrix}.$$

Then the claim is that the matrix representation of A^* with respect to this basis is

49 $$\Phi_0 = \text{Block Diag } \{F_0(\lambda_1), \cdots, F_0(\lambda_m)\}.$$

To prove the claim, let v_{kl} be any basis function of the form (46). We must show that the matrix (49) gives the right answer for $A^* v_{kl}$. Now the coordinate vector of v_{kl} is $[0 \cdots 1 \cdots 0]'$, where the 1 is in the position $(k-1)s+l$. Applying the matrix Φ_0 to this vector gives $[0 \cdots F_0(\lambda_k)e_l \cdots 0]^*$ as the coordinate vector of $A^* v_{kl}$; in other words,

50 $$A^* v_{kl} = \sum_{p=1}^{s} f_{0lp}^*(\lambda_k) v_{kp}.$$

To show that (50) is correct, it is enough to show that it gives the right answer for the inner products $<v_{ij}, A^* v_{kl}>$ for all i, j. By definition,

51 $$\overline{<v_{ij}, A^* v_{kl}>} = <A^* v_{kl}, v_{ij}> = <v_{kl}, A v_{ij}>$$
$$= <v_{kl}, \Pi T_{F_0} v_{ij}> = <v_{kl}, F_0 v_{ij}>$$
$$= e_l^* F_0(\lambda_k) v_{ij}(\lambda_k)$$
$$= f_{0lj}(\lambda_k) \frac{1}{1-\bar{\lambda}_i \lambda_k},$$

where the last two steps follow from Lemma (6.4.33) and (46), respectively. On the other hand, letting u_{kl} denote the function in (50) leads to

52 $$\langle v_{ij}, u_{kl} \rangle = \sum_{p=1}^{s} f^{*}_{0lp}(\lambda_k) \langle v_{ij}, v_{kp} \rangle .$$

Note that

53 $$\langle v_{ij}, v_{kp} \rangle = \frac{1}{1-\bar{\lambda}_k \lambda_i} \delta_{jp},$$

where δ is the Kronecker delta. Hence (52) becomes

54 $$\langle v_{ij}, u_{kl} \rangle = f^{*}_{0lj}(\lambda_k) \cdot \frac{1}{1-\bar{\lambda}_k \lambda_j},$$

which agrees with (51).

By previous discussion, $\| A^* \|$ is given by

55 $$\| A^* \| = \inf \{\lambda : \lambda \Gamma_0 - \Phi_0{}^* \Phi_0 \text{ is positive definite} \},$$

where Γ_0 is the Grammian matrix of the basis in (46). One can readily compute that $\Gamma_0 = (\Gamma_{0ij})_{i,j=1}^{m}$ where

56 $$\Gamma_{0ij} = \frac{1}{1-\bar{\lambda}_j \lambda_i} \cdot I_s .$$

Now the only difference between Γ and Γ_0 is in their dimensions. Similarly, if we define Φ as in (42), then Φ is a submatrix of Φ_0, since F is a submatrix of F_0. If $r = s$ then we are done, since $\Gamma = \Gamma_0$, $\Phi = \Phi_0$ and (55) is the same as (45). Otherwise, by symmetric row and column permutations, one can transform $\lambda \Gamma_0 - \Phi_0{}^* \Phi_0$ into the form

57 $$\lambda \Gamma_0 \rightarrow \begin{bmatrix} \lambda \Psi & 0 \\ 0 & \lambda \Gamma - \Phi^* \Phi \end{bmatrix},$$

where $\Psi = (\Psi_{ij})_{i,j=1}^{m}$, and

58 $$\Psi_{ij} = \frac{1}{1-\bar{\lambda}_j \lambda_i} \cdot I_{s-r} .$$

It is clear from (57) that $\lambda \Gamma_0 - \Phi_0{}^* \Phi_0$ is positive definite if and only if

$\lambda\Gamma - \Phi^*\Phi$ is; hence (55) reduces to (45). □

Now that an expression is available for the infimum of $\| F - bR \|_\infty$, the next step is to give a procedure for computing an R that attains the infimum. In the case where the inner function b has only simple zeros, such a procedure is provided by the *matrix Nevanlinna-Pick algorithm.*[10] This algorithm is also applicable to the scalar problem studied in Section 6.4, and forms an alternative to the procedure of Fact (6.4.27). However, the latter is preferrable, since it leads to an optimal interpolating function in one step, where as the Nevanlinna-Pick algorithm is iterative, and requires as many iterations as there are zeros of b.

The main result in the sequel is Theorem (71), which overlaps somewhat with Lemma (41). However, an independent proof is given for this theorem, as the iterative interpolation algorithm is really a consequence of the method of proof.

To set up the procedure, define

59 $$\mu = \inf_{R \in \mathbf{H}_\infty^{r \times s}} \| F - bR \|_\infty,$$

and let $\lambda_1, \cdots, \lambda_n$ denote the (simple) zeros of b. Then the problem is to find a matrix $G \in \mathbf{H}_\infty^{r \times s}$ such that

60 $$G(\lambda_i) = F(\lambda_i),\ i = 1, \cdots, n, \text{ and } \| G \|_\infty = \mu.$$

(Note that $G \in \mathbf{H}_\infty^{r \times s}$ is of the form $F - bR$ for some $R \in \mathbf{H}_\infty^{s \times s}$ if and only if $G(\lambda_i) = F(\lambda_i)\ \forall i$.) Equivalently, if we define

61 $$F_i = [F(\lambda_i)]/\mu,$$

then the problem is one of finding a matrix $\Phi \in \mathbf{H}_\infty^{r \times s}$ such that $\| \Phi \|_\infty = 1$ and $\Phi(\lambda_i) = F_i\ \forall i$.

[10] Strictly speaking, the Pick problem only makes sense for square matrices, so it is probably more accurate to speak of the matrix Nevanlinna algorithm.

The remainder of this section is devoted to the matrix version of the so-called *Nevanlinna-Pick problem,* which can be stated as follows: Given complex numbers $\lambda_1, \cdots, \lambda_n$ with $|\lambda_i| < 1 \; \forall i$, together with complex matrices $F_1, \cdots, F_n$ with $\| F_i \| \leq 1 \; \forall i$, find all matrix-valued functions $\mathbf{\Phi}$ (if any) such that $\| \Phi \|_\infty \leq 1$ and $\Phi(\lambda_i) = F_i \; \forall i$. The relevance of this problem to sensitivity minimization is clear.

In the sequel, it is important note that $\| \cdot \|$ denotes the norm of a complex matrix whereas $\| \cdot \|_\infty$ denotes the norm of a matrix-valued function.

Before proceeding to the interpolation algorithm, it is established that one may assume that $\| F_i \| < 1 \; \forall i$, without any loss of generality. Consider first the sclar case, where one is given scalars $f_1, \cdots f_n$ with $|f_i| \leq 1 \; \forall i$, and is asked to find a $\phi \in \mathbf{H}_\infty$ such that $\| \phi \|_\infty \leq 1$ and $\phi(\lambda_i) = f_i \; \forall i$. Suppose one of the constants has magnitude exactly equal to 1, and renumber the constants if necessary so that $|f_1| = 1$. Since λ_1 belongs to the *open* unit disc, it follows from the maximum modulus principle that the only $\mathbf{H}_\infty$-function with norm 1 that equals f_1 at λ_1 is the *constant* function $\phi(z) \equiv 1 \; \forall z$. Hence the interpolation problem does not have a solution unless $f_i = f_1 \; \forall i$. In the latter case, the solution is the obvious one, namely $\phi(z) \equiv f_1$. Thus one may as well assume at the outset that $|f_i| < 1 \; \forall i$, because the problem is trivial otherwise. But this reasoning breaks down in the matrix case, because there exist nonconstant matrices $\mathbf{\Phi}$ such that $\Phi(z) \equiv 1 \; \forall z \in \mathbf{D}$; for example, let

62 $$\Phi(z) = \begin{bmatrix} 1 & 0 \\ 0 & z \end{bmatrix}.$$

However, Lemma (63) below shows that all such functions are essentially of the form (62). One can think of Lemma (63) as providing a matrix version of the maximum modulus principle.

63 *Lemma* Suppose $\Phi \in \mathbf{H}_\infty^{k \times l}$ and that $\| \Phi \|_\infty \leq 1$. Suppose $\| \Phi(z_0) \| = 1$ for some z_0 in the open unit disc, and let r denote the multiplicity of 1 as a singular value of $\Phi(z_0)$ (i.e. as an eigenvalue of $\Phi^*(z_0)\Phi(z_0)$). Then $\| \Phi(z) \| = 1 \; \forall z \in \mathbf{D}$. Moreover, there exist unitary

matrices $U \in C^{k \times k}$, $V \in C^{l \times l}$ such that

$$64 \qquad U\Phi(z)V = \begin{bmatrix} I_r & 0_{l-r \times r} \\ 0_{r \times k-r} & \Psi(z) \end{bmatrix},$$

where $\| \Psi(z) \| < 1 \ \forall z \in \mathbf{D}$.

The proof of Lemma (63) requires the following observation on complex matrices.

65 *Lemma* Suppose $M \in C^{k \times l}$ has norm 1. Then

(i) $| m_{ij} | \leq 1 \ \forall i,j$,

(ii) If $| m_{ij} | = 1$ for some i,j, then all other elements of row i and column j are zero.

Proof Temporarily let $\| \cdot \|_2$ denote the Euclidean vector norm on both C^k and C^l. Then

$$66 \qquad \| Mx \|_2 \leq \| x \|_2 \ \forall x \in C^l,$$

since $\| M \| = 1$. To prove (i), suppose $| m_{ij} | > 1$ for some i,j, and let $x = e_j$, the j-th elementary unit vector in C^l. Then $\| Mx \|_2 > 1$, which is a contradiction. To prove (ii), suppose $| m_{ij} | = 1$ for some i,j, and again let $x = e_j$. Then $\| Me_j \|_2 = 1$ implies that all other elements of the j-th column of M are zero. The same reasoning applied to M^* shows that all other elements of the i-th row of M are zero. □

Proof of Lemma (63) Suppose 1 is an r-fold eigenvalue of $\Phi^*(z_0)$, so that 1 is an r-fold singular value of $\Phi(z_0)$. Thus there exist unitary matrices $U \in C^{k \times k}$, $V \in C^{l \times l}$ such that

$$67 \qquad U\Phi(z_0)V = \begin{bmatrix} I_r & 0_{r \times l-r} \\ 0_{k-r \times r} & M \end{bmatrix},$$

where $\| M \| < 1$. Define $y(z) = (U\Phi(z)V)_{11}$, i.e. y is the corner element of $U\Phi(z)V$. Then y is analytic, satisfies $| y(z) | \leq 1 \ \forall z \in \mathbf{D}$ (by Lemma (65i)), and $y(z_0) = 1$ (by (67)). Hence, by the maximum modulus theorem, $y(z) = 1 \ \forall z \in \mathbf{D}$, since z_0 is in the open unit disc. Now by Lemma

(65ii), all other elements in the first row and first column of $U\Phi(z)V$ are identically zero. The same reasoning applies to *all* elements of the form $(U\Phi(z)V)_{ii}$ where $i \leq r$. Hence $U\Phi(z)V$ is of the form (64), and $\| \Phi(z) \| = 1 \; \forall z \in \mathbf{D}$. It only remains to show that $\| \Psi(z) \| < 1 \; \forall z \in \mathbf{D}$. Suppose to the contrary that $\| \Psi(z_1) \| = 1$ for some $z_1 \in \mathbf{D}$. Then, by previous reasoning, $\| \Psi(z) \| = 1 \; \forall z \in \mathbf{D}$. In particular, $\| M \| = \| \Psi(z_0) \| = 1$, which contradicts the assumption that 1 is an r-fold singular value of $\Phi(z0)$. Hence $\| \Psi(z) \| < 1 \; \forall z \in \mathbf{D}$. □

Let us now return to the interpolation problem, where one is given points $\lambda_1, \cdots, \lambda_n$ and matrices $F_1, \cdots, F_m$, with $| \lambda_i | < 1$ and $\| F_i \| \leq 1 \; \forall i$. Recall that the objective is to show that one may assume $\| F_i \| < 1 \; \forall i$ without sacrificing generality. Assume to the contrary that $\| F_i \| = 1$ for some values of i. Among these select one for which the multiplicity of 1 as a singular value is the largest, and renumber it as F_1. Select unitary matrices U, V such that

68 $$UF_1V = \begin{bmatrix} I_r & 0 \\ 0 & G_1 \end{bmatrix},$$

where r is the multiplicity of 1 as a singular value of F, and $\| G_1 \| < 1$. Now, by Lemma (63), any Φ such that $\| \Phi \|_\infty \leq 1$ and $\Phi(\lambda_i) = F_i$ must satisfy

69 $$U\Phi(z)V = \begin{bmatrix} I_r & 0 \\ 0 & \Psi(z) \end{bmatrix} \; \forall z \in \mathbf{D}.$$

In particular, $F_2, \cdots, F_n$ must satisfy

70 $$UF_iV = \begin{bmatrix} I_r & 0 \\ 0 & G_i \end{bmatrix}, \; i = 2, \cdots, n.$$

Otherwise the interpolation problem has no solution. On the other hand, suppose (70) holds. Then $\| G_i \| < 1 \; \forall i$; otherwise, if $\| G_i \| = 1$ for some i, then F_i has 1 as a singular value of multiplicity larger than r, contradicting the manner in which F_1 was selected. Now replace the original interpolation problem by the following modified problem: Find a Ψ such that $\| \Psi \|_\infty \leq 1$ and $\Psi(\lambda_i) = G_i \ \forall i$. Then there is a one-to-one correspondence between solutions Φ of the original problem and Ψ of the modified problem, via (69). But in the modified problem all the matrices G_i have norms strictly less than one.

We are now ready for the main result on the *matrix Nevanlinna problem.*

71 *Theorem* Suppose $\lambda_1, \cdots, \lambda_n$ are distinct complex numbers and $F_1, \cdots, F_n$ are complex matrices with $| \lambda_i | < 1$ and $\| F_i \| < 1 \ \forall i$. Define a partitioned matrix

72 $$P = \begin{bmatrix} P_{11} & \cdots & P_{1n} \\ \vdots & & \vdots \\ P_{n_1} & \cdots & P_{nn} \end{bmatrix},$$

where

73 $$P_{ij} = \frac{1}{1-\bar{\lambda}_i \lambda_j} \cdot (I - F_i^* F_j).$$

Then there exists a $\Phi \in \mathbf{M}(\mathbf{H}_\infty)$ such that $\| \Phi \|_\infty < 1$ and $\Phi(\lambda_i) = F_i \ \forall i$, if and only if the matrix P is positive definite.

In the current set-up Lemma (41) leads to the following result: There exists a $\Phi \in \mathbf{M}(\mathbf{H}_\infty)$ such that $\| \Phi \|_\infty \leq 1$ and $\Phi(\lambda_i) = F_i$, if and only if the matrix P is nonnegative definite.

The proof of Theorem (71) is rather long and is based on a preliminary lemma. We shall see that an itereative algorithm for computing Φ will emerge from the proof.

74 *Lemma* Suppose $\| E \| < 1$, and define

75 $$L(E) = \begin{bmatrix} A & B \\ C & D \end{bmatrix},$$

where

76 $$A = (I - EE^*)^{-1/2},\ B = -(I - EE^*)^{-1/2}E,$$
$$C = -(I - E^*E)^{-1/2}E^*,\ D = (I - E^*E)^{-1/2}.$$

(Note that $M^{1/2}$ denotes the Hermitian square root of M.) Then for each matrix X with $\| X \| \leq 1$, the matrix

77 $$T_E(X) = (AX + B)(CX + D)^{-1}$$

is well-defined. Moreover,

78 $$\| T_E(X) \| \leq 1 \text{ (resp. } < 1) \Longleftrightarrow \| X \| \leq 1 \text{ (resp. } < 1).$$

Proof First of all, $CX + D = (I - E^*E)^{-1/2}(E^*X + I)$. Since $\| E \| < 1$, it follows that, whenever $\| X \| \leq 1$, the matrix E^*X also has norm less than one, so that $E^*X + I$ has an inverse. Hence $T_E(X)$ is well-defined.

Now define

79 $$J = \begin{bmatrix} I & 0 \\ 0 & -I \end{bmatrix}.$$

Then routine calculations show that $L^*JL = J$; these calculations make use of the matrix identities

80 $$(I - MN)^{-1}M = M(I - NM)^{-1},$$

81 $$(I - MN)^{-1} = I + M(I - NM)^{-1}N.$$

Note that for any matrix M, we have that $\| M \| \leq 1$ (resp. < 1) if and only if $M^*M - I \leq 0$ (resp. < 0). Hence (78) is equivalent to

82 $$T^*T - I \leq 0 \text{ (resp. } < 0) \Longleftrightarrow X^*X - I \leq 0 \text{ (resp. } < 0),$$

where T is a shorthand for $T_E(X)$. But since $CX + D$ is nonsingular,

83 $$T^*T - I \leq 0 \text{ (resp. } < 0) \Longleftrightarrow$$

$$(CX+D)^*(T^*T-I)(CX+D)\leq 0 \text{ (resp. } <0).$$

The latter quantity equals

84 $$(AX+B)^*(AX+B) - (CX+D)^*(CX+D)$$

$$= [X^* \quad I]\, L^*JL \begin{bmatrix} X \\ I \end{bmatrix}$$

$$= [X^* \quad I]\, J \begin{bmatrix} X \\ I \end{bmatrix}$$

$$= X^*X - I.$$

Hence (78) follows from (83) and (84). □

Proof of Theorem (71) "if" The proof is by induction on the number n of points to be interpolated. If $n = 1$, the theorem is clearly true: Take $\Phi(z) = F_1$. To establish the inductive step, suppose the theorem is true for $n-1$ points, and suppose that P is positive definite; it is shown below that this implies the existence of a $\Phi \in \mathbf{M}(\mathbf{H}_\infty)$ satisfying $\|\Phi\|_\infty < 1$ and $\Phi(\lambda_i) = F_i \; \forall i$.

Let y be the inner function[11]

85 $$y(z) = \frac{|\lambda_1|\,(z-\lambda_1)}{\lambda_1\,(1-\bar{\lambda}_1 z)},$$

and define

86 $$T_i = T_{F_1}(F_i),\; i = 2, \cdots, n,$$

87 $$G_i = \frac{1}{y(\lambda_i)}\, T_i,\; i = 2, \cdots, n,$$

where $T_{F_1}(\cdot)$ is defined in (77). Now define a partitioned matrix

[11] Note that $y(z) = z$ if $\lambda_1 = 0$.

88 $$Q = \begin{bmatrix} Q_{22} & \cdots & Q_{2n} \\ \vdots & & \vdots \\ Q_{n2} & \cdots & Q_{nn} \end{bmatrix},$$

where

89 $$Q_{ij} = \frac{1}{1-\bar{\lambda}_i\lambda_j} \cdot (I - G_i{}^*G_j)$$

It is claimed that Q is positive definite. To prove the claim, compute Q_{ij} from (89), and make liberal use of the identities (80) and (81). This leads to

90 $$Q_{ij} = \frac{1}{1-\bar{\lambda}_i\lambda_j}\left[I - \frac{1}{1-\bar{y}(\lambda_i)y(\lambda_j)} T_i^* T_j\right]$$

$$= \frac{1}{1-\bar{\lambda}_i\lambda_j}\left[(1 - \frac{1}{1-\bar{y}(\lambda_i)y(\lambda_j)}) \cdot I + \frac{1}{1-\bar{y}(\lambda_i)y(\lambda_j)}(I - T_i^* T_j)\right],$$

where, from (86),

91 $$I - T_i^* T_j = (I-F_1^*F_1)^{1/2}(I-F_i^*F_1)^{-1}(I-F_i^*F_j)$$
$$(I-F_1^*F_j)^{-1}(I-F_1^*F_1)^{1/2}.$$

For clarity, define

92 $$R_i^{-1} = (I-F_1^*F_i)^{-1}(I-F_1^*F_1)^{1/2},\ i = 1, \cdots, n,$$

and note that

93 $$\frac{1}{1-\bar{\lambda}_i\lambda_j} \cdot \left[1 - \frac{1}{1-\bar{y}(\lambda_i)y(\lambda_j)}\right] = \frac{\bar{\lambda}_1\lambda_1 - 1}{\bar{\lambda}_i - \bar{\lambda}_1)(\lambda_j - \lambda_1)}.$$

Then (90) becomes

94 $$Q_{ij} = \frac{\bar{\lambda}_1\lambda_1 - 1}{\bar{\lambda}_i - \bar{\lambda}_1)(\lambda_j - \lambda_1)} \cdot I + \frac{R^{*}{}_i^{-1}}{y(\lambda_i)} P_{ij} \frac{R_j^{-1}}{y(\lambda_j)},$$

or, upon rearrangement,

95 $$P_{ij} = R_i^* \left[\frac{1 - \bar{\lambda}_1\lambda_1}{(1 - \bar{\lambda}_i\lambda_1)(1 - \lambda_j\bar{\lambda}_1)} I + \bar{y}(\lambda_i) Q_{ij} y(\lambda_j) \right] R_j.$$

Now (95) implies that

96 $$P = R^* \begin{bmatrix} c_1 I & 0 \\ 0 & Q \end{bmatrix} R,$$

where

97 $$c_i = \frac{1}{1 - \bar{\lambda}_1\lambda_i} \text{ for } i = 1, \cdots, n,$$

and R is a partitioned matrix of the form

98 $$R = \begin{bmatrix} c_1R_1 & c_2R_2 & c_3R_3 & \cdots & c_nR_n \\ 0 & y(\lambda_2)c_2R_2 & 0 & \cdots & 0 \\ 0 & 0 & y(\lambda_3)c_3R_3 & \cdots & 0 \\ \vdots & \vdots & \vdots & & \vdots \\ 0 & 0 & 0 & \cdots & y(\lambda_n)c_nR_n \end{bmatrix}.$$

Since R is nonsingular, $c_1 > 0$ and P is positive definite, it follows from (96) that Q is also positive definite. In particular, Q_{ii} is positive definite for all i, and $\| G_i \| < 1$ for all i. By the inductive hypothesis, there exists a function $\Psi \in \mathbf{M}(\mathbf{H}_\infty)$ such that $\| \Psi \|_\infty < 1$ satisfying $\Psi(\lambda_i) = G_i$ for $i = 2, \cdots, n$. Now define

99 $$\Phi(z) = (A - y(z)\Psi(z)C)^{-1}(y(z)\Psi(z)D - B),$$

where

100 $$A = (I - F_1 F_1^*)^{-1/2},\ B = (I - F_1 F_1^*)^{-1/2} F_1,$$
$$C = (I - F_1^* F_1)^{-1/2} F_1^*,\ D = (I - F_1^* F_1)^{-1/2}.$$

Note that (99) is the inverse of the transformation T_{F_1} defined in (77). Hence, from (86), (87) and (99),

101 $$\Phi(\lambda_i) = F_i \text{ for } i = 2, \cdots, n.$$

Since $y(\lambda_1) = 0$, (99) shows that

102 $$\Phi(\lambda_1) = A^{-1} B = F_1.$$

Finally, by Lemma (74), $\| \Psi(z) \| < 1$ for all z implies that $\| \Phi(z) \| < 1$ for all z. Hence Φ is a solution to the interpolation problem with n points. This completes the proof by induction.

"only if" One needs to prove the following statement: "Suppose $\Phi \in \mathbf{M}(\mathbf{H}_\infty)$ and that $\| \Phi \|_\infty < 1$. Let $\lambda_1, \cdots, \lambda_n$ be any n distinct points in the open unit disc, let $F_i = \Phi(\lambda_i)$, and define P as in (62),(63). Then P is positive definite."

The proof is again by induction on the number n of points. The statement is clearly true if $n = 1$. Suppose it is true for $n-1$ points. Given the n points $\lambda_1, \cdots, \lambda_n$ and the function Φ, let $F_i = \Phi(\lambda_i)$ and define the function $\Psi \in \mathbf{M}(\mathbf{H}_\infty)$ by

103 $$\Psi(z) = \frac{1}{y(z)} (A\Phi(z) + B)(C\Phi(z) + D)^{-1},$$

where y is as in (85) and A, B, C, D are as in (100). Then $\| \Psi \|_\infty < 1$ by Lemma (74). Moreover, $\Psi(\lambda_i) = G_i$ where G_i is defined in (87). By the inductive hypothesis, the matrix Q of (88) is positive definite. Since (96) holds, P is also positive definite. Hence the statement is true for n points. □

The proof of the "if" part of the theorem suggests an iterative algorithm for computing Φ. By the transformation (86)-(87) one can replace the given interpolation problem involving n points by another involving only $n-1$ points. Applying the transformation repeatedly reduces the number of interpolation constraints by one each time, until eventually one is left with a *single* matrix H_n such that $\| H_n \| < 1$, and is faced with the problem of finding a matrix $\Gamma(\cdot)\in\mathbf{M}(\mathbf{H}_\infty)$ such that $\Gamma(\lambda_n) = H_n$ and $\| \Gamma \|_\infty < 1$. An obvious solution is $\Gamma(z)\equiv H_n \ \forall z$. By applying the inverse transformation (99) $n-1$ times one can find a solution to the original problem. Moreover, there is a one-to-one correspondence between the set of *all* solutions $\Gamma(\cdot)$ to the above problem and the set of *all* solutions to the original interpolation problem.

The fact that the matrix R in (96) is upper-triangular suggests that (96) is the first step in a block Cholesky decomposition of P. This fact can be used to generate Q quickly and reliably.

Suppose now that it is desired to find a function $\Phi(\cdot)\in\mathbf{M}(\mathbf{H}_\infty)$ satisfying $\Phi(\lambda_i) = F_i$ and such that $\| \Phi \|_\infty \leq 1$ (as opposed to $\| \Phi \|_\infty < 1$). In this case, a necessary and sufficient condition that such a Φ exist is that the test matrix P of (72) be nonnegative definite (as opposed to positive definite as in Theorem (71)). To construct such a Φ, one can use the iterative procedure described above, but with a few modifications. As mentioned before Theorem (71), one can assume at the outset that each matrix F_i has a norm strictly less than one, by applying the reduction procedure of Lemma (63) if needed. Then, by applying the transformation (86)-(89), one can replace the original problem involving the n matrices $F_1, \cdots ,F_n$ by a reduced interpolation problem involving the $n-1$ matrices $G_2, \cdots ,G_n$. Moreover, if P is nonnegative definite, so is Q, because of (96). However, even if $\| F_i \| < 1 \ \forall i$, it can happen that $\| G_i \| = 1$ for some i. In such a case, it is necessary first to reduce the size of the matrices by first using Lemma (63) before carrying out another step of the Nevanlinna iteration.

This section concludes with an example that illustrates the use of the matrix Nevanlinna algorithm.

104 *Example* Suppose it is desired to find a matrix $\Phi\in\mathbf{H}_\infty^{1\times 2}$ that has norm less than one and satisfies the interpolation constraints

$$\Phi(0) = [0.2 \ \ 0.4], \Phi(0.5) = [0.3 \ \ 0.7].$$

The test matrix P of (72) is of order 4×4 in this case. To save some labor, let us consider the equivalent problem of finding a $\Phi\in\mathbf{H}_\infty^{2\times1}$ that has norm less than one and satisfies the interpolation constraints

$$\Phi(0) = \begin{bmatrix}0.2\\0.4\end{bmatrix}, \Phi(0.5) = \begin{bmatrix}0.3\\0.7\end{bmatrix}.$$

In this case $\lambda_1 = 0, \lambda_2 = 0.5, F_1, F_2$ are self-evident, and

$$P = \begin{bmatrix}0.8 & 0.66\\0.66 & 0.56\end{bmatrix},$$

which is positive definite. Hence the problem has a solution. From (76),

$$L(F_1) = \begin{bmatrix}1.0236 & 0.0472 & -0.2236\\0.0472 & 1.0944 & -0.4472\\-0.2236 & -0.4472 & 1.118\end{bmatrix}.$$

In this case $y(z) = z, y(\lambda_2) = 0.5$. Hence, from (86) and (87),

$$T_2 = \begin{bmatrix}0.11652\\0.33304\end{bmatrix}\cdot\frac{1}{0.73788} = \begin{bmatrix}0.15791\\0.45135\end{bmatrix},$$

$$G_2 = \frac{1}{0.5}\,T_2 = \begin{bmatrix}0.31582\\0.9027\end{bmatrix}.$$

One can as well choose $\Psi(z) = G_2\ \forall z$. Then from (99),

$$\Phi(z) = \begin{bmatrix}1.0236+0.0706z & 0.0472+0.1412z\\0.0472+0.2018z & 1.0944+0.4037z\end{bmatrix}^{-1}$$

$$\begin{bmatrix}0.2236+0.3531z\\0.4472+1.0092z\end{bmatrix}$$

$$= \frac{1}{1.118+0.4743z} \begin{bmatrix} 0.2236+0.3659z \\ 0.4472+1.0028z \end{bmatrix}$$

$$= \frac{1}{1+0.4242z} \begin{bmatrix} 0.2+0.3273z \\ 0.4+0.8969z \end{bmatrix}.$$

6.6 SENSITIVITY MINIMIZATION: GENERAL CASE

In this section, the sensitivity minimization problem is studied in full generality, i.e. without any simplifying assumptions on the relative numbers of plant inputs and outputs, and on the dimensions of the weighting matrices. The filtering problem in the general case is not discussed separately in the interests of brevity, but reasoning similar to that given here applies to this problem as well. As shown in earlier sections, the problem is one of minimizing a cost function

1 $\quad J(R) = \| F - GRH \|_\infty$

by a suitable choice of $R \in \mathbf{M}(\mathbf{H}_\infty)$, where $F, G, H \in \mathbf{M}(\mathbf{H}_\infty)$ are given rational matrices. It is assumed throughout that $G(e^{j\theta})$, $H(e^{j\theta})$ have full rank (either row or column, as appropriate) for all $\theta \in [0, 2\pi]$. If this is not so, then it is necessary to extend Lemma (6.5.9) to the present case. This extension is straight forward and is left to the reader.

Some preliminary results are presented before we proceed to the main solution algorithm. In the sequel, the notation $A > 0$ (resp. $A \geq 0$) means that the Hermitian matrix A is positive definite (resp. nonnegative definite). The notation $A > B$ (resp. $A \geq B$) denotes that A and B are Hermitian, and that $A - B > 0$ (resp. $A - B \geq 0$). The notations $A < B$ and $A \leq B$ are defined analogously.

2 $\quad$ *Lemma* Suppose $A \in \mathbf{M}(\mathbb{R}(z))$ and satisfies two conditions:

3 $\quad A'(z^{-1}) = A(z) \; \forall z$, and

4 $\quad A(e^{j\theta}) \geq 0 \ \forall \theta \in [0,2\pi]$.

Then there exists an outer matrix $U \in \mathbf{M}(\mathbf{R}_\infty)$ such that

5 $\quad A(z) = U'(z^{-1})U(z) \ \forall z.$

Further, if (4) is strengthened to

6 $\quad A(e^{j\theta}) > 0 \ \forall \theta \in [0,2\pi],$

then U can be chosen to a unit of $\mathbf{M}(\mathbf{R}_\infty)$. In this case, U is unique to within left multiplication by an orthogonal matrix.

A matrix U satisfying (5) is called a *spectral factor* of A and is denoted by $A^{1/2}$.

Proof See [3, p.240].

7 *Lemma* Suppose $A \in \mathbf{R}_\infty^{n \times n}$ where $n > m$, and that A is inner. Then there exists an inner matrix $B \in \mathbf{R}_\infty^{n \times (n-m)}$ such that $\Phi = [A \quad B] \in \mathbf{R}_\infty^{n \times n}$ is also inner.

The matrix B is sometimes referred to as a *complementary inner matrix* of A.

Proof Since A is inner, it has full column rank in $\mathbf{M}(\mathbf{R}_\infty)$. By permuting rows if necessary, which does not affect the validity of the lemma, it can be assumed that

8 $$A = \begin{bmatrix} D \\ N \end{bmatrix},$$

where $|D| \neq 0$ as an element of $\mathbf{R}_\infty$. Note that $D \in \mathbf{R}_\infty^{m \times m}$, $N \in \mathbf{R}_\infty^{n-m \times m}$. Now define

9 $\quad P(z) = -[D'(z^{-1})]^{-1} N'(z^{-1}).$

Then $P \in (\mathbb{R}(z))^{m \times (n-m)}$ and therefore has an r.c.f. (U,V) over $\mathbf{M}(\mathbf{R}_\infty)$. Specifically, $U \in \mathbf{R}_\infty^{m \times (n-m)}$, $V \in \mathbf{R}_\infty^{(n-m) \times (n-m)}$, and

10 $$P(z) = U(z)[V(z)]^{-1} \;\forall z.$$

Combining (10) and (11) shows that

11 $$D'(z^{-1})U(z) + N'(z^{-1})V(z) = 0 \;\forall z.$$

Now define

12 $$W = \begin{bmatrix} U \\ V \end{bmatrix} \in \mathbf{R}_{\infty}^{n \times (n-m)}$$

and express (11) in the form

13 $$A'(z^{-1})W(z) = 0 \;\forall z.$$

Clearly W has full column rank as an element of $\mathbf{M}(\mathbf{R}_{\infty})$. Hence, by Fact (6.2.17), W has an inner-outer factorization $W_i W_o$, where $W_i \in \mathbf{R}_{\infty}^{n \times (n-m)}$ and $W_o \in \mathbf{R}_{\infty}^{(n-m) \times (n-m)}$. Multiplying (13) by $[W_o(z)]^{-1}$ gives

14 $$A'(z^{-1})W_i(z) = \;\forall z.$$

It is claimed that the choice $B = W_i$ makes $\Phi = [A \;\; B]$ inner (and of course square). Note that

15 $$\Phi'(z^{-1})\Phi(z) = \begin{bmatrix} A'(z^{-1})A(z) & A'(z^{-1})B(z) \\ B'(z^{-1})A(z) & B'(z^{-1})B(z) \end{bmatrix} = I,$$

since A, B are both inner and satisfy (14). □

Now we come to the main result of this section. To be specific, suppose $F \in \mathbf{R}_{\infty}^{n \times m}$, $G \in \mathbf{R}_{\infty}^{n \times l}$, $H \in \mathbf{R}_{\infty}^{k \times m}$, and that $n > l$, $k < m$; the cases $n \leq l$ and/or $k \geq m$ are simpler and the requisite modifications will be obvious. Recall that the problem is to select $R \in \mathbf{R}_{\infty}^{l \times k}$ so as to minimize

16 $$J(R) = \| F - GRH \|_{\infty}.$$

The first step in the solution procedure is to show that the above problem can be reduced to that of minimizing

$$17 \quad \| \begin{bmatrix} A-Q & B \\ C & D \end{bmatrix} \|_\infty$$

by a suitable choice of $Q \in \mathbf{R}_\infty^{l \times k}$, where A,B,C,D are $\mathbf{L}_\infty$ (not $\mathbf{H}_\infty$!) matrices of appropriate dimensions. For this purpose, first factor G as $G_i G_o$, where $G_i \in \mathbf{R}_\infty^{n \times l}$ is inner and $G_o \in \mathbf{R}_\infty^{l \times l}$ is outer. Then, using Lemma (7), construct a *square* inner matrix $\Phi \in \mathbf{R}_\infty^{n \times n}$ containing G_i as its first l columns. Then

$$18 \quad G = \Phi \begin{bmatrix} G_o \\ 0_{(n-l) \times l} \end{bmatrix}.$$

Similar reasoning applied to $H' \in \mathbf{R}_\infty^{m \times k}$ shows that there exist an outer matrix $H_o \in \mathbf{R}_\infty^{k \times k}$ and a *square* inner matrix $\Psi \in \mathbf{R}_\infty^{m \times m}$ such that

$$19 \quad H = [H_o \;\; 0_{k \times (m-k)}]\Psi.$$

Hence

$$20 \quad J(R) = \| F - \Phi \begin{bmatrix} G_o \\ 0 \end{bmatrix} R \, [H_o \;\; 0]\Psi \|_\infty$$

$$= \| \Phi^* F \Psi^* - \begin{bmatrix} G_o \\ 0 \end{bmatrix} R \, [H_o \;\; 0] \|_\infty.$$

Now define

$$21 \quad \begin{bmatrix} A & B \\ C & D \end{bmatrix} = \Phi^* F \Psi^* \in \mathbf{M}(\mathbf{L}_\infty),$$

and let Q denote the "free" parameter $G_o R H_o$ (recall that both G_o and H_o are units of $\mathbf{M}(\mathbf{H}_\infty)$ by virtue of the rank conditions on G and H). Then the cost function becomes

$$22 \quad \| \begin{bmatrix} A-Q & B \\ C & D \end{bmatrix} \|_\infty.$$

To proceed further, we require another very simple result:

23 *Lemma* Suppose $W = [X \;\; Y] \in \mathbf{M}(\mathbf{L}_\infty)$ is rational. Then $\| W \|_\infty \geq \| Y \|_\infty$. Further, if $\gamma > \| Y \|_\infty$, then

$$\| W \|_\infty \leq \gamma \Longleftrightarrow \| (\gamma^2 I - YY^*)^{-1/2} X \|_\infty \leq 1, \tag{24}$$

where $Y^*(z) = Y'(z^{-1})$, and $(\gamma^2 I - YY^*)^{-1/2}$ denotes the spectral factor of $\gamma^2 I - YY^*$.

Proof It is obvious that $\| W \|_\infty \geq \| Y \|_\infty$. Moreover, if $\gamma > \| Y \|_\infty$, then $(\gamma^2 I - YY^*)^{1/2}$ exists and is a unit of $\mathbf{M}(\mathbf{R}_\infty)$, by Lemma (2); hence $(\gamma^2 I - YY^*)^{-1/2}$ is well-defined.

By definition,

$$\| W \|_\infty = \max_{\theta \in [0, 2\pi]} \bar{\sigma}(W(e^{j\theta})). \tag{25}$$

Hence

$$\begin{aligned} \| W \|_\infty \leq \gamma &\Longleftrightarrow \bar{\sigma}(W(e^{j\theta})) \leq \gamma \;\; \forall \theta \in [0, 2\pi] \\ &\Longleftrightarrow W(e^{j\theta}) W^*(e^{j\theta}) \leq \gamma^2 I \;\; \forall \theta \\ &\Longleftrightarrow [XX^* + YY^*](e^{j\theta}) \leq \gamma^2 I \;\; \forall \theta \\ &\Longleftrightarrow (XX^*)(e^{j\theta}) \leq \gamma^2 I - (YY^*)(e^{j\theta}) \;\; \forall \theta. \end{aligned} \tag{26}$$

This leads immediately to (24). □

A parallel result to Lemma (23) is the following.

27 *Lemma* Suppose

$$W = \begin{bmatrix} X \\ Y \end{bmatrix} \in \mathbf{M}(\mathbf{L}_\infty) \tag{28}$$

is rational. Then $\| W \|_\infty \geq \| Y \|_\infty$. Further, if $\gamma > \| Y \|_\infty$, then

29 $$\| W \|_\infty \le \gamma \Longleftrightarrow \| X(\gamma^2 I - Y^* Y)^{-1/2} \|_\infty \le 1.$$

Now let us return to the problem of minimizing the cost function in (22). Let γ be a current guess of the minimum, and let K denote $[C \;\; D]$. Then, from Lemma (27),

30 $$\min_{Q \in \mathbf{M}(\mathbf{H}_\infty)} \left\| \begin{bmatrix} A-Q & B \\ C & D \end{bmatrix} \right\|_\infty \le \gamma$$

if and only if

31 $$\min_{Q \in \mathbf{M}(\mathbf{R}_\infty)} \| [A-Q \;\; B](\gamma^2 I - K^* K)^{-1/2} \|_\infty \le 1.$$

Now note that $(\gamma^2 I - K^* K)^{-1/2}$ is a unit of $\mathbf{M}(\mathbf{R}_\infty)$. Hence the product $Q(\gamma^2 I - K^* K)^{-1/2}$ can be replaced by another free parameter R. If we define

32 $$A_1 = A(\gamma^2 I - K^* K)^{-1/2} \in \mathbf{M}(\mathbf{L}_\infty),$$

33 $$B_1 = B(\gamma^2 I - K^* K)^{-1/2} \in \mathbf{M}(\mathbf{L}_\infty),$$

then (31) is equivalent to

34 $$\min_{R \in \mathbf{M}(\mathbf{H}_\infty)} \| [A_1 - R \;\; B_1] \|_\infty \le 1.$$

Next, from Lemma (23), (34) holds if and only if

35 $$\min_{R \in \mathbf{M}(\mathbf{H}_\infty)} \| (I - B_1 B^*_1)^{-1/2} (A_1 - R) \|_\infty \le 1.$$

Once again, it is possible to replace $(I - B_1 B^*_1)^{-1/2} R$ by another free parameter $S \in \mathbf{M}(\mathbf{H}_\infty)$. If we define

36 $$A_2 = (I - B_1 B^*_1)^{-1/2} A_1 \in \mathbf{M}(\mathbf{L}_\infty),$$

then (35) is equivalent to

37 $$\min_{S \in \mathbf{M}(\mathbf{H}_\infty)} \| A_2 - S \|_\infty \leq 1.$$

The problem of finding the minimum in (37) is one of finding the closest matrix in $\mathbf{M}(\mathbf{H}_\infty)$ to a given matrix in $\mathbf{M}(\mathbf{L}_\infty)$. This can be reduced to the problem studied in Section 6.5, namely minimizing a cost function of the form (6.5.25). This can be accomplished as follows: Since $A_2 \in \mathbf{M}(\mathbf{L}_\infty)$ and is rational, it has an l.c.f. (T, U) over $\mathbf{M}(\mathbf{R}_\infty)$. Now T is square and full rank in $\mathbf{M}(\mathbf{R}_\infty)$, and therefore has a factorization $T_o T_i$ where T_o is a unit of $\mathbf{M}(\mathbf{R}_\infty)$ and T_i is inner. Hence

38 $$\begin{aligned} A_2(z) &= [T(z)]^{-1} U(z) \\ &= [T_i(z)]^{-1} [T_o(z)]^{-1} U(z) \\ &= [T_i(z)]^{-1} V(z) \end{aligned}$$

where $V = T_o^{-1} U \in \mathbf{M}(\mathbf{R}_\infty)$. Now

39 $$\begin{aligned} \| A_2 - S \|_\infty &= \| T_i^{-1} V - S \|_\infty \\ &= \| T_i (T_i^{-1} V - S) \|_\infty \\ &= \| V - T_i S \|_\infty \\ &= \| T_i^{adj} V - bS \|_\infty \end{aligned}$$

where $b = | T_i |$. This is the type of objective function covered by Theorem (6.5.23).

The preceding discussion can be summarized in the following iterative procedure for computing the minimum of the function in (22).

Step 1 Choose

40 $$\gamma \geq \max\{ \| [C \;\; D] \|_\infty, \| \begin{bmatrix} B \\ D \end{bmatrix} \|_\infty \}.$$

Step 2 Form

41 $$L = (\gamma^2 I - K^* K)^{-1/2},$$

where $K = [C \;\; D]$.

Step 3 Form

42 $$M = (I - BLL^* B^*)^{-1/2}.$$

If $\| BL \|_\infty \geq 1$, increase γ and go to Step 2; otherwise go on to Step 4.

Step 4 Compute

43 $$J^* = \min_{S \in \mathbf{M}(\mathbf{H}_\infty)} \| MAL - S \|_\infty.$$

If $J^* < 1$, decrease γ and go to Step 2. If $J^* > 1$, increase γ and go to Step 2. If $\gamma = 1$, go on to Step 5.

Step 5 Set $Q = M^{-1} S L^{-1}$, where S achieves the minimum in (43). This choice of Q minimizes the original objective function in (22).

6.7 TWO-PARAMETER COMPENSATORS

The objective of this section is to discuss the relative merits of a two-parameter compensator in filtering and sensitivity minimization applications, compared to a one-parameter compensator. The conclusions can be briefly stated as follows: In problems of disturbance rejection, both schemes are equivalent. In tracking problems involving a stable plant, again both schemes are equivalent. Finally, in tracking problems with an unstable plant, whether one uses the $\mathbf{H}_2$-norm or the $\mathbf{H}_\infty$-norm, there exist plants for which the optimum achievable with a two-parameter scheme is as small as one wishes compared to the one-parameter optimum.

To avoid purely technical details caused by the nonexistence of optimal compensators in the continuous-time case, the discussion below is addressed to the discrete-time case. However, with the aid of Lemmas (6.3.10), (6.4.11) and (6.5.9), this discussion can be translated to the continuous-time case.

Accordingly, suppose P is a given multivariable plant with transfer matrix $P(z)$, where z is the transfer function of a unit delay. Suppose u_1 is a reference input to be tracked, while u_2 and u_3 are sensor noises to be rejected. Let W_1, W_2, W_3, W_4 be given weighting matrices in $\mathbf{M}(\mathbf{H}_\infty)$. It is assumed that (i) P has no poles on the unit circle, (ii) W_1, W_3 are outer, (iii) W_2, W_4 are square outer, and (iv) $P(e^{j\theta}), W_1(e^{j\theta}), \cdots, W_4(e^{j\theta})$ all have full row rank at all $\theta \in [0,2\pi]$.

For convenience, a two-parameter compensation scheme is reproduced in Figure 6.2. The *optimal tracking problem* is to minimize either

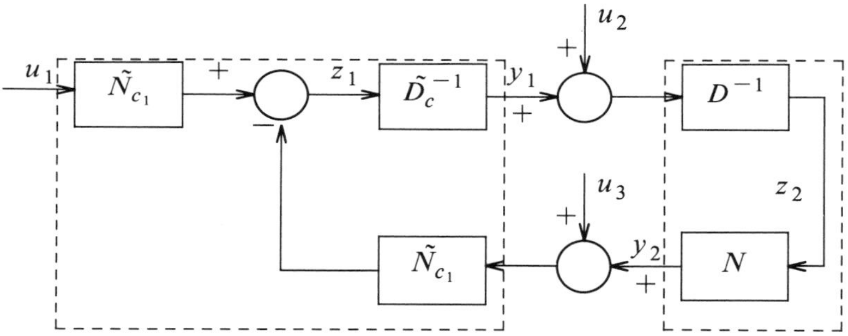

Figure 6.2. Two-Parameter Compensation Scheme.

1 $$J_2 = \| W_1(I - H_{yu_1})W_2 \|_2$$

or

2 $$J_\infty = \| W_1(I - H_{yu_1})W_2 \|_\infty$$

by means of a suitable stabilizing compensator. The *optimal disturbance rejection problem* is to select a stabilizing compensator that minimizes either

3 $$K_{21} = \| W_3 H_{yu_2} W_4 \|_2 \text{ or } K_{22} = \| W_3 H_{vu_3} W_4 \|_2,$$

or

4 $$K_{\infty 1} = \| W_3 H_{yu_2} W_4 \|_\infty \text{ or } K_{\infty 2} = \| W_3 H_{vu_3} W_4 \|_\infty.$$

From the discussion in Section 5.6, it is at once clear that, as far as minimizing K_{21}, K_{22}, $K_{\infty 1}$, $K_{\infty 2}$, there is no advantage to using a two-parameter scheme instead of a one-parameter scheme. This is because the set of achievable stable transfer marices H_{yu_2}, H_{vu_3} are exactly the same with either scheme. But with the optimal tracking problem, the situation is quite different. The minimum J_2 or J_∞ achievable with a two-parameter scheme is no larger than with a one-parameter scheme, and can be arbitrarily smaller, depending on the plant. Moreover, with the two-parameter scheme, it is possible *simultaneously* to minimize J_2 and K_{21}, (or J_2 and K_{22}, or J_∞ and $K_{\infty 1}$ or any other combination). This is the overwhelming advantage of the two-parameter scheme. To justify the latter statement first, recall from (5.6.17) that the set of achievable stable transfer matrices from (u_1, u_2, u_3) to y are parametrized by

5 $$[NQ \quad N(Y - R\tilde{N}) \quad -N(X + R\tilde{D})],$$

where (N, D), $(\tilde{D}, \tilde{N})$ are any r.c.f. and l.c.f. of P, $X, Y \in \mathbf{M}(\mathbf{H}_\infty)$ satisfy $XN + YD = I$, and $Q, R \in \mathbf{M}(\mathbf{H}_\infty)$ are *arbitrary*. Thus to minimize J_∞ and K_{22} simultaneously, for example, one chooses Q to minimize

6 $$J_\infty = \| W_1 (I - NQ) W_2 \|_\infty,$$

and R to minimize

7 $$K_{22} = \| W_3 (I - N(X + R\tilde{D})) W_4 \|_2.$$

The associated compensator

8 $$\tilde{N}_{c_1} = Q, \tilde{N}_{c_2} = X + R\tilde{D}, \tilde{D}_c = Y - R\tilde{N}$$

minimizes the weighted tracking error *and* optimally rejects the weighted disturbance.

From (5.6.18), the set of stable transfer matrices H_{yu_1} that can be achieved using a one-parameter compensator is $\{N(X+R\tilde{D}): R \in \mathbf{M}(\mathbf{H}_\infty)\}$. Let us define

9 $$J_\infty^* = \min_{Q \in \mathbf{M}(\mathbf{H}_\infty)} \| W_1(I-NQ)W_2 \|_\infty,$$

10 $$J_2^* = \min_{Q \in \mathbf{M}(\mathbf{H}_\infty)} \| W_1(I-NQ)W_2 \|_2,$$

11 $$h_\infty^* = \min_{R \in \mathbf{M}(\mathbf{H}_\infty)} \| W_1(I-N(X+R\tilde{D}))W_2 \|_\infty,$$

12 $$h_2^* = \min_{R \in \mathbf{M}(\mathbf{H}_\infty)} \| W_1(I-N(X+R\tilde{D}))W_2 \|_2.$$

Then J_∞^*, J_2^* (resp. h_∞^*, h_2^*) are the best weighted tracking errors achievable with a two-parameter (resp. one-parameter) compensation scheme. Clearly $J_\infty^* \le h_\infty^*$, $J_2^* \le h_2^*$. If P is a *stable* plant, then one can choose $N = P, X = 0, \tilde{D} = I$, in which case it is immediate that $J_2^* = h_2^*, J_\infty^* = h_\infty^*$. On the other hand, there exist plants for which the ratios J_∞^*/h_∞^* or J_2^*/h_2^* are as small as one wishes. To set up these examples, let us study the computation of the various minima in (9)-(12). Consider first the constants J_∞^*, h_∞^*, and restrict attention to the scalar case for ease of discussion. In this case we can define a single weighting function $w = w_1 w_2$. Also, factor n as $n_i n_o$, where n_i, n_o are respectively inner and outer functions. Then

13 $$J_\infty^* = \min_{q \in \mathbf{H}_\infty} \| (1-nq)w \|_\infty.$$

At this stage it can be assumed without loss of generality that w is outer: If $w = w_i w_o$ is an inner-outer factorization of w, then multiplication by w_i does not affect the norm. Also, since $w(e^{j\theta}) \ne 0 \ \forall \theta \in [0,2\pi]$, w is actually a *unit* of $\mathbf{H}_\infty$. Therefore

14 $$J_\infty^* = \min_{q \in \mathbf{H}_\infty} \| w - n_i n_o q w \|_\infty$$

$$= \min_{q \in \mathbf{H}_\infty} \| w - n_i q \|_\infty,$$

after absorbing the units n_o and w into q. Thus the problem is to find a function f of minimum $\mathbf{H}_\infty$-norm such that f interpolates the values of w at the zeros of n_i (i.e. at the plant zeros inside the unit disc).

Now consider the computation of h^*_∞. By definition,

15 $$h^*_\infty = \min_{r \in \mathbf{H}_\infty} \| [1 - n(x + rd)]w \|_\infty$$

$$= \| (y - rn)dw \|_\infty.$$

Factor d as $d_i d_o$ where d_i, d_o are respectively inner and outer, and absorb all units into the free parameter r. This leads to

16 $$h^*_\infty = \min_{r \in \mathbf{H}_\infty} \| [yd_o w - n_i r]d_i \|_\infty$$

$$= \min_{r \in \mathbf{H}_\infty} \| yd_o w - n_i r \|_\infty,$$

since multiplication by d_i preserves norms. Thus the problem is to find a function g of minimum $\mathbf{H}_\infty$-norm that interpolates the values of $yd_o w$ at the zeros of n_i. This can be restated in a more transparent way. Suppose λ is a zero of n_i. Then

17 $$n_i(\lambda) = 0 \Rightarrow (nx)(\lambda) = 0 \Rightarrow (yd)(\lambda) = 1$$

$$\Rightarrow (yd_o)(\lambda) = 1/d_i(\lambda).$$

Thus g must interpolate the values of w/d_i at the zeros of n_i, while f interpolates the values of w at the zeros of n_i. Since $| d_i(z) | < 1$ for all z in the open unit disc, we see that h^*_∞ is in general larger than J^*_∞. In fact, if p has a pole-zero pair close together inside the unit disc, then $| d_i(\lambda) |$ will be very small at some zero λ of n_i, and h^*_∞ will be considerably larger than J^*_∞.

18 *Example* Let $w = 1$, and let

$$p(z) = \frac{z-0.5-\varepsilon}{z-0.5}.$$

Then

$$n = z-0.5-\varepsilon,\ n_i = \frac{z-0.5-\varepsilon}{1-(0.5+\varepsilon)z},$$

$$d = z-0.5,\ d_i = \frac{z-0.5}{1-0.5z}.$$

Since n_i has only one zero, at $z = 0.5+\varepsilon$,

$$J_\infty^* = |w(0.5+\varepsilon)| = 1,$$

$$h_\infty^* = |(w/d_i)(0.5+\varepsilon)| = 1/|d_i(0.5+\varepsilon)|$$

$$= \frac{0.75-0.5\varepsilon}{\varepsilon} \approx \frac{0.75}{\varepsilon}.$$

By choosing ε sufficiently small, the ratio J_∞^*/h_∞^* can be made as small as one desires.

Now consider the computation of J_2^* and h_2^*. Recall from Section 6.2 that the space $\mathbf{L}_2[0,2\pi]$ can be decomposed as the direct sum of the orthogonal subspaces $\mathbf{H}_2^+$ and $\mathbf{H}_2^-$, where $\mathbf{H}_2^+$ consists of the boundary values of $\mathbf{H}_2$-functions. Given $f \in \mathbf{L}_2[0,2\pi]$, let f_+ (resp. f_-) denote its projection onto $\mathbf{H}_2^+$ (resp. $\mathbf{H}_2^-$). Finally, if $f \in \mathbf{H}_2^-$, then f^* denotes the function $z \rightarrow z^{-1} f(z^{-1})$, while if $F \in \mathbf{M}(\mathbf{H}_\infty)$, then F_o (resp. F_i) denotes the outer (resp. inner) part of F.

For ease of discussion, suppose $W_1 = I$, and denote W_2 by W. Also, suppose W is a unit of $\mathbf{M}(\mathbf{H}_\infty)$, without loss of generality. Then

19 $$J_2^* = \min_{Q \in \mathbf{M}(\mathbf{H}_\infty)} \| (I-NQ)W \|_2$$

$$= \min_{Q \in \mathbf{M}(\mathbf{H}_\infty)} \| W - N_i Q \|_2, \text{ after absorbing } N_o, W \text{ into } Q$$

$$= \min_{Q \in \mathbf{M}(\mathbf{H}_\infty)} \| N_i^* W - Q \|_2$$

$$= \| (N_i^* W)_- \|_2,$$

corresponding to the choice $Q = (N_i^* W)_+$. On the other hand,

20 $$h_2^* = \min_{R \in \mathbf{M}(\mathbf{H}_\infty)} \| (I - N(X + R\tilde{D}))W \|_2.$$

To bring out the connection between h_2^* and J_2^* more clearly, select $\tilde{X}, \tilde{Y} \in \mathbf{M}(\mathbf{H}_\infty)$ such that

21 $$\begin{bmatrix} Y & X \\ -\tilde{N} & \tilde{D} \end{bmatrix} \begin{bmatrix} D & -\tilde{X} \\ N & \tilde{Y} \end{bmatrix} = \begin{bmatrix} D & -\tilde{X} \\ N & \tilde{Y} \end{bmatrix} \begin{bmatrix} Y & X \\ -\tilde{N} & \tilde{D} \end{bmatrix} = I.$$

This is always possible, by Theorem (4.1.60). If we let $(\tilde{D}W)_o$ denote the outer part of $\tilde{D}W$, then

22 $$h_2^* = \min_{R \in \mathbf{M}(\mathbf{H}_\infty)} \| (\tilde{Y} - NR)\tilde{D}W \|_2$$

$$= \min_{R \in \mathbf{M}(\mathbf{H}_\infty)} \| \tilde{Y}(\tilde{D}W)_o - NR(\tilde{D}W)_o \|_2$$

$$= \min_{R \in \mathbf{M}(\mathbf{H}_\infty)} \| N_i^* \tilde{Y}(\tilde{D}W)_o - N_o R(\tilde{D}W)_o \|_2$$

$$= \min_{R \in \mathbf{M}(\mathbf{H}_\infty)} \| N_i^* \tilde{Y}(\tilde{D}W)_o - R \|_2$$

$$= \| [N_i^* \tilde{Y}(\tilde{D}W)_o]_- \|_2.$$

To compare (19) and (22), note that

23 $$NX + \tilde{Y}\tilde{D} = I,$$

24 $$N_i^* W = N_i^*(NX + \tilde{Y}\tilde{D}) = N_o XW + N_i^* \tilde{Y}\tilde{D}W.$$

However, since $N_o XW \in \mathbf{M}(\mathbf{H}_2)$,

25 $(N_i^* W)_- = [N_i^* \tilde{Y}\tilde{D}W]_-.$

So if we define $F = N_i^* \tilde{Y}\tilde{D}W \in \mathbf{M}(\mathbf{L}_2)$, then

26 $J_2^* = \| F_- \|_2, h_2^* = \| [F(\tilde{D}W)_i^*]_- \|_2.$

This implies that h_2^* is in general larger than J_2^*.

27 *Example* Consider the scalar case; let $w = 1$, and let

$$p(z) = \frac{z^m}{1-2z},$$

where m is an integer. Then we can choose

$$n = z^m, d = 1-2z,$$

$$x = 2^m, y = 1+2z+4z^2+ \cdots +2^{m-1}z^{m-1}.$$

In this case $n = n_i$ since n is already inner. Hence we have

$$n_i = z^m, n_i^* = z^{-m}.$$

$$d_i = \frac{z-0.5}{1-0.5z}, d_i^* = \frac{1-0.5z}{z-0.5}.$$

$$f = n_i^* yd = n_i^*(1-nx) = n_i^* - x = z^{-m} - 2^m.$$

$$J_2^* = \| f_- \|_2 = \| z^{-m} \|_2 = 1.$$

$$fd_i^* = (z^{-m} - 2^m)d_i^*,$$

$$h_2^* = \| (fd_i^*)_- \|_2 = \| (z^{-m} d_i^*)_- - (2^m d_i^*)_- \|_2$$

$$\geq 2^m \| (d_i^*)_- \|_2 - \| (z^{-m} d_i^*)_- \|_2.$$

Now expand $d_i(z)$ in a Fourier series

$$d_i(z) = \sum_{i=0}^{\infty} c_i z^i,$$

where

$$c_0 = -0.5,\ c_i = 1.5(0.5)^i \text{ for } i \geq 1.$$

Then, since d_i is inner,

$$1 = \| d_i \|_2^2 = \sum_{i=0}^{\infty} c_i^2.$$

Therefore

$$\sum_{i=1}^{\infty} c_i^2 = 0.75.$$

Since

$$d_i^*(z) = \sum_{i=0}^{\infty} c_i z^{-i},$$

$$(d_i^*)_- = \sum_{i=1}^{\infty} c_i z^{-i},$$

we see that

$$\| (d_i^*)_- \|_2^2 = \sum_{i=1}^{\infty} c_i^2 = 0.75.$$

On the other hand, if $m \geq 1$, then we have

$$(z^{-m} d_i^*)_- = z^{-m} d_i^* = \sum_{i=0}^{\infty} c_i z^{-(i+m)},$$

$$\| (z^{-m} d_i^*)_- \|_2 = \| d_i^* \|_2 = 1.$$

Hence

$$h_2^* \geq 2^{m-1} \cdot \sqrt{3} - 1.$$

By choosing the integer m sufficiently large, the ratio J_2^*/h_2^* can be made as small as one desires.

Notes and References

Section 6.2 contains only the barest minimum of the theory of Hardy spaces. For excellent treatments of this topic, see Duren [32], Koosis [55] or Sz. Nagy and Foias [65].

Solving the filtering problem of Section 6.3 was the original motivation of Youla, Jabr and Bongiorno [107] in parametrizing the set of all compensators that stabilize a given plant. Kucera [56] gives a solution to the discrete-time filtering problem that is similar in its approach to that of [107]. The treatment given here follows Francis [34].

The sensitivity minimization problem was formulated by Zames [109] and solved in the scalar case by Zames and Francis [37, 111]. The treatment here is similar to that in [F4]. In the multivariable case, the fat plant case was solved by Chang and Pearson [12] and Francis, Helton and Zames [35]. Our treatment is adapted from the former. The general sensitivity minimization problem was solved by Doyle [29]. Note that it is possible to carry out all of the steps of Sections 6.4 to 6.6 using purely state-space methods; see [16, 29, 41]. The paper of Sarason [82] contains some mathematical results pertinent to the sensitivity minimization problem. The Nevanlinna-Pick algorithm used in Section 6.5 is taken from the paper by Delsarte, Genin and Kamp [18], but this paper treats only the square matrix case. Finally, Potapov [71] contains several useful results on matrices of analytic functions, such as Lemma (6.5.44).

Chapter 7. ROBUSTNESS

This chapter begins with a detailed study of the issue of the robustness of feedback stability in the presence of plant and/or compensator perturbations. Both the nature and the extent of the perturbations that maintain feedback stability are discussed. Then the problem of robust regulation (i.e. stabilization plus tracking and/or disturbance rejection) is examined, for both the one-parameter as well as the two-parameter compensation schemes. Finally, various generic properties of control systems are studied.

7.1. INTRODUCTION

Consider the standard feedback system shown in Figure 7.1, and suppose the compensator C stabilizes the plant P, i.e. that the closed-loop transfer matrix

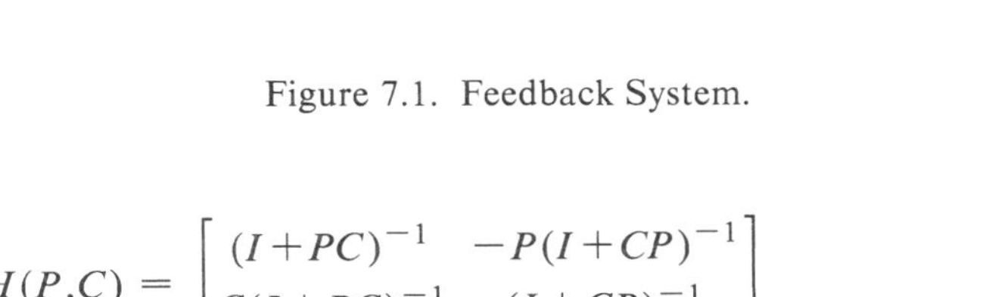

Figure 7.1. Feedback System.

1 $$H(P,C) = \begin{bmatrix} (I+PC)^{-1} & -P(I+CP)^{-1} \\ C(I+PC)^{-1} & (I+CP)^{-1} \end{bmatrix}$$

belongs to $\mathbf{M}(\mathbf{S})$. The central question studied in this chapter is: What sort of perturbations can be permitted in P and/or C without destroying feedback stability? This question is very relevant in any practical application of the controller synthesis theory developed in the previous two chapters. The first step in designing a compensator for any given plant is to obtain a mathematical model for it. As this process always involves some modeling errors, simplifying assumptions, etc., the "true" plant is in general different from the nominal model used in designing a compensator. In the same vein, one can normally expect some errors or simplifications in implementing the compensator, so that the "true" compensator is not necessarily the same as the designed compensator. These phenomena can be captured by supposing that the true plant and compensator are perturbed versions of the nominal plant and compensator respectively. Hence there is interest in obtaining characterizations of both the nature and the extent of the permissible perturbations.

Before attempting to characterize permissible perturbations, we should first decide on a reasonable definition of what constitutes a permissible perturbation. Suppose (P_0,C_0) is a nominal plant-compensator pair which is stable, so that $H(P_0,C_0) \in \mathbf{M}(\mathbf{S})$, and suppose P_0, C_0 are perturbed to P, C, respectively. In order for (P,C) to be a permissible perturbation of (P_0,C_0), we require as a minimum that (P,C) also be a stable pair. But this is not enough. The objective of compensator design is not merely to stabilize a plant but also to improve its response (this is why compensation is applied even to stable plants in some instances). Hence (P,C) is deemed to be a permissible perturbation of (P_0,C_0) if (P,C) is stable, and in addition the closed-loop transfer matrix $H(P,C)$ is "close" to $H(P_0,C_0)$.

There are two aspects to the study of robustness. The first is *qualitative*, and pertains to the *nature* of the permissible perturbations. Mathematically, this is the problem of defining a *topology* on the set of plant-compensator pairs. The second is *quantitative*, and is concerned with the *extent* of the permissible perturbations. This is the problem of defining a *metric* on the set of plant-compensator pairs. Each has its own significance, which is discussed at the appropriate place.

In the study of robustness, a very helpful feature is that the map from the pair (P,C) to the closed-loop transfer matrix $H(P,C)$ is *one-to-one*. That is, given $H(P,C) \in \mathbf{M}(\mathbf{S})$, one can *uniquely* determine P and C, by proceeding as follows: Define

2 $$G = \begin{bmatrix} C & 0 \\ 0 & P \end{bmatrix}, F = \begin{bmatrix} 0 & I \\ -I & 0 \end{bmatrix}.$$

Then it is easy to verify (as in Section 5.1) that $H(P,C) = (I+FG)^{-1}$. Solving this relationship for G in terms of $H(P,C)$ gives

3 $$G = \begin{bmatrix} C & 0 \\ 0 & P \end{bmatrix} = F[I-(H(P,C))^{-1}],$$

where we use the fact that $F^{-1} = -F$. Now suppose that (P_0,C_0) is a given stable pair corresponding to the nominal plant and compensator, and let $\mathbf{B}$ denote a subset of $\mathbf{M}(\mathbf{S})$ containing $H(P_0,C_0)$, representing the acceptable perturbations of the closed-loop transfer matrix resulting from changes in the plant and/or compensator. Now the set

4 $$\{G \in \mathbf{M}(\mathbb{R}(s)): G = F(I-H)^{-1} \text{ for some } H \in \mathbf{B}\}$$

certainly contains *all* pairs (P,C) such that $H(P,C) \in \mathbf{B}$. But there is a flaw in this argument: If H is an *arbitrary* element of $\mathbf{B}$, then $G = F(I-H)^{-1}$ need not be block-diagonal and therefore need not correspond to a plant-compensator pair. Of course, the set of *block-diagonal* matrices of the form (4) is precisely the set of permissible plant and compensator perturbations. However, this set is very difficult to characterize explicitly. Hence we are forced to seek alternate approaches. This is what is done in the next few sections.

7.2 GRAPH TOPOLOGY

Suppose it is desired to design a stabilizing compensator for a plant that is only specified imprecisely. Suppose further that the plant uncertainty can be modeled by considering a *family* of plants P_λ, where the uncertainty parameter λ assumes values in some topological space Λ. Typically λ represents some physical parameters of the plant, such as component values, and the function $\lambda \rightarrow P_\lambda$ captures the manner in which this uncertainty in the physical parameters is reflected in the plant description. The parameter space Λ may be $\mathbb{R}^n$ for some integer n, or a more general normed space, or any topological space. An example of the latter situation arises in the design of compensators for singularly perturbed systems, where $\Lambda = [0,\infty)$. For technical reasons, it is assumed that Λ is a first-countable topological space (see Section C.1 for the definition). As all metric spaces are first-countable, this poses no practical restriction. Let P_{λ_0} be the nominal plant description on which the compensator design is based. The question is: When is it possible to design a stabilizing compensator for P_{λ_0} that is "robust" against the plant uncertainty? To answer this question, it is necessary first to define robust stabilization. The definition adopted here is that a compensator[1] $C \in S(P_{\lambda_0})$ *robustly stabilizes* the family of plants P_λ if (i) there exists a neighborhood $\mathbf{N}$ of λ in Λ such that $C \in S(P_\lambda)\ \forall \lambda \in \mathbf{N}$, and (ii) the stable closed-loop transfer matrix $H(P_\lambda, C)$ is continuous in λ at λ_0. The first condition means that C stabilizes not only the nominal plant P_{λ_0} but also all "perturbed" plants P_λ for all λ sufficiently near λ_0. But this alone is not enough: The objective of compensator design is not merely to stabilize a plant but also to shape the stable closed-loop response. Thus the second condition is imposed, requiring that the perturbed (stable) closed-loop response $H(P_\lambda, C)$ be close to the nominal (stable) closed-loop response $H(P_{\lambda_0}, C)$ whenever λ_0 is near λ. Note that the question as posed above is purely qualitative, in that one is not concerned about the "size" of the neighborhood $\mathbf{N}$, but merely about its existence. The quantitative aspects of robustness are studied in the next two sections.

The question studied (and answered) in this section is: When does there exist a robustly stabilizing compensator for a given family P_λ of plants? The answer involves the introduction of an appropriate topology on the set $\mathbf{M}(\mathbb{R}(s))$ of possibly unstable plants, called the *graph topology*. In terms of the graph topology, the answer to the above question turns out to be exceedingly simple and satisfactory: *Robust stabilization is possible if and only if the function* $\lambda \rightarrow P_\lambda$ *is continuous in the graph topology at* $\lambda = \lambda_0$. In

---

[1] Recall that $S(P)$ is the set of compensators that stabilize P.

other words, robust stabilization is possible if and only if any variations in the parameter λ from its nominal value λ_0 affect the plant description in a *continuous* manner, where the continuity of P_λ is judged in the context of the graph topology.

In this section, the graph topology is defined, and some of its properties are proved, including the above-stated result on robust stabililzability. Then several examples are given with a view to developing some insight into the graph topology. One of these "examples" contains a fairly substantial application of these ideas to the design of compensators for singularly perturbed systems.

As the name implies, the graph topology has to do with the graph of a plant, and we begin with this notion.

Suppose $P \in \mathbf{M}(\mathbb{R}(s))$. Then P is a possibly unstable plant; that is, not all bounded inputs applied to P need result in bounded outputs. But undoubtedly *some* (though not necessarily all) bounded inputs applied to P *will* produce bounded outputs. For example, suppose $p(s) = 1/(s-1)$; then it is intuitively clear that a bounded input $u(\cdot)$ whose Laplace transform has a zero at $s = 1$ will produce a bounded output from p. The *graph* of a plant P is just the set of bounded input-output pairs corresponding to P. Precisely, if $P \in \mathbb{R}(s)^{n \times m}$, then

1 $$\mathbf{G_S}(P) = \{(u,y) \in \mathbf{S}^{n+m} : y = Pu\}$$

is the *graph of* P with respect to $\mathbf{S}$.

2 *Lemma* Suppose $P \in \mathbb{R}(s)^{n \times m}$, and let (N,D) be any r.c.f. of P. Then

3 $$\mathbf{G_S}(P) = \{(Dz, Nz) : z \in \mathbf{S}^m\}.$$

Remarks This lemma shows that the graph of a plant P can be very simply parametrized in terms of any r.c.f. (N,D) of P. As the free parameter z varies over $\mathbf{S}^m$, the pair (Dz, Nz) generates the set of all "bounded" input-output pairs corresponding to P.

As in [26, 77], let $\mathbf{L}_\infty$ denote the set of all measurable essentially bounded functions mapping $[0,\infty)$ into the real numbers, and let $\mathbf{L}_2$ denote the set of all measurable square-integrable functions mapping $[0,\infty)$ into $\mathbb{R}$. Then one can also define the graph of P with respect to the spaces $\mathbf{L}_\infty$ and $\mathbf{L}_2$ in a manner entirely analogous with (1), namely

4 $$G_{\mathbf{L}_\infty}(P) = \{(u,y)\in\mathbf{L}_\infty^{n+m}: \hat{y} = P\hat{u}\},$$

5 $$\mathbf{G}_{\mathbf{L}_2}(P) = \{(u,y)\in\mathbf{L}_2^{n+m}: \hat{y} = P\hat{u}\},$$

where the hat denotes Laplace transformation. It can be shown [V5] that these graphs can also be simply characterized in terms of an r.c.f. of P. Specifically, if (N,D) is any r.c.f. of P, then

6 $$\mathbf{G}_{\mathbf{L}_\infty}(P) = \{(Dz,Nz): z\in\mathbf{L}_\infty^m\},$$

7 $$\mathbf{G}_{\mathbf{L}_2}(P) = \{(Dz,Nz): z\in\mathbf{L}_2^m\},$$

where D and N are now interpreted as bounded operators on $\mathbf{L}_\infty$ or $\mathbf{L}_2$. But technically Lemma (2) is easier to prove than (6) or (7).

Proof Suppose $u = Dz$ for some $z\in\mathbf{S}^m$. Then $Pu = ND^{-1}Dz = Nz\in\mathbf{S}^n$. Hence, for all $z\in\mathbf{S}^m$, the pair (Dz,Nz) belongs to $\mathbf{G}_\mathbf{S}(P)$. It remains to show that every point $(u,y)\in\mathbf{G}_\mathbf{S}(P)$ is of the form (Dz,Nz) for some $z\in\mathbf{S}^m$. Since N and D are right-coprime, one can select $X,Y\in\mathbf{M}(\mathbf{S})$ such that $XN + YD = I$. Now suppose $(u,y)\in\mathbf{G}_\mathbf{S}(P)$, i.e. suppose $u\in\mathbf{S}^m$, $y\in\mathbf{S}^n$ and $y = Pu$. Define $z = D^{-1}u\in\mathbb{R}(s)^m$. Then clearly $(u,y) = (Dz,Nz)$. The proof is complete if it can be shown that z actually belongs to $\mathbf{S}^m$. Note that

8 $$z = D^{-1}u = (XN+YD)D^{-1}u = XND^{-1}u+Yu$$
$$= Xy+Yu\in\mathbf{S}^m,$$

since all quantities in the last expression belong to $\mathbf{M}(\mathbf{S})$. □

9 *Corollary* Suppose $P \in \mathbf{S}^{n \times m}$. Then

10 $$\mathbf{G}_{\mathbf{S}}(P) = \{(z, Pz) \colon z \in \mathbf{S}^m\}.$$

Proof Note that (P, I) is an r.c.f. of P. □

Now we are ready to define the graph topology. Suppose (N,D) is a right-coprime pair, and that $|D| \neq 0$. Then there exists a constant $\mu = \mu(N,D)$ such that (N_1, D_1) is also a right-coprime pair and $|D_1| \neq 0$ whenever $\| [(N_1 - N)' \ (D_1 - D)']' \| < \mu(N,D)$. To see this, first select any $X, Y \in \mathbf{M}(\mathbf{S})$ such that $XN + YD = I$, and let $\mu_1 = 1 / \| [X \ \ Y] \|$. Then

11 $$\| XN_1 + YD_1 - I \| < 1 \text{ whenever } \left\| \begin{bmatrix} N_1 - N \\ D_1 - D \end{bmatrix} \right\| < \mu_1.$$

This shows that $XN_1 + YD_1 \in \mathbf{U}(\mathbf{S})$, whence N_1 and D_1 are right-coprime. Next, since $|D| \neq 0$, the ball $\mathbf{B}(|D|, r)$ does not contain 0 whenever $r \leq \| \, |D| \, \|$. Let $r = \| \, |D| \, \|$. Since the map $D_1 \rightarrow |D_1|$ is continuous, the set of matrices D_1 such that $|D_1| \in \mathbf{B}(|D|, r)$ contains a ball $\mathbf{B}(D, \mu_2)$ centered at D. Now let $\mu = \min\{\mu_1, \mu_2\}$. Then the preceding arguments show that N_1, D_1 are coprime and $|D_1| \neq 0$ whenever $(N_1, D_1) \in \mathbf{B}((N,D), \mu)$. Of course, the bound μ is not uniquely specified by the above conditions, but this does not matter.

Given a plant $P \in \mathbf{M}(\mathbb{R}(s))$, a *basic neighborhood of* P is defined as follows: Let (N,D) be any r.c.f. of P, and let ε be any positive number less than $\mu(N,D)$. Then the set

12 $$\mathbf{N}(N,D;\varepsilon) = \{P_1 = N_1 D_1^{-1} \colon \left\| \begin{bmatrix} N_1 - N \\ D_1 - D \end{bmatrix} \right\| < \varepsilon\}$$

is a basic neighborhood of P. Thus a basic neighborhood of P merely consists of all plants P_1 which have an r.c.f. (N_1, D_1) that is "close" to an r.c.f. of P. To put it another way, a basic neighborhood of P is obtained by starting with some r.c.f. (N,D) of P and making a "small" perturbation of (N,D) to (N_1, D_1), and then taking the ratio $N_1 D_1^{-1}$. Now by varying ε over all positive numbers less than $\mu(N,D)$, by varying (N,D) over all r.c.f.'s of P, and by varying P over all elements of $\mathbf{M}(\mathbb{R}(s))$, we obtain a collection of basic neighborhoods.

13 *Lemma* The collection of basic neighborhoods forms a base for a topology on $\mathbf{M}(\mathbb{R}(s))$.

Proof In accordance with the contents of Section C.1, it is necessary to prove that axioms (B1) and (B2) are satisfied. For convenience, these axioms are recalled below:

(B1) The collection of basic neighborhoods covers $\mathbf{M}(\mathbb{R}(s))$.

(B2) If $\mathbf{N}_1$, $\mathbf{N}_2$ are basic neighborhoods and $P \in \mathbf{N}_1 \cap \mathbf{N}_2$, then there is a basic neighborhood $\mathbf{N}$ such that $P \in \mathbf{N} \subseteq (\mathbf{N}_1 \cap \mathbf{N}_2)$.

Of these, it is at once obvious that (B1) holds, since every $P \in \mathbf{M}(\mathbb{R}(s))$ belongs to some basic neighborhood. To prove (B2), suppose

14 $$\mathbf{N}_1 = \{\overline{N}\overline{D}^{-1}: \left\| \begin{bmatrix} \overline{N} - N_1 \\ \overline{D} - D_1 \end{bmatrix} \right\| < \varepsilon_1\},$$

15 $$\mathbf{N}_2 = \{\overline{N}\overline{D}^{-1}: \left\| \begin{bmatrix} \overline{N} - N_2 \\ \overline{D} - D_2 \end{bmatrix} \right\| < \varepsilon_2\},$$

and suppose $P \in \mathbf{N}_1 \cap \mathbf{N}_2$. Then P has r.c.f.'s $(\overline{N}_1, \overline{D}_1)$, $(\overline{N}_2, \overline{D}_2)$ such that

16 $$\left\| \begin{bmatrix} \overline{N}_1 - N_1 \\ \overline{D}_1 - D_1 \end{bmatrix} \right\| =: r_1 < \varepsilon_1, \quad \left\| \begin{bmatrix} \overline{N}_2 - N_2 \\ \overline{D}_2 - D_2 \end{bmatrix} \right\| =: r_2 < \varepsilon_2.$$

Moreover, there exists a $U \in \mathbf{U}(\mathbf{S})$ such that $\overline{N}_2 = \overline{N}_1 U$, $\overline{D}_2 = \overline{D}_1 U$. Now select a positive number r such that $r \le \min\{\varepsilon_1 - r_1, \varepsilon_2 - r_2 / \| U \| \}$, and define

17 $$\mathbf{N} = \{ND^{-1}: \| \begin{bmatrix} N - \bar{N}_1 \\ D - \bar{D}_1 \end{bmatrix} \| < r\}.$$

Then $\mathbf{N}$ is certainly a basic neighborhood containing P. It only remains to show that $\mathbf{N} \subseteq \mathbf{N}_1 \cap \mathbf{N}_2$. This follows readily: First,

18 $$\| \begin{bmatrix} N - \bar{N}_1 \\ D - \bar{D}_1 \end{bmatrix} \| < r \Rightarrow \| \begin{bmatrix} N - N_1 \\ D - D_1 \end{bmatrix} \| < r_1 + r < \varepsilon_1,$$

which shows that $ND^{-1} \in \mathbf{N}_1$. Next,

19 $$\| \begin{bmatrix} N - \bar{N}_1 \\ D - \bar{D}_1 \end{bmatrix} \| < r \Rightarrow \| \begin{bmatrix} NU - \bar{N}_2 \\ DU - \bar{D}_2 \end{bmatrix} \| < r \| U \|$$

$$\Rightarrow \| \begin{bmatrix} NU - N_2 \\ DU - D_2 \end{bmatrix} \| < r_2 + r \| U \| < \varepsilon_2,$$

which shows that $(NU)(DU)^{-1} = ND^{-1} \in \mathbf{N}_2$. Hence $\mathbf{N} \subseteq \mathbf{N}_1 \cap \mathbf{N}_2$. □

The topology on $\mathbf{M}(\mathbb{R}(s))$ defined by the collection of sets of the form (12) as base is called the *graph topology*. In it, two plants P_1 and P_2 are "close" if they have r.c.f.'s (N_1, D_1), (N_2, D_2) such that $\| (N_1, D_1) - (N_2, D_2) \|$ is "small." In view of Lemma (2), this means that P_1 and P_2 are "close" if their graphs are "close." This is the reason for the name "graph topology."

The central result of this section is that the graph topology on $\mathbf{M}(\mathbb{R}(s))$ is the weakest topology in which feedback stability is a robust property. In order to state this result precisely, it is necessary first to characterize the convergence of sequences and continuity of functions in the graph topology. Suppose $\{P_i\}$ is a sequence in $\mathbf{M}(\mathbb{R}(s))$ and that $P \in \mathbf{M}(\mathbb{R}(s))$. Then by definition (Section C.1), $\{P_i\}$ converges to P if and only if every neighborhood of P contains all but a finite number of terms of the sequence $\{P_i\}$. Since every neighborhood of P contains a basic neighborhood of P, it is easy to see that $\{P_i\}$ converges to P if and only if every *basic* neighborhood of P contains all but a finite number of terms of $\{P_i\}$. All this is a ready consequence of the definition of the graph topology and of the definition of convergence in a topological space. The next lemma gives several equivalent characterizations of convergence in the graph topology.

20 *Lemma* Suppose $\{P_i\}$ is a sequence in $\mathbf{M}(\mathbb{R}(s))$ and that $P \in \mathbf{M}(\mathbb{R}(s))$. Then the following statements are equivalent:

(i) $\{P_i\}$ converges to P in the graph topology.

(ii) For *every* r.c.f. (N,D) of P, there exist r.c.f.'s (N_i,D_i) of P_i such that $N_i \to N$, $D_i \to D$ in $\mathbf{M}(\mathbf{S})$.

(iii) There exist *an* r.c.f. (N,D) of P and a sequence $\{(N_i,D_i)\}$ of r.c.f.'s of P_i such that $N_i \to N$, $D_i \to D$ in $\mathbf{M}(\mathbf{S})$.

(iv) For *every* l.c.f. $(\tilde{D},\tilde{N})$ of P, there exists a sequence $\{(\tilde{D}_i,\tilde{N}_i)\}$ of l.c.f.'s of P_i such that $\tilde{N}_i \to \tilde{N}$, $\tilde{D}_i \to \tilde{D}$ in $\mathbf{M}(\mathbf{S})$.

(v) There exist *an* l.c.f. $(\tilde{D},\tilde{N})$ of P and a sequence $\{(\tilde{D}_i,\tilde{N}_i)\}$ of l.c.f.'s of P_i such that $\tilde{N}_i \to \tilde{N}$, $\tilde{D}_i \to \tilde{D}$ in $\mathbf{M}(\mathbf{S})$.

Proof (i) $\Longleftrightarrow$ (ii). This is a ready consequence of the definition of the graph topology and of the discussion in the paragraph preceding the lemma. Suppose $\{P_i\}$ converges to P. Then every basic neighborhood of P contains all but a finite number of terms in the sequence $\{P_i\}$. In particular, let (N,D) be any r.c.f. of P and let ε_i be any sequence of positive numbers converging to zero. (Assume without loss of generality that $\varepsilon_i < \mu(N,D)\ \forall i$.) Define

21 $$\mathbf{N}_i = \{\overline{N}\overline{D}^{-1} : \left\| \begin{bmatrix} \overline{N} - N \\ \overline{D} - D \end{bmatrix} \right\| < \varepsilon_i \}.$$

Then for each integer k there exists an integer j_k such that $P_i \in \mathbf{N}_k\ \forall i \geq j_k$. That is, there exist r.c.f.'s (N_i,D_i) of P_i such that

22 $$\left\| \begin{bmatrix} N_i - N \\ D_i - D \end{bmatrix} \right\| < \varepsilon_k \ \forall i \geq j_k.$$

From this, one can conclude that

23 $$\left\| \begin{bmatrix} N_i - N \\ D_i - D \end{bmatrix} \right\| \to 0,$$

i.e. that $N_i \to N$, $D_i \to D$ in $\mathbf{M}(\mathbf{S})$. This shows that (i) $\Rightarrow$ (ii). The converse implication follows just as easily. Suppose (ii) is true and let $\mathbf{N}$ be any basic neighborhood of P. Then $\mathbf{N}$ is of the form (12) for some r.c.f. (N,D) of P and some $\varepsilon > 0$. By (ii), there exists a sequence $\{(N_i,D_i)\}$ of r.c.f.'s of P_i such that $(N_i,D_i) \to (N,D)$ in $\mathbf{M}(\mathbf{S})$. Hence $\| [N_i' \ D_i']' - [N' \ D']' \| < \varepsilon$ for large enough i, i.e. $P_i \in \mathbf{N}$ for large enough i. Since this is true for *all* basic neighborhoods of P, it follows that $\{P_i\}$ converges to P.

(ii) $\Rightarrow$ (iii). Obvious.

(iii) $\Rightarrow$ (ii). Suppose (N,D) is an r.c.f. of P with the property that there exists a sequence $\{(N_i,D_i)\}$ of r.c.f.'s of P_i such that $N_i \to N$, $D_i \to D$. Let $(\overline{N},\overline{D})$ be any other r.c.f. of P. It is shown that there exists a sequence $\{(\overline{N}_i,\overline{D}_i)\}$ of r.c.f.'s of P_i such that $\overline{N}_i \to \overline{N}$, $\overline{D}_i \to \overline{D}$. Since $(N,D),(\overline{N},\overline{D})$ are both r.c.f.'s of P, there exists a $U \in \mathbf{U}(\mathbf{S})$ such that $\overline{N} = NU$, $\overline{D} = DU$. Now define $\overline{N}_i = N_i U$, $\overline{D}_i = D_i U$. Then $\overline{N}_i \to \overline{N}$, $\overline{D}_i \to \overline{D}$. Thus (ii) holds.

(iv) $\Leftrightarrow$ (v). The proof is entirely analogous to that of (ii) $\Leftrightarrow$ (iii) above.

Thus far it has been shown that (i), (ii), (iii) are all equivalent, and that (iv) and (v) are equivalent. The proof of the lemma is completed by showing that (iii) $\Leftrightarrow$ (v).

(iii) $\Rightarrow$ (v). By hypothesis, there exist an r.c.f. (N,D) of P and a sequence of r.c.f.'s $\{(N_i,D_i)\}$ of P_i such that $N_i \to N$, $D_i \to D$. Now select an l.c.f. $(\tilde{D},\tilde{N})$ of P and matrices $X,Y,\tilde{X},\tilde{Y} \in \mathbf{M}(\mathbf{S})$ such that

24 $$\begin{bmatrix} Y & X \\ -\tilde{N} & \tilde{D} \end{bmatrix} \begin{bmatrix} D & -\tilde{X} \\ N & \tilde{Y} \end{bmatrix} = I.$$

This can always be done, in view of Theorem (4.1.60). Since $(N_i, D_i) \to (N, D)$ and since $\mathbf{U}(\mathbf{S})$ is an *open* subset of $\mathbf{M}(\mathbf{S})$, it follows that

25 $$F_i = \begin{bmatrix} D_i & -\tilde{X} \\ N_i & \tilde{Y} \end{bmatrix} \in \mathbf{U}(\mathbf{S}) \text{ for large enough } i.$$

Now let $G_i = F_i^{-1} \in \mathbf{M}(\mathbf{S})$ for large enough i, and partition G_i as

26 $$G_i = \begin{bmatrix} Y_i & X_i \\ -\tilde{N}_i & \tilde{D}_i \end{bmatrix}.$$

Then $\tilde{N}_i$, $\tilde{D}_i$ are left-coprime (for large enough i), since $\tilde{N}_i X_i + \tilde{D}_i Y_i = I$. Further, $\tilde{N}_i D_i + \tilde{D}_i N_i = 0$ implies that $\tilde{D}_i^{-1} \tilde{N}_i = N_i D_i^{-1} = P_i$.[2] Hence $(\tilde{D}_i, \tilde{N}_i)$ is an l.c.f. of P_i. Finally, since inversion of unimodular matrices is continuous, we have that

27 $$G_i \to (\lim_i F_i)^{-1} = \begin{bmatrix} Y & X \\ -\tilde{N} & \tilde{D} \end{bmatrix}.$$

This shows that $(\tilde{D}_i, \tilde{N}_i) \to (\tilde{D}, \tilde{N})$, which is an l.c.f. of P. This is (v).

(v) $\Rightarrow$ (iii). This is entirely analogous to the above. □

The next result characterizes the continuity of functions mapping into $\mathbf{M}(\mathbb{R}(s))$. It is a ready consequence of Lemma (20).

28 *Corollary* Suppose Λ is a first-countable topological space, and that $\lambda \to P_\lambda$ is a function mapping Λ into $\mathbf{M}(\mathbb{R}(s))$. Then the following are equivalent:

---

[2] It still has to be shown that $\tilde{D}_i$ is nonsingular for this argument to be valid. However, this is straight-forward; see the proof of Lemma (8.1.45).

(i) The function $\lambda \longrightarrow P_\lambda$ is continuous at $\lambda = \lambda_0$ when $\mathbf{M}(\mathbb{R}(s))$ is equipped with the graph topology.

(ii) For each r.c.f. (N_0, D_0) of P_{λ_0}, there exists a family of r.c.f.'s (N_λ, D_λ) of P_λ such that $(N_{\lambda_0}, D_{\lambda_0}) = (N_0, D_0)$, and such that N_λ, D_λ are continuous (in $\mathbf{M}(\mathbf{S})$) at $\lambda = \lambda_0$.

(iii) There exists a family (N_λ, D_λ) of r.c.f.'s of P_λ such that N_λ, D_λ are continuous (in $\mathbf{M}(\mathbf{S})$) at $\lambda = \lambda_0$.

(iv) For each l.c.f. $(\tilde{D}_0, \tilde{N}_0)$ of P_{λ_0}, there exists a family $(\tilde{D}_\lambda, \tilde{N}_\lambda)$ of l.c.f.'s of P_λ such that $(\tilde{D}_{\lambda_0}, \tilde{N}_{\lambda_0}) = (\tilde{D}_0, \tilde{N}_0)$, and such that $\tilde{D}_\lambda, \tilde{N}_\lambda$ are continuous (in $\mathbf{M}(\mathbf{S})$) at $\lambda = \lambda_0$.

(v) There exists a family $(\tilde{D}_\lambda, \tilde{N}_\lambda)$ of l.c.f.'s of P_λ such that $\tilde{D}_\lambda, \tilde{N}_\lambda$ are continuous (in $\mathbf{M}(\mathbf{S})$) at $\lambda = \lambda_0$.

Now we come to the main result of this section. It is rather wordy in the interests of precision, but can be loosely stated as follows: A family of plants $\{P_\lambda\}$ can be robustly stabilized by a compensator $C \in S(P_{\lambda_0})$ if and only if the function $\lambda \longrightarrow P_\lambda$ is continuous at $\lambda = \lambda_0$ in the graph topology; further, if $\lambda \longrightarrow P_\lambda$ is continuous at λ_0, then *every* $C \in S(P_{\lambda_0})$ robustly stabilizes the family $\{P_\lambda\}$.

29 *Theorem* Suppose $\lambda \longrightarrow P_\lambda$ is a function mapping a first-countable topological space Λ into $\mathbf{M}(\mathbb{R}(s))$.

(i) Suppose the function $\lambda \longrightarrow P_\lambda$ is continuous at $\lambda = \lambda_0$ when $\mathbf{M}(\mathbb{R}(s))$ is equipped with the graph topology. Let C be any element of $S(P_{\lambda_0})$. Then there exists a neighborhood $\mathbf{N}$ of λ_0 such that $C \in S(P_\lambda)\ \forall \lambda \in \mathbf{N}$, and moreover $H(P_\lambda, C)$ is continuous (in $\mathbf{M}(\mathbf{S})$) as a function of λ at λ_0.

(ii) Conversely, suppose there exists a $C \in S(P_{\lambda_0})$ such that $C \in S(P_\lambda)$ for all λ in some neighborhood $\mathbf{N}$ of λ_0, and that $H(P_\lambda, C)$ is continuous (in $\mathbf{M}(\mathbf{S})$) as a function of λ at λ_0. Then the function $\lambda \longrightarrow P_\lambda$ is continuous when $\mathbf{M}(\mathbb{R}(s))$ is equipped with the graph topology.

Proof (i) By hypothesis, the function $\lambda \rightarrow P_\lambda$ is continuous at λ_0. By Corollary (28iii), there exists a family (N_λ, D_λ) of r.c.f.'s of P_λ such that N_λ, D_λ are continuous at λ_0. Since $C \in S(P_{\lambda_0})$, there is an l.c.f. $(\tilde{D}_c, \tilde{N}_c)$ of C such that $\tilde{D}_c D_{\lambda_0} + \tilde{N}_c N_{\lambda_0} = I$ (see Corollary (5.1.30)). Moreover,

30 $$H(P_{\lambda_0}, C) = \begin{bmatrix} I - N_{\lambda_0}\tilde{N}_c & -N_{\lambda_0}\tilde{D}_c \\ D_{\lambda_0}\tilde{N}_c & D_{\lambda_0}\tilde{D}_c \end{bmatrix}.$$

Now define $\Delta_\lambda = \tilde{D}_c D_\lambda + \tilde{N}_c N_\lambda$. Then $\Delta_{\lambda_0} = I$ and $\lambda \rightarrow \Delta_\lambda$ is a continuous function from Λ into $\mathbf{M}(\mathbf{S})$. Since $\mathbf{U}(\mathbf{S})$ is an open subset of $\mathbf{M}(\mathbf{S})$, there is a neighborhood $\mathbf{N}$ of λ_0 in Λ such that $\Delta_\lambda \in \mathbf{U}(\mathbf{S})\ \forall \lambda \in \mathbf{N}$. Hence $C \in S(P_\lambda)\ \forall \lambda \in \mathbf{N}$. Moreover, since inversion of unimodular matrices is continuous, the map $\lambda \rightarrow \Delta_\lambda^{-1}$ is continuous. It now follows from

31 $$H(P_\lambda, C) = \begin{bmatrix} I - N_\lambda \Delta_\lambda^{-1}\tilde{N}_c & N_\lambda \Delta_\lambda^{-1}\tilde{D}_c \\ D_\lambda \Delta_\lambda^{-1}\tilde{N}_c & D_\lambda \Delta_\lambda^{-1}\tilde{D}_c \end{bmatrix}$$

that $H(P_\lambda, C)$ is continuous at $\lambda = \lambda_0$.

(ii) By hypothesis, there exists a $C \in S(P_{\lambda_0})$ and a neighborhood $\mathbf{N}$ of λ_0 such that $C \in S(P_\lambda)\ \forall \lambda \in \mathbf{N}$. Select any r.c.f. (N_0, D_0) of P_{λ_0}. Since $C \in S(P_{\lambda_0})$, by Corollary (5.1.30) there exists an l.c.f. $(\tilde{D}_c, \tilde{N}_c)$ of C such that $\tilde{D}_c D_0 + \tilde{N}_c N_0 = I$. Now, since $C \in S(P_\lambda)\ \forall \lambda \in \mathbf{N}$, it follows by symmetry that $P_\lambda \in S(C)\ \forall \lambda \in \mathbf{N}$. Hence, once again by Corollary (5.1.30), there exists r.c.f.'s (N_λ, D_λ) of P_λ such that $\tilde{D}_c D_\lambda + \tilde{N}_c N_\lambda = I\ \forall \lambda \in \mathbf{N}$. Moreover,

32 $$H(P_{\lambda_0}, C) = \begin{bmatrix} I - N_0\tilde{N}_c & -N_0\tilde{D}_c \\ D_0\tilde{N}_c & D_0\tilde{D}_c \end{bmatrix},$$

33 $$H(P_\lambda, C) = \begin{bmatrix} I - N_\lambda\tilde{N}_c & -N_\lambda\tilde{D}_c \\ D_\lambda\tilde{N}_c & D_\lambda\tilde{D}_c \end{bmatrix}.$$

By hypothesis $H(P_\lambda, C)$ is continuous at $\lambda = \lambda_0$. Hence so is the function

34 $$F_\lambda = \begin{bmatrix} N_\lambda \tilde{N}_c & N_\lambda \tilde{D}_c \\ D_\lambda \tilde{N}_c & D_\lambda \tilde{D}_c \end{bmatrix}.$$

Since $N_{\lambda_0}, D_{\lambda_0}$ are *fixed* elements of $\mathbf{M}(\mathbf{S})$ (i.e. independent of λ), the function

35 $$F_\lambda \begin{bmatrix} N_0 \\ D_0 \end{bmatrix} = \begin{bmatrix} N_\lambda \\ D_\lambda \end{bmatrix} \text{(since } \tilde{N}_c N_0 + \tilde{D}_c D_0 = I)$$

is also continuous at λ_0. Thus N_λ, D_λ are continuous at λ_0. Finally, since $H(P_\lambda, C)$ is continuous at λ_0, it follows that

36 $$\lim_{\lambda \to \lambda_0} H(P_\lambda, C) = H(P_{\lambda_0}, C).$$

Then from (35) it follows that

37 $$\lim_{\lambda \to \lambda_0} \begin{bmatrix} N_\lambda \\ D_\lambda \end{bmatrix} = \begin{bmatrix} N_0 \\ D_0 \end{bmatrix}.$$

Now by Corollary (28ii), the map $\lambda \to P_\lambda$ is continuous at λ_0. □

In the proof of Theorem (29), the only properties of $\mathbf{M}(\mathbf{S})$ used were (i) the set of units is open, and (ii) inversion of a unit is continuous. Thus Theorem (29) holds under extremely general conditions. Specifically, one can replace $\mathbf{S}$ by any topological ring with the above two properties, plus the obvious ones (namely $\mathbf{S}$ is a commutative domain with identity), and Theorem (29) remains valid. The same remark applies to Theorem (38) and Corollary (39) below.

Theorem (29) is addressed to the robustness of feedback stability in the case where the plant is perturbed but the compensator remains fixed. The next result is addressed to the case where *both* the plant and the compensator may be perturbed. Suppose a family of plants $\{P_\lambda\}$ and a family of compensators $\{C_\lambda\}$ is specified, and consider the function $\lambda \to (P_\lambda, C_\lambda)$. Theorem (38) below relates the behavior of this function to the robustness of feedback stability. Before the presentation of the theorem, however, a word is needed about the problem formulation. Modeling the plant and compensator perturbations by a family $\{P_\lambda, C_\lambda\}$ where *both* P and C are functions of the *same* parameter λ gives the misleading impression that plant and compensator perturbations are in some way "coupled"" But this is not so. Suppose plant uncertainty is modeled by specifying a family of plants P_α where α belongs to a topological space A, and compensator uncertainty similarly leads to a family of compensators C_β where β belongs to another topological space B. In such a case, define the parameter space Λ to be the cartesian product of A and B equipped with the product topology, and define $P_{(\alpha,\beta)} = P_\alpha$, $C_{(\alpha,\beta)} = C_\beta$. In this way, "independent" plant and compensator uncertainty can also be treated by Theorem (38).

38 *Theorem* Suppose $\lambda \rightarrow P_\lambda$, $\lambda \rightarrow C_\lambda$ are functions mapping a first-countable topological space Λ into $\mathbf{M}(\mathbb{R}(s))$, and suppose that the pair $(P_{\lambda_0}, C_{\lambda_0})$ is stable.

(i) Suppose the functions $\lambda \rightarrow P_\lambda$, $\lambda \rightarrow C_\lambda$ are continuous at $\lambda = \lambda_0$ in the graph topology. Then there exists a neighborhood $\mathbf{N}$ of λ_0 such that (P_λ, C_λ) is stable for all $\lambda \in \mathbf{N}$, and in addition $H(P_\lambda, C_\lambda)$ is continuous at $\lambda = \lambda_0$ in the norm topology on $\mathbf{M}(\mathbf{S})$.

(ii) Conversely, suppose there is a neighborhood $\mathbf{N}$ of λ_0 such that (P_λ, C_λ) is stable for all $\lambda \in \mathbf{N}$, and such that $H(P_\lambda, C_\lambda)$ is continuous at $\lambda = \lambda_0$ in the norm topology on $\mathbf{M}(\mathbf{S})$. Then the functions $\lambda \rightarrow P_\lambda$, $\lambda \rightarrow C_\lambda$ are continuous at $\lambda = \lambda_0$ in the graph topology.

The proof of this theorem is deferred to Section 7.3. However, combining Theorems (38) and (29) leads to the following result, which is relevant in the context of singularly perturbed systems.

39 *Corollary* Suppose $\lambda \rightarrow P_\lambda$ is a function mapping a first-countable topological space Λ into $\mathbf{M}(\mathbb{R}(s))$, and suppose a point $\lambda_0 \in \Lambda$ is specified. If there exist a neighborhood $\mathbf{N}$ of λ_0 and a family $\{C_\lambda, \lambda \in \mathbf{N}\}$ in $\mathbf{M}(\mathbb{R}(s))$ of compensators such that (P_λ, C_λ) is stable for all $\lambda \in \mathbf{N}$ and such that $H(P_\lambda, C_\lambda)$ is continuous at $\lambda = \lambda_0$, then there exists a $C \in \mathbf{M}(\mathbb{R}(s))$ and a neighborhood $\mathbf{B}$ of λ_0 such that (P_λ, C) is stable for all $\lambda \in \mathbf{B}$ and $H(P_\lambda, C)$ is

continuous at $\lambda = \lambda_0$.

Remarks The point of the corollary is this: Suppose a family of plants $\{P_\lambda\}$ is specified, together with a nominal plant P_{λ_0}. If it is possible to design a *family* of compensators C_λ such that C_λ robustly stabilizes the family of plants $\{P_\lambda\}$, then the same can be accomplished using a *fixed* (i.e. λ-independent) compensator.

Proof The hypotheses imply, by Theorem (38ii), that both $\lambda \rightarrow P_\lambda$, $\lambda \rightarrow C_\lambda$ are continuous in the graph topology at $\lambda = \lambda_0$. Now let $C = \lim_{\lambda \rightarrow \lambda_0} C_\lambda$, or any other element of $S(P_{\lambda_0})$. Then by Theorem (29i), C robustly stabilizes the family $\{P_\lambda\}$. □

The preceding results show that the graph topology is of fundamental importance in the study of the robustness of feedback stability. Hence it is worthwhile to develop a further understanding of it, with particular emphasis on what continuity in this topology means. The next several propositions are directed towards this aim.

40 *Proposition* On the set $\mathbf{M}(\mathbf{S})$ viewed as a subset of $\mathbf{M}(\mathbb{R}(s))$, the norm topology and the graph topology are the same. $\mathbf{M}(\mathbf{S})$ is an open subset of $\mathbf{M}(\mathbb{R}(s))$ in the graph topology.

Proof To prove the first sentence, it is enough to show that convergence in the norm topology is equivalent to convergence in the graph topology in $\mathbf{M}(\mathbf{S})$. Accordingly, suppose $\{P_i\}$ is a sequence[3] in $\mathbf{M}(\mathbf{S})$ and that $P \in \mathbf{M}(\mathbf{S})$. Suppose first that $\| P_i - P \| \rightarrow 0$. Then $C = 0$ stabilizes P as well as P_i, since $P, P_i \in \mathbf{M}(\mathbf{S})$. Further, since $\| P_i - P \| \rightarrow 0$, it follows from (7.1.1) that $H(P_i, 0) \rightarrow H(P, 0)$ in $\mathbf{M}(\mathbf{S})$. Hence by Theorem (29ii), we conclude that $\{P_i\}$ converges to P in the graph topology.

[3] Strictly speaking, one should examine the convergence of nets rather than sequences. However, it is shown in Section 7.3 that the graph topology is metrizable and hence first-countable; so it is enough to examine only sequences.

To prove the converse, and incidentally to prove the second sentence, suppose $\{P_i\}$ is a sequence in $\mathbf{M}(\mathbb{R}(s))$ (not necessarily in $\mathbf{M}(\mathbf{S})$) converging in the graph topology to $P \in \mathbf{M}(\mathbf{S})$. Now (P,I) is an r.c.f. of P. Hence by Lemma (20ii) there exist r.c.f.'s (N_i,D_i) of P_i such that $N_i \rightarrow P$, $D_i \rightarrow I$ in $\mathbf{M}(\mathbf{S})$. Now I is a unimodular matrix. Since $\mathbf{U}(\mathbf{S})$ is an open subset of $\mathbf{M}(\mathbf{S})$ and since the inversion of unimodular matrices is continuous, this implies that D_i is unimodular for large enough i and that $D_i^{-1} \rightarrow I$ in $\mathbf{M}(\mathbf{S})$. Hence $P_i = N_i D_i^{-1} \in \mathbf{M}(\mathbf{S})$ for large enough i; moreover, since $\| N_i - P \| \rightarrow 0$ and $\| D_i^{-1} - I \| \rightarrow 0$, it follows that $\| P_i - P \| = \| N_i D_i^{-1} - I \| \rightarrow 0$. As it has been shown during the preceding argument that every sequence in $\mathbf{M}(\mathbb{R}(s))$ that converges to a limit in $\mathbf{M}(\mathbf{S})$ must eventually lie in in $\mathbf{M}(\mathbf{S})$, we conclude that $\mathbf{M}(\mathbf{S})$ is an open subset of $\mathbf{M}(\mathbb{R}(s))$. □

Proposition (40) shows that, in the case of functions mapping into $\mathbf{M}(\mathbf{S})$, continuity in the graph topology is the same as continuity in the usual norm topology. In the case of unstable plant sequences, however, several strange things can happen. Before illustrating these difficulties, a preliminary result is presented that is often helpful in concluding *lack* of continuity.

41 *Proposition* Suppose $\lambda \rightarrow P_\lambda$ is a function from a first-countable topological space Λ into $\mathbf{M}(\mathbb{R}(s))$, and suppose this function is continuous in the graph topology at $\lambda = \lambda_0$. Let $s_1, \cdots, s_l$ denote the poles of P_{λ_0} in the closed RHP.

(i) Let Ω be any compact subset of the open RHP such that none of the s_i lies on the boundary of Ω, and let γ denote the number of poles of P_{λ_0} inside Ω, counted according to their McMillan degrees. Then there is a neighborhood $\mathbf{N}$ of λ_0 such that P_λ has exactly γ poles inside Ω for all $\lambda \in \mathbf{N}$.

(ii) Let Ω be any closed subset of the closed RHP that does not contain $s_1, \cdots, s_l$; if P_{λ_0} has a pole at infinity, suppose in addition that Ω is bounded. Then $P_\lambda(s) \rightarrow P_{\lambda_0}(s)$ as $\lambda \rightarrow \lambda_0$, uniformly for all $s \in \Omega$.

Remarks In general, the number of *closed* RHP poles of P_λ and of P_{λ_0} need not be equal. For example, let $\Lambda = [0,\infty)$, $\lambda_0 = 0$, and $P_\lambda(s) = 1/(s+\lambda)$.

Proof Since the function $\lambda \rightarrow P_\lambda$ is continuous in the graph topology at $\lambda = \lambda_0$, it follows from Corollary (28iii) that there exists a family (N_λ, D_λ) of r.c.f.'s of P_λ such that N_λ, D_λ are continuous at λ_0. To prove (i), observe that $|D_\lambda|$ is continuous at λ_0; hence $|D_\lambda|(s) \rightarrow |D_{\lambda_0}|(s)$ uniformly in s as $\lambda \rightarrow \lambda_0$. Since $|D_\lambda|$ and $|D_{\lambda_0}|$ are both analytic in the open RHP, and since $|D_{\lambda_0}|$ has no zeros on the boundary of Ω, it follows by applying the principle of the argument to $|D_\lambda|$ and to $|D_{\lambda_0}|$ that, for some neighborhood $\mathbf{N}$ of λ_0, the number of zeros of $|D_\lambda|$ inside Ω is the same for all $\lambda \in \mathbf{N}$. Since the RHP poles of P_λ are in one-to-one correspondence with the zeros of $|D_\lambda|$, and since the McMillan degree of an RHP pole of P_λ is precisely its multiplicity as a zero of $|D_\lambda|$, (i) follows.

To prove (ii), observe that since $D_\lambda(s) \rightarrow D_{\lambda_0}(s)$ uniformly over the closed RHP, it follows that $[D_\lambda(s)]^{-1} \rightarrow [D_{\lambda_0}(s)]^{-1}$ wherever the latter is well-defined, i.e. wherever $|D_{\lambda_0}(s)| \neq 0$. Moreover, if we exclude a neighborhood of every closed RHP zero of $|D_{\lambda_0}|$, the convergence is uniform with respect to s. Since $P_\lambda(s) = N_\lambda(s)[D_\lambda(s)]^{-1}$, the same is true of P_λ.

□

One of the nice features of the norm topology in $\mathbf{M}(\mathbf{S})$ is that it is a *product* topology: That is, a sequence of matrices $\{P^{(i)}\}$ in $\mathbf{M}(\mathbf{S})$ converges to $P \in \mathbf{M}(\mathbf{S})$ if and only if each component sequence $\{p_{jk}^{(i)}\}$ converges to p_{jk} for all j, k. Consequently, a function $\lambda \rightarrow P_\lambda$ mapping Λ into $\mathbf{M}(\mathbf{S})$ is continuous at λ_0 if and only if each component of P_λ is continuous at λ_0. Now Proposition (40) shows that the graph topology and the norm topology coincide on $\mathbf{M}(\mathbf{S})$. Hence, *when restricted to* $\mathbf{M}(\mathbf{S})$, the graph topology is also a product topology. However, in the case where a function $\lambda \rightarrow P_\lambda$ maps Λ into *unstable multivariable plants*, the continuity of the matrix-valued map P_λ at λ_0 *neither implies nor is implied by* the continuity at λ_0 of the scalar-valued maps $\lambda \rightarrow p_{ij\lambda}$ for all i, j. This is illustrated through some examples. First a scalar example is presented to set the stage.

42 *Example* Let $\Lambda = \mathbb{R}, \lambda_0 = 0$, and consider the function

43 $$p_\lambda(s) = \frac{s-1-\lambda}{s-1}.$$

Then at $\lambda = 0$ the function p_λ has a pole-zero cancellation in the open right half-plane, namely at $s = 1$. This causes p_λ to be discontinuous at $\lambda = 0$, as can be seen by the following reasoning. Now $p_0 = 1$, which belongs to $\mathbf{S}$. If p_λ were to be continuous at $\lambda = 0$, it would follow from Proposition (40) that $p_\lambda \in \mathbf{S}$ for all small enough λ. Since clearly $p_\lambda \notin \mathbf{S}\ \forall \lambda \neq 0$, the conclusion is that p is *not* continuous at $\lambda = 0$.

44 *Example* Again let $\Lambda = \mathbb{R}, \lambda_0 = 0$ and consider the function

45 $$P_\lambda(s) = \begin{bmatrix} \dfrac{-\lambda s}{s-1} & \dfrac{s-1-\lambda}{s-1} \\ \dfrac{2s^2-1}{s^2-1} & \dfrac{-s^2+s+1}{s^2-1} \end{bmatrix}.$$

Then, from Example (42), the (1,2)-component of P_λ is not continuous in the graph topology at $\lambda = 0$. Nevertheless, the function $\lambda \rightarrow P_\lambda$ mapping $\mathbb{R}$ into $\mathbf{M}(\mathbb{R}(s))$ *is* continuous in the graph topology at $\lambda = 0$. To see this, define

46 $$N_\lambda(s) = \begin{bmatrix} z+\lambda & z-1 \\ z-2 & 1 \end{bmatrix}, D_\lambda(s) = \begin{bmatrix} z-1 & z \\ z & z-1 \end{bmatrix},$$

where $z = s/(s+1) \in \mathbf{S}$. Then it is easy to see that N_λ is continuous at $\lambda = 0$, and D_λ is independent of λ. Moreover, $P_\lambda = N_\lambda D_\lambda^{-1}$, as is easily checked. Hence by Corollary (28iii), the function P_λ is continuous in the graph topology at $\lambda = 0$. This example shows that the continuity of a matrix-valued function does *not* imply the continuity of its components.

47 *Example* Once again let $\Lambda = \mathbb{R}, \lambda_0 = 0$, and consider

48 $$P_\lambda(s) = \left[\frac{s+1}{s-1-\lambda} \quad \frac{s+1}{s-1+\lambda} \right].$$

Then

49 $$P_0(s) = \left[\frac{s+1}{s-1} \quad \frac{s+1}{s-1} \right].$$

It is immediate from (48) and (49) that each component of P_λ is continuous in the graph topology at $\lambda = 0$. Nevertheless, P_λ itself is *not* continuous at $\lambda = 0$. To see why this is so, note that P_0 has a pole at $s = 1$ of McMillan degree one, and this is the only pole of P_0. On the other hand, for *each nonzero* $\lambda \in (-1,1)$, P_λ has *two* poles in the closed RHP, each of McMillan degree one. Since Proposition (41i) fails to hold, the conclusion is that P_λ is not continuous in the graph topology at $\lambda = 0$.

The next result shows that, while the graph topology may be the "right" topology for studying the robustness of feedback stability, addition and multiplication in $\mathbb{R}(s)$ (or $\mathbf{M}(\mathbb{R}(s))$) are not necessarily continuous operations in this topology.

50 *Proposition* $\mathbb{R}(s)$ equipped with the graph topology is not a topological ring.

Proof It is necessary to demonstrate that either addition or multiplication fails to be continuous. In actuality, *both* operations are not continuous in general. First, let $\{\varepsilon_i\}$ be any sequence converging to zero, and let

51 $$p_i(s) = \frac{s-3-\varepsilon_i}{s-2}, \; q_i(s) = \frac{1}{s-2},$$

$$p(s) = \frac{s-3}{s-2}, \; q(s) = \frac{1}{s-2}.$$

Then $p_i \rightarrow p$, $q_i \rightarrow q$ in the graph topology, but

52 $$(p_i + q_i)(s) = \frac{s-2-\varepsilon_i}{s-2}$$

does *not* approach $p+q = 1$ in the graph topology (see Example (42)). Hence addition is not continuous in general. An example of the noncontinuity of multiplication is provided by

53 $$p_i(s) = \frac{s-1-\varepsilon_i}{s+1}, \; q_i(s) = \frac{s+1}{s-1+\varepsilon_i},$$

$$p(s) = \frac{s-1}{s+1}, \; q(s) = \frac{s+1}{s-1}.$$

The details are left to the reader. □

Using entirely analogous reasoning, one can show that the matrix ring $\mathbb{R}(s)^{n \times n}$ equipped with the graph topology fails to be a topological ring for every integer $n \geq 1$.

Up to now we have seen a spate of negative results, including the fact that the continuity of a matrix-valued function cannot be related in a simple way to the continuity of its components, and that addition and multiplication need not be continuous. These results are presented primarily to impress upon the reader the need to proceed with caution, and to demonstrate that "obvious" statements such as "$\lim_{\lambda \to \lambda_0} (P_\lambda + Q_\lambda) = \lim_{\lambda \to \lambda_0} P_\lambda + \lim_{\lambda \to \lambda_0} Q_\lambda$" need not always hold true. Happily, there are also several *positive* statements that one can make about the graph topology.

54 *Proposition* Suppose Λ is a first-countable topological space, and that $P, Q : \Lambda \to \mathbf{M}(\mathbb{R}(s))$, $R : \Lambda \to \mathbf{M}(\mathbf{S})$ satisfy $P_\lambda = Q_\lambda + R_\lambda \; \forall \lambda \in \Lambda$. Suppose in addition that R is continuous at λ_0. Then P is continuous at λ_0 if and only if Q is continuous at λ_0.

Remarks As we have seen in the proof of Proposition (50), the statement

55 $$\lim_{\lambda \to \lambda_0} P_\lambda = \lim_{\lambda \to \lambda_0} Q_\lambda + \lim_{\lambda \to \lambda_0} R_\lambda$$

is not valild in general. Thus the point of Proposition (54) is that (55) *is* valid provided one of the two limits on the right side of (55) belongs to $\mathbf{M}(\mathbf{S})$.

Proof "if" Since Q is continuous at λ_0, by Corollary (28iii) it follows that there exists a family (N_λ, D_λ) of r.c.f.'s of Q_λ such that N_λ, D_λ are continuous at λ_0. Since $R_\lambda \in \mathbf{M}(\mathbf{S})$, the pair $(N_\lambda + R_\lambda D_\lambda, D_\lambda)$ is an r.c.f. of P_λ. Finally, since $\mathbf{M}(\mathbf{S})$ is a topological ring, the functions $\lambda \rightarrow N_\lambda + R_\lambda D_\lambda$ and $\lambda \rightarrow D_\lambda$ are both continuous at λ_0. This implies, by Corollary (28), that P_λ is continuous at λ_0.

"only if" Write $Q_\lambda = P_\lambda - R_\lambda$ and apply the above reasoning. □

The multiplicative analog of Proposition (54) is stated next.

56 *Proposition* Suppose Λ is a first-countable topological space, and that $P, Q : \Lambda \rightarrow \mathbf{M}(\mathbb{R}(s))$, $R : \Lambda \rightarrow \mathbf{U}(\mathbf{S})$ satisfy $P_\lambda = Q_\lambda R_\lambda \; \forall \lambda \in \Lambda$. Suppose in addition that R is continuous at λ_0. Then P is continuous at λ_0 if and only if Q is continuous at λ_0.

It is left to the reader to prove Proposition (56), by exploiting the fact that R_λ is unimodular for all λ, and to extend Proposition (56) to the case where $P_\lambda = R_\lambda Q_\lambda$.

The final result along these lines shows that the graph topology on $\mathbf{M}(\mathbb{R}(s))$ *is* a product topology under certain circumstances.

57 *Proposition* Suppose P maps a first-countable topological space Λ into $\mathbf{M}(\mathbb{R}(s))$, and that P is block-diagonal of the form $P = \text{Block Diag } \{P_1, \cdots, P_n\}$ where the dimensions of the blocks are the same for all λ. Then P is continuous at $\lambda_0 \in \Lambda$ if and only if each function P_i is continuous at λ_0.

The proof of this proposition is given in Section 7.3. One can generate some more results by combining the preceding three propositions in various ways.

Consider a function $\lambda \rightarrow P_\lambda$ mapping Λ into $\mathbf{M}(\mathbb{R}(s))$. In view of Examples (44) and (47), one can ask whether there is a simple test to determine whether or not P_λ is continuous at λ_0. Such a test is provided by

Theorem (29), and can be stated as follows: Find *any* $C \in S(P_{\lambda_0})$; then P_λ is continuous at λ_0 if and only if C robustly stabilizes the family $\{P_\lambda\}$. Of course, this test begs the question, since the issue of continuity arises only in conjunction with the existence of a robustly stabilizing compensator. Thus the next several propositions are addressed to some *very specific* situations of practical interest.

58 *Proposition* Let $\Lambda = \mathbb{R}, \lambda_0 = 0$, and suppose $P_\lambda(s)$ is the transfer matrix of the system

59 $$\dot{x}(t) = A_\lambda x(t) + B_\lambda u(t), y(t) = C_\lambda x(t).$$

Suppose the triple (C_0, A_0, B_0) is stabilizable and detectable, and that $(C_\lambda, A_\lambda, B_\lambda)$ is continuous in λ at $\lambda = 0$ in the usual norm topology on Euclidean space. Then the map $\lambda \rightarrow P_\lambda$ is continuous in the graph topology at $\lambda = 0$.

Proof This is a straight-forward application of Lemma (4.2.8). Select a matrix F such that $A_0 - FC_0$ is Hurwitz, i.e. such that all eigenvalues of $A_0 - FC_0$ have negative real parts. By continuity, $A_\lambda - FC_\lambda$ is also Hurwitz for sufficiently small λ. From Lemma (4.2.8), an l.c.f. of P_λ is given by

60 $$\tilde{D}_\lambda(s) = I - C_\lambda(sI - A_\lambda + FC_\lambda)^{-1}F,$$

61 $$\tilde{N}_\lambda(s) = C_\lambda(sI - A_\lambda FC_\lambda)^{-1}B_\lambda,$$

for sufficiently small λ. Since $\tilde{D}_\lambda$, $\tilde{N}_\lambda$ are continuous at $\lambda = 0$ in the norm topology on $\mathbf{M}(\mathbf{S})$, it follows from Corollary (28iii) that P_λ is continuous in the graph topology at $\lambda = 0$. □

A consequence of Proposition (58) is that any compensator C that stabilizes the nominal plant P_0, which is described by (59) with $\lambda = 0$, also stabilizes P_λ for all sufficiently small λ; moreover, the closed-loop transfer matrix $H(P_\lambda, C)$ is continuous at $\lambda = 0$.

In Proposition (58), each of the plants P_λ has the same dynamic order (i.e. the same state-space dimension). Proposition (62) below is addressed to the so-called *singular perturbation* case, where the plant P_0 has a lower dynamic order than the plants P_λ, $\lambda > 0$.

62 *Proposition* Suppose $\Lambda = [0,\infty)$, and for each $\lambda > 0$ let $P_\lambda(s)$ be the transfer matrix of the system

63 $$\begin{bmatrix} \dot{x}(t) \\ \lambda \dot{z}(t) \end{bmatrix} = \begin{bmatrix} A_{11} & A_{12} \\ A_{21} & A_{22} \end{bmatrix} \begin{bmatrix} x(t) \\ z(t) \end{bmatrix} + \begin{bmatrix} B_1 \\ B_2 \end{bmatrix} u(t),$$

64 $$y(t) = [C_1 \;\; C_2] \begin{bmatrix} x(t) \\ z(t) \end{bmatrix} + D_1 u(t).$$

Suppose A_{22} is a Hurwitz (and hence nonsingular) matrix, and let $P_0(s)$ be the transfer matrix of the system resulting from substituting $\lambda = 0$ in (63) and (64), namely

65 $$\dot{x}(t) = Ax(t) + Bu(t), y(t) = Cx(t) + D_0 u(t),$$

where

66 $$A = A_{11} - A_{12}A_{22}^{-1}A_{21}, B = B_1 - A_{12}A_{22}^{-1}B_2,$$

$$C = C_1 - C_2A_{22}^{-1}A_{21}, D_0 = D_1 - C_2A_{22}^{-1}B_2.$$

Suppose the triple (C,A,B) is stabilizable and detectable. Then P_λ is continuous at $\lambda = 0$ in the graph topology if $C_2(sI - A_{22})^{-1}B_2 \equiv 0$.

Remarks Equations of the form (63)-(64) are usually used to model systems with both fast and slow dynamical variables. The state variables x (resp. z) are referred to as the slow (resp. fast) modes. In this context, P_0 represents a simplified model of P_λ obtained by neglecting the fast modes. A relevant question for such systems is the following: When does a compensator designed on the basis of the simplified system P_0 work for the true system P_λ? More precisely, suppose one has found a stabilizing compensator C_0 for the simplified system P_0. The issue is whether C_0 also stabilizes P_λ for all sufficiently small positive λ, and if so whether $H(P_\lambda, C_0)$ approaches

$H(P_0, C_0)$ as $\lambda \to 0^+$. On the basis of Theorem (29), we know that this is so if and only if P_λ is continuous in the graph topology at $\lambda = 0$. Now Proposition (62) states that continuity holds if $C_2(sI - A_{22})^{-1} B_2 \equiv 0$.

Proposition (62) has several interesting consequences. For instance, suppose the simplified system P_0 is stabilized by state feedback. This corresponds to the case $y(t) = x(t)$ (i.e. $C_1 = I$, $C_2 = 0$), and $C_0 = K$, a constant matrix. Since the requisite condition holds in this case, it follows that the control law $u(t) = -Kx(t)$ also stabilizes P_λ for all sufficiently small positive λ, and that $H(P_\lambda, K)$ converges to $H(P_0, K)$ as $\lambda \to 0^+$. Similarly, if the control input u does not directly affect the fast modes, (i.e. $B_2 = 0$), then it is again immediate that any compensator that stabilizes the reduced order model also robustly stabilizes the singularly perturbed system for all sufficiently small positive λ.

Proof Suppose $C_2(sI - A_{22})^{-1} B_2 = 0$. The continuity of P_λ at $\lambda = 0$ is established using Corollary (28iii), i.e. by constructing a family $(\tilde{D}_\lambda, \tilde{N}_\lambda)$ of l.c.f.'s of P_λ, $\lambda \geq 0$, such that both $\tilde{D}_\lambda$ and $\tilde{N}_\lambda$ are continuous at $\lambda = 0$ in the norm topology on $\mathbf{M}(\mathbf{S})$.

Select a matrix F such that $A - FC$ is Hurwitz; this can always be done since (C, A) is detectable. Define

67 $$\tilde{D}_0(s) = I - C(sI - A + FC)^{-1} F,$$

68 $$\tilde{N}_0(s) = C(sI - A + FC)^{-1}(B - FD_0) + D_0.$$

Then, by Theorem (4.2.4), $(\tilde{D}_0, \tilde{N}_0)$ is an l.c.f. of P_0. To obtain an l.c.f. of P_λ, let $F_0 = [F' \ \ 0]'$ and define

69 $$\bar{A}_\lambda = \begin{bmatrix} A_{11} & A_{12} \\ A_{21}/\lambda & A_{22}/\lambda \end{bmatrix} - \begin{bmatrix} F \\ 0 \end{bmatrix} [C_1 \ \ C_2]$$
$$= \begin{bmatrix} A_{11} - FC_1 & A_{12} - FC_2 \\ A_{21}/\lambda & A_{22}/\lambda \end{bmatrix}.$$

From [15], the eigenvalues of $\bar{A}_\lambda$ are asymptotically equal to those of A_{22}/λ

plus those of

70 $$A_{11}-FC_1-(A_{12}-FC_2)A_{22}^{-1}A_{21} = A-FC.$$

Hence $\bar{A}_\lambda$ is Hurwitz for all sufficiently small positive λ. Thus, once again by Theorem (4.2.4), an l.c.f. of P_λ is given by

71 $$\tilde{D}_\lambda(s) = I-C_0(sI-\bar{A}_\lambda)^{-1}F_0,$$

72 $$\tilde{N}_\lambda(s) = C_0(sI-\bar{A}_\lambda)^{-1}(B_\lambda-F_0D_1) + D_1,$$

where

73 $$B_\lambda = \begin{bmatrix} B_1 \\ B_2/\lambda \end{bmatrix}, C_0 = [C_1 \ \ C_2].$$

The remainder of the proof consists of studying $\tilde{N}_\lambda$ and $\tilde{D}_\lambda$ in detail and showing that $\tilde{N}_\lambda \to \tilde{N}_0$, $\tilde{D}_\lambda \to \tilde{D}_0$ as $\lambda \to 0^+$. Recall that the inverse of a partitioned matrix is given by

74 $$\begin{bmatrix} X & Y \\ W & V \end{bmatrix}^{-1} = \begin{bmatrix} \Delta^{-1} & -\Delta^{-1}YV^{-1} \\ -V^{-1}W\Delta^{-1} & V^{-1}+V^{-1}W\Delta^{-1}YV^{-1} \end{bmatrix},$$

where

75 $$\Delta = X-YV^{-1}W.$$

Hence

76 $$\tilde{D}_\lambda = I-[C_1 \ \ C_2]\begin{bmatrix} sI-A_{11}+FC_1 & -A_{12}+FC_2 \\ -A_{21}/\lambda & sI-A_{22}/\lambda \end{bmatrix}^{-1}\begin{bmatrix} F \\ 0 \end{bmatrix}$$

$$= I-C_1\Delta^{-1}F+C_2V^{-1}W\Delta^{-1}F,$$

where

77 $$\Delta = (sI-A_{11}+FC_1)-(A_{12}-FC_2)(sI-A_{22}/\lambda)^{-1}A_{21}/\lambda$$

$$= (sI - A_{11} + FC_1) - (A_{12} - FC_2)(\lambda sI - A_{22})^{-1} A_{21},$$

78 $$V = sI - A_{22}/\lambda, \; W = -A_{21}/\lambda.$$

Simplification of (76) yields

79 $$\tilde{D}_\lambda = I - M \Delta^{-1} F,$$

where

80 $$M = C_1 - C_2 V^{-1} W = C_1 + C_2(\lambda sI - A_{22})^{-1} A_{21}.$$

Now, letting $\bar{A}$ denote the Hurwitz matrix $A - FC$, we get

81 $$\tilde{D}_\lambda = I - C(sI - \bar{A})^{-1} F$$
$$= I - C_1(sI - \bar{A})^{-1} F + C_2 A_{22}^{-1} A_{21}(sI - \bar{A})^{-1} F.$$

Hence it can be concluded that $\tilde{D}_\lambda \to \tilde{D}_0$ if it can be shown that

82 $$\Delta^{-1} \to (sI - \bar{A})^{-1},$$

83 $$(\lambda sI - A_{22})^{-1} A_{21} \Delta^{-1} \to -A_{22}^{-1} A_{21}(sI - \bar{A})^{-1}.$$

Note that $(\lambda sI - A_{22})^{-1}$ does *not* approach $-A_{22}^{-1}$.[4] Hence (83) does not automatically follow from (82).

To prove (82), let $\bar{A}_{11}$ denote $A_{11} - FC_1$, and note that

84 $$\Delta^{-1} = [sI - \bar{A}_{11} - (A_{12} - FC_2)(\lambda sI - A_{22})^{-1} A_{21}]^{-1}$$
$$= \{sI - \bar{A}_{11} - (A_{12} - FC_2) A_{22}^{-1} A_{21}$$
$$- (A_{12} - FC_2)[(\lambda sI - A_{22})^{-1} - A_{22}^{-1}] A_{21}\}^{-1}$$
$$= (I - R)^{-1}(sI - \bar{A})^{-1},$$

where

[4] Observe that $(\lambda sI - A_{22})^{-1} \to 0$ as $s \to \infty$ whenever $\lambda > 0$. Hence the norm of the difference between the functions $(\lambda sI - A_{22})^{-1}$ and A_{22}^{-1} is at least $\| A_{22}^{-1} \|$, no matter how small λ is.

85 $$R = (A_{12}-FC_2)[(\lambda sI-A_{22})^{-1}-A_{22}^{-1}](sI-\bar{A})^{-1}.$$

Hence

86 $$\Delta^{-1}-(sI-\bar{A})^{-1} = [(I-R)^{-1}-I](sI-\bar{A})^{-1}.$$

Now $(I-R)(s)$ approaches I uniformly on every *compact* subset of C_{+e} as $\lambda \to 0$, and $(sI-\bar{A})^{-1} \to 0$ as $|s| \to \infty$. Hence $\Delta^{-1} \to (sI-\bar{A})^{-1}$ and (82) is proved. To establish (83), observe that $(\lambda sI-A_{22})^{-1}$ approaches $-A_{22}^{-1}$ uniformly on all compact subsets of C_{+e} as $\lambda \to 0$, and that both Δ^{-1} and $(sI-\bar{A})^{-1}$ are strictly proper. Thus (83) follows from (82) *plus* the strict properness of the quantities on both sides of (82). Finally, (82) and (83) together imply that $\tilde{D}_\lambda \to \tilde{D}_0$ as $\lambda \to 0$.

Next, we examine the "numerator" matrix $\tilde{N}_\lambda$. From (72)

87 $$\tilde{N}_\lambda = [C_1 \;\; C_2] \begin{bmatrix} sI-A_{11}+FC_1 & -A_{12}+FC_2 \\ -A_{21}/\lambda & sI-A_{22}/\lambda \end{bmatrix} \begin{bmatrix} B_1-FD_1 \\ B_2/\lambda \end{bmatrix} + D_1.$$

Expanding the above equation results in an expression for $\tilde{N}_\lambda$ as a sum of five terms, the first four of which arise from the matrix product and the last is D_1. From (74), the first of these terms is $C_1\Delta^{-1}(B_1-FD_1)$. Now (82) implies that this term converges to $C_1(sI-\bar{A})^{-1}(B_1-FD_1)$. The second term in the expansion is

88 $$\begin{aligned} -C_2V^{-1}W\Delta^{-1}(B_1-FD_1) \\ &= C_2(sI-A_{22}/\lambda)^{-1}(A_{21}/\lambda)\Delta^{-1}(B_1-FD_1) \\ &= C_2(\lambda sI-A_{22})^{-1}A_{21}\Delta^{-1}(B_1-FD_1). \end{aligned}$$

From reasoning analogous to that used to establish (83), it follows that this term converges to $-C_2A_{22}^{-1}A_{21}(sI-A_{22})^{-1}(B_1-FD_1)$. The third term is

89 $$-C_1\Delta^{-1}YV^{-1}B_2/\lambda$$
$$= C_1\Delta^{-1}(A_{12}-FC_2)(sI-A_{22}/\lambda)^{-1}B_2/\lambda$$
$$= C_1\Delta^{-1}(A_{12}-FC_2)(\lambda sI-A_{22})^{-1}B_2.$$

As before, this converges to

90 $$-C_1(sI-\bar{A})^{-1}(A_{12}-FC_2)A_{22}^{-1}B_2.$$

However, note that $C_2A_{22}^{-1}B_2 = 0$. Hence the above limit simplifies to

91 $$-C_1(sI-\bar{A})^{-1}A_{12}A_{22}^{-1}B_2.$$

The fourth term is the most complicated one; it equals

92 $$C_2V^{-1}(I+W\Delta^{-1}YV^{-1})B_2/\lambda$$
$$= C_2(sI-A_{22}/\lambda)^{-1}$$
$$\cdot\left[I+\frac{A_{21}}{\lambda}\Delta^{-1}(A_{12}-FC_2)\left(sI-\frac{A_{22}}{\lambda}\right)^{-1}\right]\cdot\frac{B_2}{\lambda}$$
$$= C_2(\lambda sI-A_{22})^{-1}[I+A_{22}\Delta^{-1}(A_{12}-FC_2)(\lambda sI-A_{22})^{-1}]B_2$$
$$=C_2(\lambda sI-A_{22})^{-1}B_2 + C_2(\lambda sI-A_{22})^{-1}A_{21}\Delta^{-1}$$
$$\cdot(A_{12}-FC_2)(\lambda sI-A_{22})^{-1}B_2.$$

Now the first term on the right side is identically zero by hypothesis, while the second term is strictly proper. As a result, the second term converges without dificulty to $C_2A_{22}^{-1}A_{21}(sI-\bar{A})^{-1}A_{12}A_{22}^{-1}B_2$. Of course the final term is the constant D_1. Combining the five limits, we see that $\tilde{N}_\lambda$ approaches

93 $$(C_1-C_2A_{22}^{-1}A_{21})(sI-\bar{A})^{-1}(B_1-FD_1-A_{12}A_{22}^{-1}B_2)+D_1$$

Finally, using the relations (66), and noting that $D_1 = D_0$, since $C_2A_{22}^{-1}B_2 = 0$, the above limit simplifies to

94 $$C(sI-\bar{A})^{-1}(B-FD_0)+D_0 = \tilde{N}_0.$$

From Corollary (28), it now follows that P_λ converges to P_0 in the graph topology. □

While the result contained in Proposition (62) is interesting, its use is limited by the fact that the condition it requires is quite restrictive. The origin of the restrictiveness is the definition of robust stabilizability. Recall that a controller $C \in S(P_0)$ was not only required to stabilize P_λ for all sufficiently small positive λ, but the resulting stable family of transfer matrices $H(P_\lambda, C)$ was required to converge *uniformly* over C_{+e} to $H(P_0, C)$ as $\lambda \to 0$. The next proposition is devoted to a study of the situation where a weaker form of convergence of the family $H(P_\lambda, C)$ is acceptable.

95 *Proposition* Consider the family $\{P_\lambda\}$ defined in (63)-(64). Suppose $C \in (P_0)$ is strictly proper. Then $C \in S(P_\lambda)$ for all sufficiently small positive λ. Moreover, $[H(P_\lambda, C)](s)$ converges to $[H(P_0, C)](s)$ uniformly over all compact subsets of C_{+e} as $\lambda \to 0$.

Proof In the interests of brevity, let us say that a family $\{F_\lambda\}$ *converges weakly* to $F \in \mathbf{M}(\mathbf{S})$ if $F_\lambda(s) \to F(s)$ uniformly over all compact subsets of C_{+e} as $\lambda \to 0$. In contrast, we say simply that $\{F_\lambda\}$ *converges* to F if the convergence of $F_\lambda(s)$ to $F(s)$ is uniform over the entire C_{+e}. Also note that, if $\{F_\lambda\}$ converges weakly to F and $\{G_\lambda\}$ converges to G, then the product $\{G_\lambda F_\lambda\}$ converges to GF provided G is strictly proper.

Since $C \in S(P_0)$, it has an r.c.f. (N_c, D_c) such that $\tilde{N}_0 N_c + \tilde{D}_0 D_c = I$. Moreover, since C is strictly causal, so is N_c. Now consider the return difference matrix

96 $$\Delta_\lambda = \tilde{N}_\lambda N_c + \tilde{D}_\lambda D_c.$$

In the proof of Proposition (62), it was shown that $\{\tilde{N}_\lambda\}$ converges to $\tilde{N}_0$ provided $C_2(sI-A_{22})^{-1}B_2 \equiv 0$. However, from (92) one can easily verify that, even if the above condition is not satisfied, the family $\{\tilde{N}_\lambda\}$ still converges *weakly* to $\tilde{N}_0$. Since N_c is strictly proper, it follows that $\tilde{N}_\lambda N_c \to \tilde{N}_0 N_c$. Also $\tilde{D}_\lambda \to \tilde{D}_0$, since the condition was not used in the proof of this result (see (82-86)). Hence $\Delta_\lambda \to \tilde{N}_0 N_c + \tilde{D}_0 D_c = I$, which implies that Δ_λ is unimodular for sufficiently small (positive) λ. By Theorem (5.1.25), this means that $C \in S(P_\lambda)$ for small enough λ.

Next, the twin formula to (5.1.28) shows that

97 $$H(P_\lambda, C) = \begin{bmatrix} D_c \Delta_\lambda^{-1} \tilde{D}_\lambda & -D_c \Delta_\lambda^{-1} \tilde{N}_\lambda \\ N_c \Delta_\lambda^{-1} \tilde{D}_\lambda & I - N_c \Delta_\lambda^{-1} \tilde{N}_\lambda \end{bmatrix}.$$

All terms except $-D_c \Delta_\lambda^{-1} \tilde{N}_\lambda$ converge to their corresponding values when $\lambda \neq 0$. But this term only converges weakly to $-D_c \tilde{N}_0$, since neither D_c nor the limit of Δ_λ^{-1} is strictly proper. However, this is enough to conclude that $\{H(P_\lambda, C)\}$ converges weakly to $H(P_0, C)$. □

Note that the set of all strictly proper compensators in $S(P_0)$ can be parametrized using Proposition (5.2.27).

If $C \in S(P_0)$ is *not* strictly proper, then it is possible to construct a family $\{P_\lambda\}$ such that (66) holds, but $C \notin S(P_\lambda)$ for all $\lambda > 0$. In other words, if P_0 is stabilized using a nonstrictly proper compensator, then there *always* exists a singular perturbation of P_0 that is not stabilized by C for each $\lambda > 0$.

In closing, let us discuss the graph topology in the case where the set of "stable" transfer matrices is something other than the set **S**. A careful examination of the various proofs in this section reveals that the actual nature of the elements of the set **S** was not used anywhere: Rather, the pertinent properties of **S** were (i) the set of units in **S** is open and (ii) inversion is continuous. Thus the graph topology can be defined over the set of fractions associated with an *arbitrary* ring **R**, provided only that **R** has the two properties mentioned above. In particular, the graph topology for discrete-time systems is self-evident, as it is in the case where the set of desirable transfer functions is taken to be the set $\mathbf{S_D}$ where **D** is some prespecified region in the complex plane. Finally, note that in [110] a so-called "gap metric" is defined which is nothing but a metrization of the graph topology corresponding to a particular choice of "stable" transfer functions, namely the set of bounded linear operators mapping the Hilbert space $\mathbf{H}_2$ into itself.

Such a choice of "stable" transfer functions takes into account the fact that BIBO-stability is a desirable attribute of a system, but does not take into account the requirement that a stable system should be causal in order to be physically useful.

7.3 GRAPH METRIC

The previous section discusses the *qualitative* aspects of the robustness of feedback stability. Specifically, the set $\mathbf{M}(\mathbb{R}(s))$ of all (possibly unstable) plants is topologized by means of the *graph topology*, and it is shown that continuity in the graph topology is a necessary and sufficient condition for feedback stability to be robust (see Theorem (7.2.29)). In order to obtain *quantitative* estimates of robustness, it is essential to metrize the graph topology, i.e. to define a metric on $\mathbf{M}(\mathbb{R}(s))$ such that convergence (and hence continuity) in this metric is equivalent to convergence in the graph topology. Such a metric, called the *graph metric*, is defined in this section. Its applications to deriving quantative robustness estimates are given in the next section.

At this stage, it should be noted that the graph metric presented in this section is only one possible means of metrizing the graph topology. It is undoubtedly possible to define alternate metrics on $\mathbf{M}(\mathbb{R}(s))$ that induce exactly the same topology. Each of these metrics would lead to a different *quantitative* estimate for the robustness of feedback stability, but would leave the *qualitative* result of Theorem (7.2.29) unchanged. Thus in some sense the graph topology is more fundamental than the graph metric. defined below.

The graph metric is defined using the concept of a normalized r.c.f. of a rational matrix, and this is introduced first. Suppose $P \in \mathbf{M}(\mathbb{R}(s))$. An r.c.f. (N,D) of P is *normalized* if

1 $$N^*(s)N(s) + D^*(s)D(s) = I, \forall s,$$

where $N^*(s) = N'(-s)$.

2 *Lemma* Every $P \in \mathbf{M}(\mathbb{R}(s))$ has a normalized r.c.f. which is unique to within right multiplication by an orthogonal matrix.

Proof Let (N_1, D_1) be any r.c.f. of P, and define

3 $$\Phi(s) = N_1^*(s) N_1(s) + D_1^*(s) D_1(s).$$

Then $\Phi^*(s) = \Phi(s)\ \forall s$. Moreover, by the right-coprimeness of N_1 and D_1, there exist positive constants α and β such that $\alpha I \leq \Phi(j\omega) \leq \beta I\ \forall \omega$. Hence, by [3, p.240], one can do a spectral factorization of Φ as $\Phi(s) = R^*(s) R(s)$, where $R \in \mathbf{U}(\mathbf{S})$. Now let $N = N_1 R^{-1}$, $D = D_1 R^{-1}$. Then (N, D) is also an r.c.f. of P by Theorem (4.1.43), and (1) holds.

To prove the assertion regarding uniqueness, let (N_2, D_2) be another normalized r.c.f. of P. Then by definition

4 $$N_2^*(s) N_2(s) + D_2^*(s) D_2(s) = I,\ \forall s.$$

Moreover, by Theorem (4.1.43), there is a unimodular matrix $U \in \mathbf{U}(\mathbf{S})$ such that $N_2 = NU$, $D_2 = DU$. Substituting this into (4) and using (1) shows that $U^*(s) U(s) = I$. In other words, U is a spectral factor of the identity matrix. Hence $U(s)$ must be constant as a function of s and must equal an orthogonal matrix. □

5 *Definition* Suppose $P_1, P_2 \in \mathbf{M}(\mathbb{R}(s))$ have the same dimensions, and let (N_i, D_i) be a normalized r.c.f. of P_i for $i = 1,2$. Let

6 $$A_i = \begin{bmatrix} D_i \\ N_i \end{bmatrix},\ i = 1,2,$$

and define

7 $$\delta(P_1, P_2) = \inf_{U \in \mathbf{M}(\mathbf{S}),\ \| U \| \leq 1} \| A_1 - A_2 U \|,$$

8 $$d(P_1, P_2) = \max \{ \delta(P_1, P_2), \delta(P_2, P_1) \}.$$

Then $d(P_1, P_2)$ is the *graph metric* on $\mathbf{M}(\mathbb{R}(s))$.

It is left to the reader to verify that $d(P_1,P_2)$ is a well-defined quantity depending only on P_1 and P_2, even though A_1 and A_2 are only unique to within right multiplication by an orthogonal matrix. The reason is that multiplication by an orthogonal matrix does not change the norm.

9 *Lemma* d is a metric on $\mathbf{M}(\mathbb{R}(s))$ assuming values in the interval $[0,1]$.

Proof If $U = 0$, then $\| A_1 - A_2 U \| = \| A_1 \| = 1$. Hence $\delta(P_1,P_2) \leq 1$ and $d(P_1,P_2) \leq 1$ for all plants P_1, P_2. In proving that d is a metric, the only nonobvious part is the proof of the traingle inequality. Accordingly, let $P_1,P_2,P_3 \in \mathbf{M}(\mathbb{R}(s))$, and select $U,V \in \mathbf{M}(\mathbf{S})$ such that $\| U \| \leq 1$, $\| V \| \leq 1$, and

10 $$\| A_1 - A_2 U \| \leq \delta(P_1,P_2) + \varepsilon,$$

11 $$\| A_2 - A_3 U \| \leq \delta(P_2,P_3) + \varepsilon,$$

where ε is some positive number. Then $VU \in \mathbf{M}(\mathbf{S})$ and $\| VU \| \leq 1$; moreover,

12 $$\begin{aligned} \| A_1 - A_3 VU \| &\leq \| A_1 - A_2 U \| + \| A_2 U - A_3 VU \| \\ &\leq \| A_1 - A_2 U \| + \| A_2 - A_3 V \| \, \| U \| \\ &\leq \delta(P_1,P_2) + \delta(P_2,P_3) + 2\varepsilon \end{aligned}$$

where in the last step we use the fact that $\| U \| \leq 1$. Now (12) shows that

13 $$\delta(P_1,P_3) \leq \delta(P_1,P_2) + \delta(P_2,P_3) + 2\varepsilon.$$

Since (13) holds for *every* $\varepsilon > 0$, it follows that

14 $$\delta(P_1,P_3) \leq \delta(P_1,P_2) + \delta(P_2,P_3).$$

By symmetry,

15 $$\delta(P_3,P_1) \leq \delta(P_3,P_2) + \delta(P_2,P_1)$$

The triangle inequality for d now follows from (8). □

The main result of this section is Theorem (18) which states that the topology on $\mathbf{M}(\mathbb{R}(s))$ induced by the graph metric is the same as the graph topology. The following technical result is needed to prove this theorem.

16 *Lemma* Suppose $P_1, P_2 \in \mathbf{M}(\mathbb{R}(s))$ have the same dimensions, and let A_1, A_2 be associated matrices in $\mathbf{M}(\mathbf{S})$ obtained from their normalized r.c.f.'s. Suppose $U, V \in \mathbf{M}(\mathbf{S})$ satisfy $\| U \| \leq 1$, and $\| A_1 - A_2 U \| + \| A_2 - A_1 V \| < 1$. Then U and V are unimodular.

Proof Note that, because of (1), we have $\| A_1 X \| = \| A_2 X \| = \| X \| \; \forall X \in \mathbf{M}(\mathbf{S})$ (in other words, multiplication by A_i preserves norms). Hence,

17 $$\begin{aligned} \| I - VU \| &= \| A_1 - A_1 VU \| \\ &\leq \| A_1 - A_2 U \| + \| A_2 U - A_1 VU \| \\ &\leq \| A_1 - A_2 U \| + \| A_2 - A_1 V \| < 1. \end{aligned}$$

Hence $VU \in \mathbf{U}(\mathbf{S})$ by Lemma (2.2.20), and $V, U \in \mathbf{U}(\mathbf{S})$ as a consequence. □

18 *Theorem* A sequence $\{P_i\}$ in $\mathbf{M}(\mathbb{R}(s))$ converges to $P \in \mathbf{M}(\mathbb{R}(s))$ in the graph topology if and only if $d(P_i, P) \rightarrow 0$.

Proof "if" Suppose $d(P_i, P) \rightarrow 0$. Let A, A_i be matrices of the form (6) obtained from normalized r.c.f.'s of P, P_i respectively. Then for each ε in the interval $(0, 1/4)$ there exists an integer n such that $d(P_i, P) \leq \varepsilon \; \forall i \geq n$. For each i, there exists U_i, V_i in the unit ball of $\mathbf{M}(\mathbf{S})$ such that $\| A - A_i U_i \| \leq d(P_i, P) + \varepsilon$. Hence, for all $i \geq n$, we have

19 $$\| A - A_i U_i \| + \| A_i - A V_i \| \leq 4\varepsilon < 1.$$

By Lemma (16), this implies that U_i, V_i are unimodular. Thus, if $A = [D' \;\; N']'$ and $A_i = [D_i' \;\; N_i']'$, then $(N_i U_i, D_i U_i)$ is an r.c.f. of P_i and $\| A - A_i U_i \| \leq 2\varepsilon \; \forall i \geq n$. It now follows from Lemma (7.2.20) that $P_i \rightarrow P$ in the graph topology.

"only if" Suppose $P_i \rightarrow P$ in the graph topology, and let (N,D) be a normalized r.c.f. of P. Then by Lemma (7.2.20) there exist r.c.f.'s (N_i,D_i) of P_i such that $N_i \rightarrow N$, $D_i \rightarrow D$. However, (N_i,D_i) need not be normalized. Let $M_i = [D_i' \ N_i']'$, and suppose $M_i = A_i R_i$ where A_i corresponds to a normalized r.c.f. of P_i and $R_i \in \mathbf{U(S)}$. Now $\| M_i - A \| \rightarrow 0$ and $\| M_i \| = \| A_i R_i \| = \| R_i \|$ since A_i is an isometry. Hence $\| R_i \| \rightarrow 1$. Define $U_i = R_i / \| R_i \|$. Then $\| U_i \| = 1$; moreover,

20 $$\begin{aligned} \| A - A_i U_i \| &\leq \| A - A_i R_i \| + \| A_i R_i - A_i U_i \| \\ &= \| A - M_i \| + \| R_i - U_i \| \\ &= \| A - M_i \| + | 1 - \| R_i \| \, | \\ &\rightarrow 0 \text{ as } i \rightarrow \infty. \end{aligned}$$

This shows that $\delta(P,P_i) \rightarrow 0$. To prove that $\delta(P_i,P) \rightarrow 0$, it is necessary to estimate $\| R_i^{-1} \|$. Let $\gamma_i = \| A - M_i \|$. Then for all vectors $x \in \mathbf{M(L_2)}$, we have $\| M_i x \| \geq \| Ax \| - \| (A - M_i)x \| \geq (1-\gamma_i) \| x \|$. Since $\| x \| = \| A_i x \| = \| M_i R_i^{-1} x \| \geq (1-\gamma_i) \| R_i^{-1} x \|$, it follows that $\| R_i^{-1} \| \leq 1/(1-\gamma_i)$ whenever $\gamma_i < 1$. In particular, $\| R_i^{-1} \| \rightarrow 1$ *as* $i \rightarrow \infty$. Let $V_i = R_i^{-1} / \| R_i^{-1} \|$. Then, as in (20), $\| V_i \| = 1$ and

21 $$\begin{aligned} \| A_i - AV_i \| &= \| M_i R_i^{-1} - AV_i \| \\ &\leq \| M_i R_i^{-1} - AR_i^{-1} \| + \| AR_i^{-1} - AV_i \| \\ &\leq \| M_i - A \| \, \| R_i^{-1} \| + \| R_i^{-1} - V_i \| \\ &\rightarrow 0 \text{ as } i \rightarrow \infty. \end{aligned}$$

Thus $\delta(P_i,P) \rightarrow 0$. □

At present, the graph metric distance between two plants cannot be computed exactly, since the problem of determining the infimum in (7) is still open. However, it is possible to give upper and lower bounds.

22 *Lemma* Suppose (N_1,D_1) is a normalized r.c.f. of P_1, (N_2,D_2) is a (not necessarily normalized) r.c.f. of P_2, and let $A_1 = [D_1{}' \ N_1{}']'$, $M_2 = [D_2{}' \ N_2{}']'$. Suppose $\| A_1 - M_2 \| =: \gamma < 1$. Then

23 $$d(P_1,P_2) \le \frac{2\gamma}{1-\gamma} .$$

Proof Suppose $M_2 = A_2 R_2$ where A_2 corresponds to a normalized r.c.f. of P_2 and $R_2 \in \mathbf{U}(\mathbf{S})$. Now $\| M_2 \| \le \| A_1 \| + \| M_2 - A_1 \| = 1+\gamma$. Hence $\| R_2 \| = \| M_2 \| \le 1+\gamma$ and $\| R_2^{-1} \| \le 1/(1-\gamma)$, as in the proof of Theorem (18). Moreover, (20) and (21) imply that

24 $$\delta(P_1,P_2) \le 2\gamma,$$

25 $$\delta(P_2,P_1) \le 2\gamma/(1-\gamma),$$

26 $$d(P_1,P_2) = \max \{\delta(P_1,P_2), \delta(P_2,P_1)\} \le 2\gamma/(1-\gamma).$$

This completes the proof. □

To compute a lower bound on the graph metric distance between two plants, note that

28 $$\delta(P_1,P_2) \ge \inf_{R \in \mathbf{M}(\mathbf{S})} \| A_1 - A_2 R \| .$$

This can be computed using the methods of Section 6.6.

We are now in a position to undertake a proof of Proposition (7.2.57). For ease of reference it is restated here in an equivalent form.

29 *Proposition* Suppose $\{P_i\}$ is a sequence in $\mathbf{M}(\mathbb{R}(s))$, $P \in \mathbf{M}(\mathbb{R}(s))$, and suppose P_i, P are block-diagonal in the form

30 $$P_i = \text{Block Diag } \{P_{i_1},\cdots,P_{i_l}\},$$

31 $$P = \text{Block Diag } \{\bar{P}_1,\cdots,\bar{P}_l\},$$

where the partitioning is commensurate. Then $d(P_i,P) \to 0$ if and only if $d(P_{i_j},\bar{P}_j) \to 0$ as $i \to \infty$, for each j in $\{1,\cdots,l\}$.

Remarks Proposition (29) is equivalent to Proposition (7.2.57) in view of Theorem (18).

Proof It is enough to prove the proposition for the case where $l = 2$, as the general case will then follow by induction. Accordingly, let us change notation, and suppose we are given a sequence $\{P_i\}$ of block-diagonal matrices and a block-diagonal matrix P having the form

32 $$P_i = \begin{bmatrix} Q_i & 0 \\ 0 & R_i \end{bmatrix}, P = \begin{bmatrix} Q & 0 \\ 0 & R \end{bmatrix}.$$

The claim is that $d(P_i,P) \to 0$ if and only if $d(Q_i,Q) \to 0$ and $d(R_i,R) \to 0$.

"if" Suppose $d(Q_i,Q) \to 0$, $d(R_i,R) \to 0$. Then, by Theorem (18), $\{Q_i\}$ converges to Q and $\{R_i\}$ converges to R in the graph topology. This implies, by Lemma (7.2.20), that there are r.c.f.'s (N_{Q_i},D_{Q_i}), (N_{R_i},D_{R_i}), (N_Q,D_Q), (N_R,D_R) of Q_i, R_i, Q, R such that $(N_{Q_i},D_{Q_i}) \to (N_Q,D_Q)$, $(N_{R_i},D_{R_i}) \to (N_R,D_R)$. Now define

33 $$N_i = \begin{bmatrix} N_{Q_i} & 0 \\ 0 & N_{R_i} \end{bmatrix}, D_i = \begin{bmatrix} D_{Q_i} & 0 \\ 0 & D_{R_i} \end{bmatrix}.$$

Then clearly (N_i,D_i) is an r.c.f. of P_i, (N,D) is an r.c.f. of P, and $(N_i,D_i) \to (N,D)$ (because the norm topology on $\mathbf{M}(\mathbf{S})$ is a product topology). Again by Lemma (7.2.20), $\{P_i\}$ converges to P in the graph topology, so that $d(P_i,P) \to 0$ by Theorem (18).

"only if" Suppose $d(P_i,P) \to 0$. Then $\{P_i\}$ converges to P in the graph topology. Let (N_Q,D_Q), (N_R,D_R) be normalized r.c.f.'s of Q, R respectively. Then (N,D) as defined in (34) is easily shown to be a normalized r.c.f. of P. Similarly, let (N_{Q_i},D_{Q_i}), (N_{R_i},D_{R_i}) be normalized r.c.f.'s of Q_i, R_i respectively. Then (N_i,D_i) as defined in (33) is a normalized r.c.f. of P_i. By hypothesis, there is a sequence $\{U_i\}$ of unimodular matrices such that

34 $$\begin{bmatrix} D_{Q_i} & 0 \\ 0 & D_{R_i} \\ N_{Q_i} & 0 \\ 0 & N_{R_i} \end{bmatrix} \begin{bmatrix} U_{1i} & U_{2i} \\ U_{3i} & U_{4i} \end{bmatrix} \rightarrow \begin{bmatrix} D_Q & 0 \\ 0 & D_R \\ N_Q & 0 \\ 0 & N_R \end{bmatrix},$$

where U_i is partitioned in the obvious way. Since the topology on $\mathbf{M(S)}$ is a product topology, each block in the partitioned matrix on the left side of (34) converges to the corresponding block on the right side of (34). In particular,

35 $$\begin{bmatrix} D_{Q_i} \\ N_{Q_i} \end{bmatrix} U_{2i} \rightarrow \begin{bmatrix} 0 \\ 0 \end{bmatrix}, \begin{bmatrix} D_{R_i} \\ N_{R_i} \end{bmatrix} U_{3i} \rightarrow \begin{bmatrix} 0 \\ 0 \end{bmatrix}.$$

Since (N_{Q_i}, D_{Q_i}), (N_{R_i}, D_{R_i}) are both *normalized* r.c.f.'s, (35) implies that $U_{2i} \rightarrow 0$, $U_{3i} \rightarrow 0$. Let A_i (resp. A) denote the block 4×2 matrix on the left (resp. right) side of (34). Since $U_{2i} \rightarrow 0$, $U_{3i} \rightarrow 0$, it follows that

36 $$A_i \begin{bmatrix} 0 & U_{2i} \\ U_{3i} & 0 \end{bmatrix} \rightarrow 0.$$

Combining (34) and (36) gives

37 $$A_i \overline{U}_i \rightarrow A,$$

where

38 $$\overline{U}_i = \begin{bmatrix} U_{1i} & 0 \\ 0 & U_{4i} \end{bmatrix}.$$

Now (37) shows that $\| \overline{U}_i \| \rightarrow 1$. Hence if we define $W_i = U_i / \| U_i \|$, then (37) shows that $\| A - A_i W_i \| \rightarrow 0$. On the other hand, since $d(P_i, P) \rightarrow 0$, there exists a sequence $\{ V_i \}$ of matrices in the unit ball of $\mathbf{M(S)}$ such that $\| A_i - A V_i \| \rightarrow 0$. Thus $\| A - A_i W_i \| + \| A_i - A V_i \| < 1$ for large enough i, and $\| W_i \| = 1$. Hence by Lemma (16), both W_i and V_i are unimodular for large enough i. In particular, U_{1i} and U_{4i} are both unimodular for large enough i. Hence $(N_{Q_i} U_{1i}, D_{Q_i} U_{1i})$ is an r.c.f. of Q_i, and $(N_{R_i}U_{4i}, D_{R_i}U_{4i})$ is an r.c.f. of R_i. Since (37) implies that

39 $$\begin{bmatrix} D_{Q_i} \\ N_{Q_i} \end{bmatrix} U_{1i} \to \begin{bmatrix} D_Q \\ N_Q \end{bmatrix}, \begin{bmatrix} D_{R_i} \\ N_{R_i} \end{bmatrix} U_{4i} \to \begin{bmatrix} D_R \\ N_R \end{bmatrix},$$

It follows from Lemma (7.2.20) that $Q_i \to Q$, $R_i \to R$ in the graph topology. Now Theorem (18) shows that $d(Q_i,Q) \to 0$, $d(R_i,R) \to 0$. □

This section is concluded with a proof of Theorem (7.2.38). It is restated here in an equivalent form for ease of reference.

40 *Proposition* Suppose $\{P_i\}$, $\{C_i\}$ are sequences in $\mathbf{M}(\mathbb{R}(s))$, and that $P_0, C_0 \in \mathbf{M}(\mathbb{R}(s))$. Then the following statements are equivalent:

(i) $d(P_i,P_0) \to 0$, $d(C_i,C_0) \to 0$ *as* $i \to \infty$.

(ii) $d(H_i,H_0) \to 0$ *as* $i \to \infty$, where $H_i = H(P_i,C_i)$, $H_0 = H(P_0,C_0)$.

Remarks If the pair (P_0,C_0) is stable, then the above result reduces to Theorem (7.2.38). In general, it shows that the map from the plant-compensator pair to the closed-loop transfer matrix is continuous if all spaces are equipped with the graph topology, even if the pair (P_0,C_0) is not stable.

Proof Let

41 $$G_0 = \begin{bmatrix} C_0 & 0 \\ 0 & P_0 \end{bmatrix}, G_i = \begin{bmatrix} C_i & 0 \\ 0 & P_i \end{bmatrix}.$$

Then Proposition (29) shows that (i) is equivalent to the following statement: (iii) $d(G_i,G_0) \to 0$ *as* $i \to \infty$. Thus the statement to be proved is: $d(G_i,G_0) \to 0$ if and only if $d(H_i,H_0) \to 0$.

Define

42 $$F = \begin{bmatrix} 0 & I \\ -I & 0 \end{bmatrix}.$$

Then (*cf.* Section 5.1)

43 $$H_i = (I+FG_i)^{-1}, H_0 = (I+FG_0)^{-1}.$$

Suppose (N_i,D_i) is an r.c.f. of G_i. Then $H_i = D_i(D_i+FN_i)^{-1}$, from (43). Moreover, (D_i,D_i+FN_i) is an r.c.f. of H_i, since

44 $$\begin{bmatrix} D_i+FN_i \\ D_i \end{bmatrix} = \begin{bmatrix} I & F \\ I & 0 \end{bmatrix} \begin{bmatrix} D_i \\ N_i \end{bmatrix},$$

and the matrix on the right side of (44) is unimodular, with the inverse

45 $$\begin{bmatrix} 0 & I \\ -F & F \end{bmatrix}.$$

(Note that $F^{-1} = -F$.) Conversely, suppose (B_i,A_i) is an r.c.f. of H_i. Then

46 $$\begin{bmatrix} D_i \\ N_i \end{bmatrix} = \begin{bmatrix} 0 & I \\ -F & F \end{bmatrix} \begin{bmatrix} A_i \\ B_i \end{bmatrix}$$

is an r.c.f. of G_i. Similar remarks apply to G_0 and H_0.

The statement to be proved is that $d(G_i,G_0) \rightarrow 0$ if and only if $d(H_i,H_0) \rightarrow 0$. Suppose first that $d(G_i,G_0) \rightarrow 0$. Then, by Theorem (18), $\{G_i\}$ converges to G_0 in the graph topology. Hence, according to Lemma (7.2.20), there exist r.c.f.'s (N_i,D_i) of G_i and (N_0,D_0) of G_0 such that $N_i \rightarrow N_0$, $D_i \rightarrow D_0$. Now define

47 $$\begin{bmatrix} A_i \\ B_i \end{bmatrix} = \begin{bmatrix} I & F \\ I & 0 \end{bmatrix} \begin{bmatrix} D_i \\ N_i \end{bmatrix}, \begin{bmatrix} A_0 \\ B_0 \end{bmatrix} = \begin{bmatrix} I & F \\ I & 0 \end{bmatrix} \begin{bmatrix} D_0 \\ N_0 \end{bmatrix}.$$

Then (B_i,A_i) is an r.c.f. of H_i, (B_0,A_0) is an r.c.f. of H_0, and $A_i \rightarrow A_0$, $B_i \rightarrow B_0$. Hence $\{H_i\}$ converges to H_0 in the graph topology and $d(H_i,H_0) \rightarrow 0$. The reverse implication follows by reversing the above steps. □

We conclude this section with a discussion of the graph metric in the case where the set of "stable" transfer functions is something other than the set **S**. First of all, the graph metric extends almost effortlessly to linear *distributed* systems. The central step in defining a graph metric is the existence of a normalized r.c.f. In the case of distributed systems, the existence of such a factorization is assured by Beurling's theorem in the scalar case and the Beurling-Lax theorem in the multivariable case, which states that a closed shift-invariant subspace of $\mathbf{H}_2^n$ is the image of an inner operator. Thus, even in the case of distributed systems, if a plant has an r.c.f., then it has a normalized r.c.f. Of course, not all distributed systems have r.c.f.'s (see Section 8.1). Thus the graph metric can only be defined between pairs of plants when each has an r.c.f. Next, in the case of discrete-time systems, the arguments given here carry over without any major modifications: One only has to replace the extended closed right half-plane by the closed unit disc. Finally, if the set of "stable" transfer functions is a set of the form $\mathbf{S}_{\mathbf{D}}$ where $\mathbf{D}$ is some prespecified region of the complex plane, or its discrete analog, then the graph metric can be defined by first conformally mapping $\mathbf{D}$ into C_{+e} (or the closed unit disc in the discrete-time case). This of course requires that the region $\mathbf{D}$ be simply connected.

7.4 ROBUSTNESS CONDITIONS

In this section, the focus is on designing stabilizing compensators for imprecisely known plants. Thus a nominal plant description is available, together with a description of the plant uncertainty, and the objective is to design a compensator that stabilizes *all* plants lying within the specified band of uncertainty. The types of plant uncertainty considered include additive perturbations, multiplicative perturbations, stable-factor perturbations, and graph-metric perturbations. The solution of the design problem is divided into two steps: The first step is the derivation of necessary and sufficient conditions that any compensator must satisfy in order to be a solution of the design problem. The second step is a further examination of these conditions to see when there actually exists a compensator satisfying these conditions. This leads naturally to the notion of optimally robust compensators. This section concludes with a discussion of the case where there are uncertainties in both the plant and compensator.

Throughout this section, several models of plant uncertainty are used; these are described next.

Suppose P_0 is a given nominal plant and that $r \in \mathbf{S}$ is a given function. An *additive perturbation* of P_0 is only defined in the case where P_0 has no poles on the imaginary axis. The class $A(P_0,r)$ consists of all $P \in \mathbf{M}(\mathbb{R}(s))$ that have the same number of open RHP poles as P_0 (counted according to McMillan degree), and satisfy

1 $$\| P(j\omega) - P_0(j\omega) \| < | r(j\omega) | \ \forall \omega \in \mathbb{R}.$$

In particular, (1) implies that if $P \in A(P_0,r)$, then P also has no poles on the $j\omega$-axis. However, $P - P_0$ need not be analytic in the open RHP, because even though P and P_0 have the same number of open RHP poles, they could be at different locations. Hence (1) does *not* imply that $\| P(s) - P_0(s) \| < | r(s) | \ \forall s \in C_{+e}$. The assumption that $r \in \mathbf{S}$ is not too restrictive. Suppose it is known that P has the same number of RHP poles as P_0, and that $\| P(j\omega) - P_0(j\omega) \| < | f(j\omega) |$ where f is some proper rational function. Then it is always possible to find a *stable* proper rational function r such that $r(j\omega) = f(j\omega) \ \forall \omega$. In fact, one can assume without loss of generality that r is outer.

The class of multiplicative perturbations of P_0 is also defined under the restriction that P_0 has no $j\omega$-axis poles. The class $M(P_0,r)$ consists of all $P \in \mathbf{M}(\mathbb{R}(s))$ that have the same number of open RHP poles as P_0 and satisfy

2 $$P(s) = (I + M(s))P_0(s), \ M \in \mathbf{M}(\mathbb{R}(s)), \text{ where}$$

$$\| M(j\omega) \| < | r(j\omega) | \ \forall \omega \in \mathbb{R}.$$

Finally, the class of stable-factor perturbations is specified not just in terms of the nominal plant description P_0, but in terms of a particular r.c.f. (N_0,D_0) of P_0 and a function $r \in \mathbf{S}$. The class $S(N_0,D_0,r)$ consists of all plants $P \in \mathbf{M}(\mathbb{R}(s))$ satisfying

3 $P(s) = N(s)[D(s)]^{-1}$, where

$$\left\| \begin{bmatrix} N(s)-N_0(s) \\ D(s)-D_0(s) \end{bmatrix} \right\| < |r(s| \;\forall s\in C_{+e}.$$

The rationale behind the above three models of plant uncertainty is fairly clear. In the case of both $A(P_0,r)$ and $M(P_0,r)$, the perturbed plant P has the same number of open RHP poles as the nominal plant P_0 (though possibly at different locations). In the case of $A(P_0,r)$, $|r(j\omega)|$ provides an upper bound on the *absolute* magnitude of the plant uncertainty at the frequency ω, while in the case of $M(P_0,r)$, $|r(j\omega)|$ provides an upper bound on the *relative* magnitude of uncertainty at the frequency ω. The class of stable-factor perturbations is free from the restructive assumption that the perturbed plant and the unperturbed plant have the same number of RHP poles. As well, P_0 is permitted to have $j\omega$-axis poles, as is the perturbed plant P. Using (3) in conjunction with the principle of the argument over various subregions of C_+, one can localize regions in C_+ where a $P\in S(N_0,D_0,r)$ can have poles.

Theorems (4) and (6) below give the main robustness results in the case of additive and multiplicative perturbations, respectively.

4 *Theorem* Suppose $P\in S(N_0,D_0,r)$, $r\in \mathbf{M}(\mathbb{R}(s))$ are specified, and suppose C stabililzes P_0. Then C stabilizes all P in the class $A(P_0,r)$ if and only if[5]

5 $$\| [C(I+P_0C)^{-1}(j\omega) \| \cdot |r(j\omega)| \leq 1 \;\forall \omega\in\mathbb{R}.$$

6 *Theorem* Suppose $P_0\in M(P_0,r)$, $r\in \mathbf{M}(\mathbb{R}(s))$ are specified, and suppose C stabilizes P_0. Then C stabilizes all P in the class $M(P_0,r)$ if and only if

[5] Recall that the norm $\| M \|$ of a matrix M equals its largest singular value.

7 $\| [P_0C(I+P_0C)^{-1}](j\omega) \| \cdot | r(j\omega) | \leq 1 \ \forall \omega \in \mathbb{R}.$

The proofs of both theorems make use of a graphical criterion for feedback stability that generalizes the Nyquist criterion. As it is of independent interest, it is stated separately.

8 *Lemma* Suppose $P, C \in \mathbf{M}(\mathbb{R}(s))$ have γ_p, γ_c poles in the open RHP, respectively, counted according to McMillan degree, and none on the $j\omega$-axis. Then C stabilizes P if and only if the plot of $|(I+PC)(j\omega)|$ as ω decreases from ∞ to $-\infty$ does not pass through the origin, and encircles the origin $\gamma_p + \gamma_c$ times in the clockwise sense.

A little notation and an appeal to the principle of the argument simplifies the proof of Lemma (8). Suppose f is a rational function that has neither poles nor zeros on the extended $j\omega$-axis. Then, as ω decreases from ∞ to $-\infty$, the graph of $f(j\omega)$ traces out a closed curve in the complex plane. Let

9 $$ind\ f = \frac{1}{2\pi} \lim_{\omega \to \infty} [arg\ f(-j\omega) - arg\ f(j\omega)]$$

denote the *index* of this closed curve with respect to the origin. In more prosaic terms, *ind* f is the number of times that the plot of $f(j\omega)$ encircles the origin in the counter-clockwise sense as ω decreases from ∞ to $-\infty$. The *principle of the argument* states that

10 $ind\ f =$ no. of zeros of f in C_+ - no. of poles of f in C_+.

Since $arg\ ab = arg\ a + arg\ b$ for any two nonzero complex numbers a and b, it follows that $ind\ fg = ind\ f + ind\ g$ for any two appropriate rational functions f and g.

Proof of Lemma (8) Let (N_p, D_p), $(\tilde{D}_c, \tilde{N}_c)$ denote any r.c.f. and any l.c.f. of P and C, respectively. Then, by Theorem (5.1.25), C stabilizes P if and only if $\tilde{D}_c D_p + \tilde{N}_c N_p$ is unimodular, or equivalently, $f = |\tilde{D}_c D_p + \tilde{N}_c N_p|$ is a unit of $\mathbf{S}$. Using the principle of the argument, this condition can be expressed in terms of the behavior of $f(j\omega)$, and ultimately, $|(I+PC)(j\omega)|$.

Note that

11 $$f = |\tilde{D}_c D_p + \tilde{N}_c N_p| = |\tilde{D}_c| \cdot |I+CP| \cdot |D_p|$$
$$= |\tilde{D}_c| \cdot |I+PC| \cdot |D_p|$$

since $|I+CP| = |I+PC|$. If $|(I+PC)(j\omega)| = 0$ for some ω (finite or infinite), it is immediate that f is not a unit. So suppose that $|(I+PC)(j\omega)| \neq 0 \ \forall \omega$ and at infinity. Then $ind\ f$ is well-defined and

12 $$ind\ f = ind\ |\tilde{D}_c| + ind\ |I+PC| + ind\ |D_p|.$$

Now for a function $g \in \mathbf{S}$ with no $j\omega$-axis zeros, $ind\ g$ equals its degree $\delta(g)$ defined in (2.1.15). Moreover, g is a unit if and only if $\delta(g) = 0$. Finally, from Theorem (4.3.22) and (10), it follows that $ind\ |\tilde{D}_c| = \gamma_c$, $ind\ |D_p| = \gamma_p$. Combining this reasoning with (12) shows that f is a unit if and only if

13 $$ind\ |I+PC| = -(\gamma_c + \gamma_p).$$

This completes the proof. □

If P and/or C has $j\omega$-axis poles, Lemma (8) has to be modified by replacing $j\omega$-axis throughout by indented $j\omega$-axis. This modification is standard (see e.g. [D1, Sec.IV.5]).

Proof of Theorem (4) For the sake of simplicity, it is assumed that C has no $j\omega$-axis poles. The modifications required to handle the case where C has $j\omega$-axis poles are obvious and left to the reader.

"only if" It is shown that if (5) does not hold, then there exists a P in the class $A(P_0,r)$ that is not stabilized by C. Accordingly, suppose (5) is not true, and let F denote $C(I+P_0C)^{-1}$. Then either $\|F(\infty)\| \cdot |r(\infty)| > 1$, or else $\|F(j\omega_0)\| \cdot |r(j\omega_0)| > 1$ for some finite ω_0. Let $\mathbb{R}_e$ denote the extended real line, i.e. the set $\mathbb{R} \cup \{\infty\}$. Then the hypothesis is that $\|F(j\omega_0)\| \cdot |r(j\omega_0)| > 1$ for some $\omega_0 \in \mathbb{R}_e$. Let F denote $F(j\omega_0)$, let $\sigma_1, \cdots, \sigma_s$ denote the nonzero singular values of F, and let Σ denote the diagonal matrix $\Sigma = \text{Diag}\ \{\sigma_1, \cdots, \sigma_s\}$. Select unitary matrices U, V such that

14 $$UF(j\omega_0)V = \begin{bmatrix} \Sigma & 0 \\ 0 & 0 \end{bmatrix},$$

Note that $\sigma_1 \cdot |r(j\omega_0)| > 1$ by assumption. There are now two cases to consider.

Case (i) $\omega_0 = 0$ *or* ∞. In this case F is a real matrix, so that U and V can also be chosen to be real (and hence orthogonal), and $r(j\omega_0)$ is real. Now define $Q \in \mathbf{M}(\mathbf{S})$ by

15 $$Q(s) = -V \begin{bmatrix} 1/\sigma_1 & 0 \\ 0 & 0 \end{bmatrix} U \frac{r(s)}{r(j\omega_0)},$$

and let $P = P_0 + Q$. Then, since Q is stable, P has the same number of (and in fact the same) RHP poles as P_0. Moreover,

16 $$\| Q(j\omega) \| = \frac{|r(j\omega)|}{\sigma_1 |r(j\omega_0)|} < |r(j\omega)| \ \forall \omega \in \mathbb{R},$$

since $\sigma_1 |r(j\omega_0)| > 1$. Hence P belongs to the class $A(P_0, r)$. However, it is claimed that C does *not* stabilize P. To prove this, it is enough, in view of Lemma (8), to show that $|(I+PC)(j\omega_0)| = 0$. Let v_1 denote the first column of V, and note that $V'v_1 = e_1 = [1\ 0 \cdots 0]'$ because V is orthogonal. Hence

17 $$\begin{aligned} (I+QF)(j\omega_0)v_1 &= v_1 - (QF)(j\omega_0)v_1 \\ &= v_1 - V \begin{bmatrix} 1/\sigma_1 & 0 \\ 0 & 0 \end{bmatrix} \begin{bmatrix} \Sigma & 0 \\ 0 & 0 \end{bmatrix} e_1 \\ &= v_1 - v_1 = 0. \end{aligned}$$

Since $v_1 \neq 0$, this implies that $|(I+QF)(j\omega_0)| = 0$. Now $|I+PC| = |I+(Q+P_0C)| = |I+QF| \cdot |I+P_0C|$, from which it follows that $|(I+PC)(j\omega_0)| = 0$.

Case (ii) ω_0 is nonzero and finite. In this case the only complication is that the matrices U, V and the constant $r(j\omega_0)$ may not be real, so that the Q defined in (15) may not be a *real* rational matrix. To handle this possibility, write $r(j\omega_0) = |r(j\omega_0)| \exp(j\delta)$ where $\delta \in [0,2\pi)$. Let v_1, u^1 denote respectively the first column of V and the first row of U. Express each component of v_1, u^1 in the form

18 $$(v_1)_i = \bar{v}_i \exp(j\phi_i), \ (u^1)_i = \bar{u}_i \exp(j\theta_i),$$

where $\bar{v}_i$, $\bar{u}_i$ are real numbers and $\phi - \delta, \theta - \delta \in (-\pi, 0] \, \forall i$. Define

19 $$Q(s) = -\frac{r(s)}{\sigma_1 \, | r(j\omega_0) |} \, \bar{v}(s) \bar{u}'(s),$$

where

20 $$\bar{v}(s) = \begin{bmatrix} \bar{v}_1 \ (s-\alpha_1)/(s+\alpha_1) \\ \vdots \\ \bar{v}_n \ (s-\alpha_n)/(s+\alpha_n) \end{bmatrix},$$

$$\bar{u}(s) = \begin{bmatrix} \bar{u}_1 \ (s-\beta_1)/(s+\beta_1) \\ \vdots \\ \bar{u}_n \ (s-\beta_n)/(s+\beta_n) \end{bmatrix},$$

and the nonnegative constants α_i, β_i are adjusted such that[6]

21 $$\arg \left\{ \frac{j\omega_0 - \alpha_i}{j\omega_0 + \alpha_i} \right\} = \phi_i - \delta, \ \arg \left\{ \frac{j\omega_0 - \beta_i}{j\omega_0 + \beta_i} \right\} = \theta_i - \delta.$$

Then Q is a real rational matrix, $Q \in \mathbf{M}(\mathbf{S})$, and

22 $$Q(j\omega_0) = -v_1 u^1 / \sigma_1.$$

Moreover, Q satisfies (16). Hence $P = P_0 + Q$ belongs to the class $A(P_0, r)$. As before, $(I+QF)(j\omega_0)$ is singular, because it maps v_1 into the zero vector. Hence C does not stabilize P.

---

[6] One can assume without loss of generality that $\omega_0 > 0$, so that such α_i, β_i exist.

"if" Suppose (5) holds. Let P be any plant in the class $A(P_0,r)$ and let Δ denote $P-P_0$. Then, by (1),

23 $$\| \Delta(j\omega) \| < | r(j\omega) | \ \forall\omega.$$

Combining (5) with (23) and letting F denote $C(I+P_0C)^{-1}$ leads to

24 $$\| \Delta(j\omega) \| \cdot \| F(j\omega) \| < 1 \ \forall\omega.$$

Hence

25 $$| I+\lambda\Delta(j\omega)F(j\omega) | \neq 0 \ \forall\omega\in\mathbb{R}, \ \forall\lambda\in[0,1].$$

Let $\mathbf{X}$ denote the topological space $C-0$ (i.e. the "punctured plane") with the topology inherited from C. Define two closed curves $f(\cdot)$ and $g(\cdot)$ in $\mathbf{X}$ by

26 $$f(j\omega) = 1 \ \forall\omega,$$

$$g(j\omega) = | I+\Delta(j\omega)F(j\omega) | \ \forall\omega.$$

Now (25) shows that the function

27 $$h(\lambda,j\omega) = | I+\lambda\Delta(j\omega)F(j\omega) |$$

continuously transforms the curve f into the curve g in $\mathbf{X}$. Hence the curves f and g are homotopic in $\mathbf{X}$, and as a result they have the same index with respect to 0. Since the index of f is clearly zero, it follows that

28 $$ind \ g = ind \ | I+\Delta F | = 0.$$

Finally, note that

29 $$| I+PC | = | I+\Delta F | \cdot | I+P_0C | .$$

Since C stabilizes P_0, it follows from Lemma (8) that

30 $$ind \ | I+P_0C | = -(\gamma_0+\gamma_c),$$

where γ_0 (resp. γ_c) is the number of RHP poles of P_0 (resp. C) counted according to McMillan degree. Now (28)-(30) imply that

31 $$ind \mid I+PC \mid = -(\gamma_0+\gamma_c).$$

Since P also has γ_0 RHP poles by assumption, (31) and Lemma (8) show that C stabilizes P. □

Suppose we define $\bar{A}(P_0,r)$ to be the set of all plants P that have the same number of RHP poles as P_0, and satisfy

32 $$\| P(j\omega)-P_0(j\omega) \| \leq \mid r(j\omega) \mid \; \forall\omega.$$

(The difference between (1) and (32) is that the $<$ is replaced by $\leq$). Then the proof of Theorem (4) can be easily adapted to prove the following result.

33 *Corollary* Suppose $C \in S(P_0)$. Then C stabilizes all plants in the class $\bar{A}(P_0,r)$ if and only if

34 $$\sup_{\omega} \| [C(I+P_0C)^{-1}](j\omega) \| \cdot \mid r(j\omega) \mid < 1.$$

Sketch of Proof of Theorem (6) Suppose $P = (I+M)P_0$. Then

35 $$\mid I+PC \mid = \mid I+(I+M)P_0C \mid = \mid I+MG \mid \cdot \mid I+P_0C \mid,$$

where $G = P_0C(I+P_0C)^{-1}$. By assumption, M satisfies $\| M(j\omega) \| < \mid r(j\omega) \mid \; \forall\omega$, and (7) states that $\| G(j\omega) \| \cdot \mid r(j\omega) \mid \leq 1 \; \forall\omega$. □

To derive a result analogous to Corollary (33), define $\bar{M}(P_0,r)$ to be the set of all plants P that have the same number of RHP poles as P_0, and satisfy $P = (I+M)P_0$ where $M \in \mathbf{M}(\mathbb{R}(s))$ and

36 $$\| M(j\omega) \| \leq \mid r(j\omega) \mid \; \forall\omega.$$

37 *Corollary* Suppose $C \in S(P_0)$. Then C stabilizes all plants in the class $\bar{M}(P_0,r)$ if and only if

38 $$\sup_{\omega} \| [P_0C(I+P_0C)^{-1}](j\omega) \| \cdot \mid r(j\omega) \mid < 1.$$

Finally, consider the case of stable-factor perturbations.

39 *Theorem* Suppose $P_0 \in \mathbf{M}(\mathbb{R}(s))$. Suppose an r.c.f. (N_0, D_0) of P_0 and a function $r \in \mathbf{S}$ are specified, and define the class $S(N_0, D_0, r)$ as in (3). Suppose $C \in S(P_0)$, and select an l.c.f. $(\tilde{D}_c, \tilde{N}_c)$ of C such that $\tilde{D}_c D_0 + \tilde{N}_c N_0 = I$. Then C stabilizes all P in the class $S(N_0, D_0, r)$ if and only if

40 $$\| [\tilde{D}_c(j\omega) \;\; \tilde{N}_c(j\omega)] \| \cdot | r(j\omega) | \leq 1 \;\forall \omega.$$

Remarks Since $\tilde{D}_c, \tilde{N}_c \in \mathbf{M}(\mathbf{S}), r \in \mathbf{S}$, the condition (40) actually implies that $\| [\tilde{D}_c(s) \;\; \tilde{N}_c(s)] r(s) \| \leq 1 \;\forall s \in C_{+e}$.

Proof For convenience, define

41 $$A_0 = \begin{bmatrix} D_0 \\ N_0 \end{bmatrix}, A_1 = \begin{bmatrix} D \\ N \end{bmatrix}, \tilde{A}_c = [\tilde{D}_c \;\; \tilde{N}_c].$$

"if" Suppose (40) holds, and suppose P is an arbitrary plant in the class $S(N_0, D_0, r)$. Then P has an r.c.f. (N, D) such that $\| A(s) - A_0(s) \| < | r(s) | \;\forall s \in C_{+e}$. Now consider the return difference matrix $\tilde{D}_c D + \tilde{N}_c N = \tilde{A}_c A$. Since $\tilde{A}_c A_0 = I$, it follows that $\tilde{A}_c A = I + \tilde{A}_c(A - A_0)$. However, from (40) we get $\| [\tilde{A}_c(A - A_0)](s) \| < \| \tilde{A}_c(s) \| \cdot \| [A - A_0](s) \| < 1 \;\forall s \in C_{+e}$. This shows that $| (\tilde{A}_c A)(s) = | I + [\tilde{A}_c(A - A_0)](s) | \neq 0 \;\forall s \in C_{+e}$, so that $\tilde{A}_c A$ is unimodular. Hence, by Theorem (5.1.25), C stabilizes P. □

"only if" Suppose (40) is false, and select an ω_0 such that $\| \tilde{A}_c(j\omega_0) \| \cdot | r(j\omega_0) | > 1$. Let $\sigma_1 \geq \sigma_2 \geq \cdots \geq \sigma_m$ denote the singular values of $\tilde{A}_c(j\omega_0)$, and note that $\sigma_1 | r(j\omega_0) | > 1$. Let $\Sigma = \text{Diag} \{\sigma_1, \cdots, \sigma_m\}$, and select unitary matrices U, V such that[7]

42 $$U\tilde{A}_c(j\omega_0)V = [\Sigma \;\; 0].$$

Since $\tilde{A}_c(j\omega_0)A_0(j\omega_0) = I$, it follows that $A_0(j\omega_0)$ must be of the form

[7] Note that $\tilde{A}_c(s)$ is a "fat" matrix and has full row rank at all $s \in C_{+e}$.

43 $$A_0(j\omega_0) = V\begin{bmatrix}\Sigma^{-1}\\ X\end{bmatrix}U.$$

By assumption, $\sigma_1 \, | \, r(j\omega_0) \, | \, > 1$. Now define $R = \text{Diag}\,\{1/\sigma_1, 0, \cdots, 0\}$, and

44 $$Q(s) = -V\begin{bmatrix}R\\ T\end{bmatrix}U\,\frac{r(s)}{r(j\omega_0)},$$

where T is a constant matrix chosen as follows: If the first column of the matrix X in (43) is nonzero, simply choose $T = 0$. Otherwise, if the first column of X is zero, choose T to be a matrix whose columns are all zero except for the first column, and whose first column is nonzero and has sufficiently small norm that $\| Q(j\omega_0) \| < 1/ | r(j\omega_0) |$. If T is so chosen, then the first column of $X-T$ is not identically zero. Now let P be the plant ND^{-1}, where $[N' \;\; D']' = A = A_0+Q$. Then P belongs to the class $S(N_0,D_0,r)$, since $\| Q(s) \| \cdot | r(s) | \le 1 \;\forall s \in C_{+e}$.[8] However,

45 $$A(j\omega_0) = A_0(j\omega_0)+Q(j\omega_0) = V\begin{bmatrix}\Sigma^{-1}-R\\ X-T\end{bmatrix}U,$$

and the first column of $\Sigma^{-1}-R$ is zero. Hence $\tilde{A}_c(jw0)A(j\omega_0)$ is singular, and $\tilde{A}_cA$ is not unimodular.

The question is: Does the singularity of $| \tilde{D}_cD + \tilde{N}_cN |$ at $j\omega_0$ imply that C fails to stabilize P? We can conclude that this is so provided it can be shown that N and D are right-coprime, or at least, that any greatest common right divisor of N and D is nonsingular at $j\omega_0$. To amplify this argument, suppose $N = N_1B$, $D = D_1B$, where B is a g.c.r.d. of N and D, and N_1 and D_1 are right-coprime. Then $| \tilde{D}_cD + \tilde{N}_cN | = | \tilde{D}_cD_1 + \tilde{N}_cN_1 | \cdot | B |$. Hence, if $| B(j\omega_0) | \neq 0$, then $| \tilde{D}_cD_1 + \tilde{N}_cN_1 |$ vanishes at $j\omega_0$, and C does not stabilize P. Thus the proof is complete if it can be shown that $| B(j\omega_0) | \neq 0$. For this it is enough to show that the matrix $[D' \;\; N']' = A_0-Q$ has full column rank at $j\omega_0$.

[8] The matrix Q may not be a real rational matrix; but this can be fixed up as in the proof of Theorem (4).

But this last fact is immediate from (45), since the first column of $X-T$ is nonzero. □

There is an interesting anamoly associated with the robustness condition (40). Comparing Theorem (4) with Corollary (33), or Theorem (6) with Corollary (37), we see that it makes very little difference, in the case of additive and multiplicative perturbations, whether the class of perturbations is defined with a strict or nonstrict inequaltiy. If the class of perturbatins is defined with a strict inequality, then the robustness condition has a nonstrict inequality, and vice versa. But this is not so in the case of stable-factor perturbations. Define the class $\bar{S}(N_0,D_0,r)$ by

46 $$\bar{S}(N_0,D_0,r) = \{P = ND^{-1}: \| \begin{bmatrix} (N-N_0)(s) \\ (D-D_0)(s) \end{bmatrix} \| \leq |r(s)| \ \forall s \in C_{+e}\}.$$

One might be tempted to conjecture the following result: $C \in S(P_0)$ stabilizes all P in the class $\bar{S}(N_0,D_0,r)$ if and only if

47 $$\sup_{\omega} \| \tilde{A}_c(j\omega) \| \cdot |r(j\omega)| < 1.$$

But this is false: (47) is certainly sufficient for robust stability, but not necessary, as the next example shows.

48 *Example* Consider

$$p_0(s) = \frac{4}{s-3} = \frac{n_0(s)}{d_0(s)}$$

where

$$n_0(s) = \frac{4(s+7)}{(s+1)^2}, \ d_0(s) = \frac{(s-3)(s+7)}{(s+1)^2}.$$

Let $c(s) = 4/(s+5)$. Then c stabilizes p_0. Moreover, $d_c d_0 + n_c n_0 = 1$ with

$$d_c(s) = \frac{s+5}{s+7}, \; n_c(s) = \frac{4}{s+7}.$$

Now consider the class $\bar{S}(n_0,d_0,r)$ with $r = 1$. Since

$$\| \begin{bmatrix} d_c(\infty) \\ n_c(\infty) \end{bmatrix} \| = \| \begin{bmatrix} 1 \\ 0 \end{bmatrix} \| = 1,$$

(47) does not hold. Nevertheless, c stabilizes every p in the class $\bar{S}(n_0,d_0,r)$, which shows that (47) is not always necessary for robust stability.

To establish the claim made in the above paragraph, suppose

$$p(s) = \frac{n(s)}{d(s)}, \text{ where } \| \begin{bmatrix} n-n_0 \\ d-d_0 \end{bmatrix}(s) \| \leq 1 \; \forall s \in C_{+e}.$$

Let $Q = [q_1 \;\; q_2]'$ denote $[d-d_0 \;\; n-n_0]'$. Then there are two cases to consider, namely: (i) $q_1(\infty)+1 \neq 0$, and (ii) $q_1(\infty)+1 = 0$. In either case we have $\| [d_c(s) \;\; n_c(s)] \| < 1 \; \forall s \in C_+$, so that

$$(d_c d + n_c n)(s) = 1 + [d_c(s) \;\; n_c(s)]Q(s) \neq 0 \; \forall s \in C_+.$$

On the other hand, $[d_c(\infty) \;\; n_c(\infty)] = [1 \;\; 0]$ so that

$$(d_c d + n_c n)(\infty) = 1 + q_1(\infty).$$

Hence, if $q_1(\infty) \neq -1$, then $d_c d + n_c n$ has no zeros in the extended RHP and is thus a unit of $\mathbf{S}$. Therefore c stabilizes p.

It only remains to show that c stabilizes p even if $q_1(\infty) = -1$. In this case $q_2(\infty) = 0$ since $\| Q(\infty) \| \leq 1$. Hence $n(\infty) = d(\infty) = 0$, and $\alpha = 1/(s+1)$ is a common divisor of n and d. Further, since $(d_c d + n_c n)(s) \neq 0 \; \forall s \in C_+$, we see that α, its powers and associates are the only possible common divisors of n and d. Now it is claimed that, whatever be q_1, the function $d_0 + q_1$ can have only a *simple* zero at infinity. Let us accept this claim for a moment; then α is a greatest common divisor of n and d, since d/α does not vanish at infinity. Let $n_1 = 1/\alpha$, $d_1 = d/\alpha$; then (n_1,d_1) is a coprime factorization of p. Now

$$(d_c d_1 + n_c n_1)(\infty) = d_1(\infty) \neq 0,$$

since d has only a simple zero at infinity. On the other hand, since it has already been established that $(d_c d + n_c n)(s) \neq 0 \ \forall s \in C_+$, it follows that

$$(d_c d_1 + n_c n_1)(s) \neq 0 \quad \forall s \in C_{+e}.$$

Hence $d_c d_1 + n_c n_1$ is a unit of $\mathbf{S}$, and c stabilizes p.

Thus the example is complete if the claim can be established. This is most easily done using the bilinear transformation $z = (s-1)/(s+1)$, which sends the function $d_0(s)$ into

$$a(z) = d_0\left(\frac{1+z}{1-z}\right) = (4-3z)(2z-1),$$

and sends $q_1(s)$ into an associated rational function $b(z)$. In this context, the claim is that $a-b$ has only a simple zero at $z=1$ whenever $b(1) = -1$ and $\| b \|_\infty \leq 1$.[9] We prove the contrapositive, namely, if g is a rational $\mathbf{H}_\infty$-function such that $g(1) = 1$ and $a-g$ has a double zero at $z = 1$, then $\| g \|_\infty > 1$. Expand g in a power series around $z = 1$, as

$$g(z) = \sum_{i=0}^{\infty} g_i (z-1)^i.$$

If $a-g$ has (at least) a double zero at $z = 1$, then

$$g(1) = a(1) = 1, \ g'(1) = a'(1) = -1.$$

Hence for $z < 1$ and sufficiently close to 1, we have $g(z) > 1$. Therefore $\| g \|_\infty > 1$. This completes the example.

Each of Theorems (4), (6) and (39) gives conditions for a particular $C \in S(P_0)$ to stabilize *all* plants in a given class of perturbations. But they do not answer the question whether, given a class of perturbations, there actually exists a $C \in S(P_0)$ that satisfies the robustness conditions. This question is addressed next. Note that the solutions given below make use of the $\mathbf{H}_\infty$-

[9] Here $\| \cdot \|_\infty$ denotes the norm in the space $\mathbf{H}_\infty$; see Section 6.2.

norm minimization results derived in Sections 6.5 and 6.6.

49 *Theorem* Suppose a nominal plant $P_0 \in \mathbf{M}(\mathbb{R}(s))$, free of poles on the extended $j\omega$-axis, and a function $r \in \mathbf{S}$ are specified. Suppose r does not have any zeros on the extended $j\omega$-axis. Let (N,D), $(\tilde{D},\tilde{N})$ be any r.c.f. and any l.c.f. of P_0, and let $\tilde{X},\tilde{Y} \in \mathbf{M}(\mathbf{S})$ be any particular solutions of the identity $\tilde{N}\tilde{X} + \tilde{D}\tilde{Y} = I$. Factor D, $\tilde{D}$ in the form

50 $$D = D_i D_o, \tilde{D} = \tilde{D}_o \tilde{D}_i,$$

where D_i, $\tilde{D}_i$ are inner and D_o, $\tilde{D}_o$ are outer, and assume without loss of generality that $|D_i| = |\tilde{D}_i|$.[10] Finally, factor r as $r_o r_i$ where r_o, r_i are outer and inner, respectively. Under these conditions, there exists a $C \in S(P_0)$ that stabilizes all plants in the class $A(P_0,r)$ if and only if there exists an $S \in \mathbf{M}(\mathbf{S})$ such that

51 $$\| D_i^{adj} \tilde{X} \tilde{D}_o r_o + \delta S \|_\infty \leq 1,$$

where $\delta = |D_i| = |\tilde{D}_i|$.

Proof By Theorem (5.2.1), the set of all $C \in S(P_0)$ is given by

52 $$S(P_0) = \{(\tilde{X}+DS)(\tilde{Y}-NS)^{-1} : S \in \mathbf{M}(\mathbf{S}),\ |\tilde{Y}-NS| \neq 0\},$$

where the nonsingularity constraint is satisfied by an open dense subset of $S \in \mathbf{M}(\mathbf{S})$ (see Lemma (5.2.11)). Moreover, from (5.2.9), if $C = (\tilde{X}+DS)(\tilde{Y}-NS)^{-1}$, then

53 $$C(I+P_0C)^{-1} = (\tilde{X}+DS)\tilde{D}.$$

Hence there exists a $C \in S(P_0)$ satisfying (5) if and only if there exists an $S \in \mathbf{M}(\mathbf{S})$ such that

54 $$\| \tilde{X}\tilde{D}r + DS\tilde{D}r \|_\infty \leq 1.$$

A few simple manipulations bring (54) into the form (51). Note that

---

[10] By Theorem (4.1.49), $|D|$ and $|\tilde{D}|$ are associates; so are $|D_o|$ and $|\tilde{D}_o|$, since both are units. Hence $|D_i|$ and $|\tilde{D}_i|$ are associates, and since both are inner, $|D_i| = \pm |\tilde{D}_i|$ in any case.

$$55 \quad \| \tilde{X}\tilde{D}r + DS\tilde{D}r \|_\infty = \| \tilde{X}\tilde{D}_o\tilde{D}_i r_o r_i + D_i D_o S\tilde{D}_o\tilde{D}_i r_o r_i \|$$

$$= \| D_i^{adj}\tilde{X}\tilde{D}_o r_o r_i \delta + \delta^2 r_i D_o S\tilde{D}_o r_o \|_\infty$$

$$= \| \delta r_i (D_i^{adj}\tilde{X}\tilde{D}_o r_o + \delta Q) \|_\infty$$

$$= \| D_i^{adj}\tilde{X}\tilde{D}_o r_o + \delta Q \|_\infty,$$

where, in successive steps, we multiply on the left by D_i^{adj} and on the right by $\tilde{D}_i^{adj}$, replace the matrix S by a new "free" parameter $Q = D_o S\tilde{D}_o r_o$, and observe that multiplication by the inner function δr_i preserves norms. Hence (54) and (51) are equivalent. □

If δ has only simple RHP zeros (or equivalently, if P_0 has only simple RHP poles), then (51) can be stated in a more readily testable form using the results of Section 6.5.

56 *Corollary* Suppose δ has only simple RHP zeros, and let $\lambda_1, \cdots, \lambda_n$ denote these zeros. Let $F_i = (D_i^{adj}\tilde{X}D_o r_o)(\lambda_j)$, $j = 1, \cdots, n$, and define

$$57 \quad Q_{jk} = \frac{I + F_j^* F_k}{\bar{\lambda}_j + \lambda_k}, \; 1 \le j, k \le n,$$

$$58 \quad Q = \begin{bmatrix} Q_{11} & \cdots & Q_{1n} \\ \vdots & & \vdots \\ Q_{n_1} & \cdots & Q_{nn} \end{bmatrix}.$$

Then there exists a $C \in S(P_0)$ that stabilizes all plants in the class $A(P_0, r)$ if and only if Q is nonnegative definite.

In (51), $\tilde{X}$ is *any* particular solution of the identity $\tilde{N}\tilde{X} + \tilde{D}\tilde{Y} = I$. Hence it is nice to know that the test matrix Q is the same no matter which $\tilde{X}$ is used. To see this, let $\tilde{X}_1$, $\tilde{Y}_1$ be another set of matrices in $\mathbf{M}(\mathbf{S})$ satisfying $\tilde{N}\tilde{X}_1 + \tilde{D}\tilde{Y}_1 = I$. Then, from the proof of Theorem (5.2.1), it follows that $\tilde{X}_1 = \tilde{X} + DR$ for some $R \in \mathbf{M}(\mathbf{S})$. Hence

59 $$D_i^{adj}\tilde{X}_1\tilde{D}_o r_o = D_i^{adj}(\tilde{X}+DR)\tilde{D}_o r_o$$
$$= D_i^{adj}\tilde{X}\tilde{D}_o r_o + \delta D_o R\tilde{D}_o r_o$$

since $D = D_i D_o$. Now the second term on the right side of (59) vanishes at all zeros of δ. Hence $F_1,\cdots,F_n$ are independent of which particular solution $\tilde{X}$ is used to compute them. Similar remarks apply even in the case where δ has repeated zeros.

If the plant P_0 is scalar, then the expression for F_j (or f_j in the scalar case) is more elegant. In this case, $D_i^{adj} = 1$ (since by convention the adjoint of a 1×1 matrix is 1) and $f_j = (xd_o r_o)(\lambda_j)$. Now, in the scalar case $\delta = d_i$; hence it follows from $d = d_i d_o = \delta d_o$ and $nx+yd = 1$ that, at any zero λ_j of δ, we have $n(\lambda_j)x(\lambda_j) = 1$. Hence

60 $$x(\lambda_j)d_o(\lambda_j) = \frac{d_o(\lambda_j)}{n(\lambda_j)}.$$

With the plant p_0, associate the *stable* plant $q_0 = p_0\delta = n/d_o$. Then

61 $$f_j = \frac{d_o(\lambda_j)}{n(\lambda_j)}\, r_o(\lambda_j) = \frac{r_o}{q_0}(\lambda_j).$$

The treatment of multiplicative perturbations is entirely similar to that of additive perturbations.

62 *Theorem* Suppose a function $r\in\mathbf{S}$ and a nominal plant $P_0\in\mathbf{M}(\mathbb{R}(s))$ are specified, and suppose that P_0 has no extended $j\omega$-axis poles. Let (N,D), $(\tilde{D},\tilde{N})$ be any r.c.f. and any l.c.f. of P_0, and let $X,Y\in\mathbf{M}(\mathbf{S})$ be any particular solutions to the identity $XN + YD = I$. Under these conditions, there exists a $C\in S(P_0)$ that stabilizes all plants in the class $M(P_0,r)$ if and only if there exists an $R\in\mathbf{M}(\mathbf{S})$ such that

63 $$\| N(X+R\tilde{D}) \|_\infty \leq 1.$$

Further, if P_0 has at least as many inputs as outputs and r has no extended $j\omega$-axis zeros, then (63) can be simplified as follows: Factor N, $\tilde{D}$ in the form

64 $$N = N_i N_o, \tilde{D} = \tilde{D}_o \tilde{D}_i,$$

where $N_i, \tilde{D}_i$ are inner and $N_o, \tilde{D}_o$ are outer. Finally, factor r as $r_i r_o$ where r_i is inner and r_o is a unit of $\mathbf{S}$. Under these conditions, there exists a $C \in S(P_0)$ that stabilizes all plants in the class $M(P_0, r)$ if and only if there exists an $R \in \mathbf{M}(\mathbf{S})$ such that

65 $$\| N_o X \tilde{D}_i^{adj} r_o + \delta R \|_\infty \leq 1,$$

where $\delta = |\tilde{D}_i|$.

Proof From Theorem (5.2.1), every $C \in S(P_0)$ is of the form $(Y - R\tilde{N})^{-1}(X + R\tilde{D})$ for some $R \in \mathbf{M}(\mathbf{S})$. Moreover,

66 $$P_0 C(I + P_0 C)^{-1} = P_0(I + CP_0)^{-1} C = N(X + R\tilde{D}).$$

Hence (7) holds for some $C \in S(P_0)$ if and only if there exists an $R \in \mathbf{M}(\mathbf{S})$ such that (63) holds. Now suppose the additional assumptions on P_0, r hold. Then, using by now familiar manipulations, we note that

67 $$\begin{aligned} \| NXr + NR\tilde{D}r \|_\infty &= \| N_i N_o X r_o r_i + N_i N_o R \tilde{D}_o r_o \tilde{D}_i r_i \|_\infty \\ &= \| N_i r_i (N_o X r_o \tilde{D}_i^{adj} + \delta R) \|_\infty \\ &= \| N_o X r_o \tilde{D}_i^{adj} + \delta R \|_\infty. \end{aligned}$$

where we use the fact that multiplication by the inner matrix $N_i r_i$ preserves norms. Now (67) is precisely (65). □

If the plant P_0 is scalar, one can again obtain a simple expression for the $N_o X r_o \tilde{D}_i^{adj}$ evaluated at the zeros of δ. Suppose $nx + dy = 1$ and suppose $\delta(\lambda) = 0$ at some point $\lambda \in C_+$. Then $d(\lambda) = 0$, $nx(\lambda) = 1$, and $(N_o X r_o \tilde{D}_i^{adj})(\lambda) = r_o(\lambda)/n_i(\lambda)$.

The foregoing results also lead very naturally to the notion of *optimally robust compensators.* Suppose a nominal plant P_0 is given, together with a function $f \in \mathbf{S}$, which represents an *uncertainty profile.* Consider now the class $A(P_0, r)$ consisting of all plants satisfying (1), where $r = \lambda f$. It is now reasonable to ask: What is the largest value of the parameter λ for which the class $A(P_0,r)$ is robustly stabilizable, and what is a corresponding robustly stabilizing compensator? This problem can be solved very easily using the results derived so far. Factor f as $f_i f_o$ where f_i, f_o are respectively inner and outer, and define

68 $$\gamma = \min_{S \in \mathbf{M}(\mathbf{S})} \| D_i^{adj} \tilde{X} \tilde{D}_o f_o + \delta S \|,$$

where all symbols are as in Theorem (49). Now, by applying Theorem (49) with $r = \lambda f$, it follows that there exists a single compensator that stabilizes all the plants in the class $A(P_0,r)$ if and only if $\gamma\lambda \leq 1$. Hence the largest value of λ for which there exists a robustly stabilizing compensator is given by $\lambda = 1/\gamma$. Further, if S is any matrix which attains the minimum in (68), then the compensator $C = (\tilde{X}+DS)(\tilde{Y}-NS)^{-1}$ is an optimally robust compensator. The case of multiplicative perturbations is entirely similar and is left to the reader.

Next, the case of stable-factor perturbations is studied.

69 *Theorem* Suppose a function $r \in \mathbf{S}$ and a nominal plant $P_0 \in \mathbf{M}(\mathbb{R}(s))$ are specified, together with an r.c.f. (N_0,D_0) of P_0. Let $X,Y \in \mathbf{M}(\mathbf{S})$ satisfy $XN_0 + YD_0 = I$, and let $(\tilde{D}_0,\tilde{N}_0)$ be any l.c.f. of P_0. Then there exists a $C \in S(P_0)$ that stabilizes all plants in the class $S(N_0,D_0,r)$ if and only if there exists a matrix $R \in \mathbf{M}(\mathbf{S})$ such that $\| A - BR \| \leq 1$, where

70 $$A = \begin{bmatrix} X \\ Y \end{bmatrix}, B = \begin{bmatrix} \tilde{D}_0 \\ -\tilde{N}_0 \end{bmatrix}.$$

The proof is obvious and is left to the reader.

This section is concluded with a discussion of the case where there are perturbations in both the plant and the compensator. The results in this case make use of the graph metric introduced in Section 7.3. A bit of notation makes the presentation smoother. Given $P,C \in \mathbf{M}(\mathbb{R}(s))$, define

71 $$T(P,C) = H(P,C) - \begin{bmatrix} I & 0 \\ 0 & 0 \end{bmatrix},$$

where $H(P,C)$ is defined in (7.1.1). Using this last equation, one can derive a few other useful representations of $T(P,C)$. For instance,

72 $$T(P,C) = \begin{bmatrix} -PC(I+PC)^{-1} & -P(I+CP)^{-1} \\ C(I+PC)^{-1} & (I+CP)^{-1} \end{bmatrix}$$

$$= \begin{bmatrix} -P \\ I \end{bmatrix} (I+CP)^{-1}[C \;\; I].$$

73 *Theorem* Suppose the pair (P_0,C_0) is stable, and that P_0, C_0 are perturbed to P, C, respectively. Then the pair (P,C) is stable provided

74 $$d(P,P_0) \| T(P_0,C_0) \| + d(C,C_0) \| T(C_0,P_0) \| < 1.$$

75 *Corollary* Under the hypotheses of Theorem (73), the pair (P,C_0) is stable provided $d(P,P_0) < 1/ \| T(P_0,C_0) \|$.

The significance of Theorem (73) and Corollary (75) is as follows: The type of plant perturbations studied in these two results is the most unstructured one considered in this section, in that (i) one is permitted to simultaneously perturb both the plant and the compensator, (ii) there is no restriction on the number of RHP poles of the perturbed and unperturbed plant being the same, and (iii) the perturbations are not couched in terms of a *particular* coprime factorization of the plant. The stability condition (74) is interesting in that the effects of the perturbations in the plant and controller enter additively. When only the plant is perturbed, the stability condition given in Corollary (75) is reminiscent of the small gain theorem (see e.g. [D1, Chapter V]). These results also bring out the utility of the graph metric in obtaining quantitative robustness estimates.

Proof Let (N_0,D_0), $(\tilde{X}_0,\tilde{Y}_0)$ be normalized r.c.f.'s of P_0, C_0, respectively. Then from the stability of the pair (P_0,C_0), and Problem 5.1.1, the matrix

76 $$U_0 = \begin{bmatrix} D_0 & -\tilde{X}_0 \\ N_0 & \tilde{Y}_0 \end{bmatrix}$$

is unimodular. Let $V_0 = U_0^{-1}$, and partition V_0 as

77 $$V_0 = \begin{bmatrix} Y_0 & X_0 \\ -\tilde{N}_0 & \tilde{D}_0 \end{bmatrix}.$$

Then $(\tilde{D}_0, \tilde{N}_0)$, (Y_0, X_0) are l.c.f.'s of P_0, C_0 respectively.

Now select real numbers $\delta_p > d(P, P_0)$, $\delta_c > d(C, C_0)$ such that

78 $$\delta_p \| T(P_0, C_0) \| + \delta_c \| T(C_0, P_0) \| < 1.$$

This is possible in view of (74). Let (N_p, D_p), (N_c, D_c) be any normalized r.c.f.'s of P, C respectively. Then, from the definition of the graph metric (see (7.3.5)), there exist matrices W_p, $W_c \in \mathbf{M}(\mathbf{S})$ with $\| W_p \| \leq 1$, $\| W_c \| \leq 1$ such that

79 $$\| \begin{bmatrix} D_0 - D_p W_p \\ N_0 - N_p W_p \end{bmatrix} \| \leq \delta_p,$$

80 $$\| \begin{bmatrix} \tilde{Y}_0 - D_c W_c \\ \tilde{X}_0 - N_c W_c \end{bmatrix} \| \leq \delta_c.$$

Define F as in (7.3.43) and observe that F is inner so that multiplication by F is norm-preserving. Thus (80) implies that

81 $$\| \begin{bmatrix} -(\tilde{X}_0 - N_c W_c) \\ \tilde{Y}_0 - D_c W_c \end{bmatrix} \| \leq \delta_c.$$

Next, define

82 $$U = \begin{bmatrix} D_p W_p & -N_c W_c \\ N_p W_p & D_c W_c \end{bmatrix}$$

$$= \begin{bmatrix} D_p & -N_c \\ N_p & D_c \end{bmatrix} \begin{bmatrix} W_p & 0 \\ 0 & W_c \end{bmatrix}.$$

It is now shown that (78) implies the unimodularity of U. This will show, *a fortiriori*, that W_p and W_c are both unimodular, and that the pair (P,C) is stable, by Problem 5.1.1. Note that

83 $$U - U_0 = [A \;\; B]$$

where

84 $$A = \begin{bmatrix} D_0 - D_p W_p \\ N_0 - N_p W_p \end{bmatrix}, \quad B = \begin{bmatrix} -(\tilde{X}_0 - N_c W_c) \\ \tilde{Y}_0 - D_c W_c \end{bmatrix}.$$

Now, if $\| [A \;\; B] U_0^{-1} \| < 1$, then (83) implies that U is unimodular. From (77),

85 $$[A \;\; B] U_0^{-1} = [A \;\; B] V_0 = A [Y_0 \;\; X_0] + B [-\tilde{N}_0 \;\; \tilde{D}_0].$$

86 $$\| [A \;\; B] U_0^{-1} \| \le \| A [Y_0 \;\; X_0] \| + \| B [-\tilde{N}_0 \;\; \tilde{D}_0] \|$$

$$\le \delta_p \| [Y_0 \;\; X_0] \| + \delta_c \| [-\tilde{N}_0 \;\; \tilde{D}_0] \|$$

where the last step follows from (79) and (81). The proof is completed by showing that

87 $$\| [Y_0 \;\; X_0] \| = \| T(P_0, C_0) \|,$$

88 $$\| [-\tilde{N}_0 \;\; \tilde{D}_0] \| = \| T(C_0, P_0) \|.$$

Then (78) and (86) will imply the unimodularity of U and the stability of (P,C).

To prove (87), recall that (N_0,D_0) is a normalized r.c.f. of P_0, and (Y_0,X_0) is the corresponding l.c.f. of C_0 such that $Y_0D_0+X_0N_0 = I$. Hence, from (5.1.32) and (71),

89 $$T(P_0,C_0) = \begin{bmatrix} -N_0X_0 & -N_0Y_0 \\ D_0X_0 & D_0Y_0 \end{bmatrix} = \begin{bmatrix} -N_0 \\ D_0 \end{bmatrix} [X_0 \;\; Y_0];$$

90 $$\| T(P_0,C_0) \| = \| [X_0 \;\; Y_0] \| \text{ since } \begin{bmatrix} -N_0 \\ D_0 \end{bmatrix} \text{ is an isometry}$$

$$= \| [Y_0 \;\; X_0] \| .$$

The proof of (88) follows essentially by symmetry arguments, after noting that $\| [-\tilde{N}_0 \;\; \tilde{D}_0] \| = \| [\tilde{N}_0 \;\; \tilde{D}_0] \|$. Recall that $(\tilde{X}_0,\tilde{Y}_0)$ is a normalized r.c.f. of C_0, that P_0 stabilizes C_0, and that $(\tilde{D}_0,\tilde{N}_0)$ is the corresponding l.c.f. of P_0 such that $\tilde{D}_0\tilde{Y}_0 + \tilde{N}_0\tilde{X}_0 = I$. □

It is not known at present how close the conditions of Theorem (73) and Corollary (75) are to being necessary. In particular, it is not known whether the condition $\| T(P_0,C) \| < r^{-1}$ is necessary for a controller C to stabilize all plants within a distance of r from the plant P_0. Nevertheless, it is reasonable to seek, among all stabilizing controllers for P_0, an optimally robust controller C_0 for which $\| T(P_0,C_0) \|$ is as small as possible. The problem of finding such a C_0 can be formulated as an $\mathbf{H}_\infty$-norm minimization problem. As C varies over $S(P_0)$, the corresponding $T(P_0,C)$ varies over all matrices of the form

91 $$T_1(R) = \begin{bmatrix} -N_0(X_0+R\tilde{D}_0) & -N_0(Y_0-R\tilde{N}_0) \\ D_0(X_0+R\tilde{D}_0) & D_0(Y_0-R\tilde{D}_0) \end{bmatrix},$$

where (N_0,D_0), $(\tilde{D}_0,\tilde{N}_0)$ are any r.c.f. and l.c.f. of P_0, $X_0,Y_0 \in \mathbf{M}(\mathbf{S})$ satisfy $X_0N_0+Y_0D_0 = I$, and $R \in \mathbf{M}(\mathbf{S})$ is a free parameter (see (5.2.8)). Thus, minimizing $\| T(P_0,C) \|$ over all $C \in S(P_0)$ is equivalent to the unconstrained minimization of $\| T_1(R) \|$ as R varies over $\mathbf{M}(\mathbf{S})$. Now note that $T_1(R)$ is of the form $U - VRW$, where

92 $$U = \begin{bmatrix} -\tilde{N}_0 \\ \tilde{D}_0 \end{bmatrix} [X_0 \;\; Y_0], \; V = \begin{bmatrix} -\tilde{N}_0 \\ \tilde{D}_0 \end{bmatrix}, \; W = [\tilde{D}_0 \;\; -\tilde{N}_0].$$

This type of minimization problem is studied in Section 6.6.

Finally, the results contained in this section carry over with obvious modifications to the study of the robustness of discrete-time systems, and to cases where the set of "stable" transfer functions is something other than $\mathbf{S}$.

7.5 ROBUST AND OPTIMAL REGULATION

In this section, several previously derived results are brought to bear to provide solutions of two problems, namely robust regulation and optimal robust regulation. The robust regulation problem encompasses two aspects, namely robust tracking and robust disturbance rejection. The robust tracking problem can be described as follows: Suppose we are given a plant P, together with an unstable reference signal generator which is factored in the form $\tilde{D}_r^{-1}\tilde{N}_r$, where $\tilde{D}_r, \tilde{N}_r$ are left-coprime. The objective is to design a compensator such that the plant output tracks the unstable reference signal, and continues to do so even as the plant is slightly perturbed. This problem is depicted in Figure 7.2, where, to achieve the greatest possible generality, a two-parameter compensation scheme is shown. Alternatively, suppose the unstable signal is a disturbance which is injected at the plant output. In this case, the objective is to design a compensator such that the plant output is insensitive to the disturbance, and remains so even as the plant is slightly perturbed. This situation is shown in Figure 7.3, and is referred to as the robust disturbance rejection problem. Note that a feedback, or one-parameter, configuration is employed in Figure 7.3 since, as shown in Section 5.6, both configurations are equivalent for the purposes of disturbance rejection.

In more precise terms, the robust disturbance rejection problem can be stated as follows: Find a compensator such that three conditions are met, namely

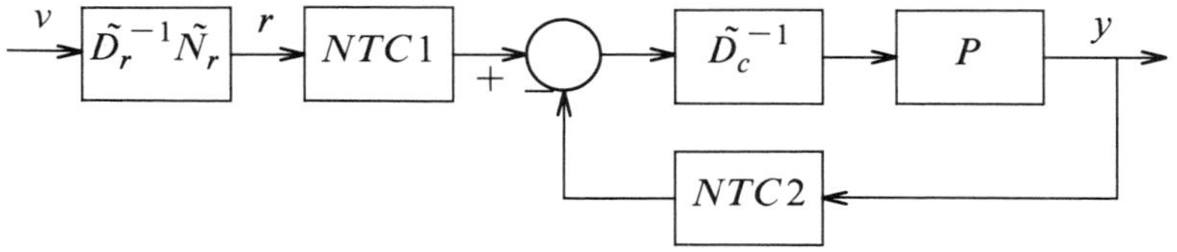

Figure 7.2. Robust Tracking Using a Two-Parameter Compensator.

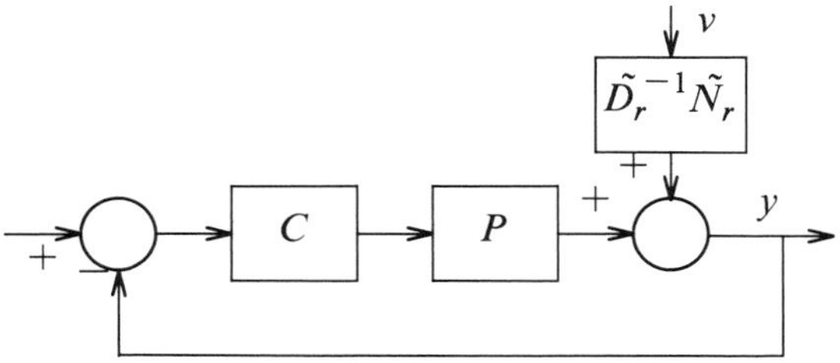

Figure 7.3. Robust Disturbance Rejection.

(i) The feedback system of Figure 7.2 is stable.

(ii) The transfer matrix from v to the tracking error $r-y$ belongs to $\mathbf{M}(\mathbf{S})$.

(iii) There exists a neighborhood $\mathbf{N}(P)$ of P in the graph topology such that conditions (i) and (ii) hold for all plants in $\mathbf{N}(P)$.

If a one-parameter or feedback compensator is used, the robust tracking problem is mathematically equivalent to the robust disturbance rejection problem, since the transfer matrix from v to the tracking error e in Figure 7.4 is exactly the same as the transfer matrix from v to the plant output y in Figure 7.3. On the other hand, if a two-parameter compensation scheme is used, there is a distinct difference between robust tracking and robust disturbance rejection. As a matter of fact, there is nothing to be gained by using a two-parameter scheme if the objective is to reject a disturbance (see

Section 5.6). For this reason, throughout this section, attention is focused on the robust tracking problem.

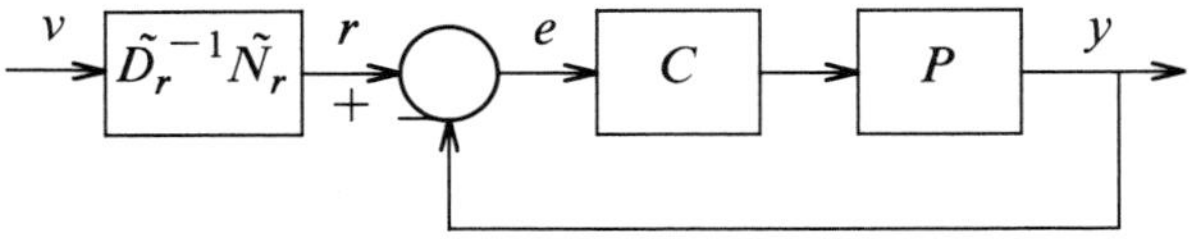

Figure 7.4. Robust Tracking Using a One-Parameter Compensator.

Let us now formulate the problem in more mathematical terms. Given the plant P, let (N,D), $(\tilde{D},\tilde{N})$ be any r.c.f. and l.c.f. of P. The objective is to find a compensator C such that three conditions are satisfied:

(R1) $C \in S(P)$.

(R2) $(I+PC)^{-1}\tilde{D}_r^{-1}\tilde{N}_r \in \mathbf{M}(\mathbf{S})$.[11]

(R3) There exists a δ such that (R1) and (R2) hold whenever P is replaced by $P_1 = \tilde{D}_1^{-1}\tilde{N}_1$ where

1 $$\| [\tilde{D}-\tilde{D}_1 \;\; \tilde{N}-\tilde{N}_1] \| < \delta.$$

The solution to the robust tracking problem is given next.

2 *Theorem* Let α_r denote the largest invariant factor of $\tilde{D}_r$.[12] Then there exists a C satisfying (R1)-(R3) if and only if N and $\alpha_r I$ are left-coprime. If N and $\alpha_r I$ are left-coprime, the set of all C that satisfy (R1)-(R3) is given by

3 $$\{C_1/\alpha_r \colon C_1 \in S(P/\alpha_r)\}.$$

Checking whether or not the problem at hand has a solution is made easier by the next result.

[11] Note that (R2) is redundant unless $\tilde{D}_r$ is non-unimodular, i.e. the reference signal generator is unstable.

[12] Note that α_r is a nonunit since $\tilde{D}_r$ is not unimodular.

4 *Lemma* The following statements are equivalent:

(i) N and $\alpha_r I$ are left-coprime.

(ii) $N(s)$ has full row rank at all C_{+e}-zeros of α_r.

(iii) N has full row rank, and the largest invariant factor of N and α_r are coprime.

Thus Lemma (4) shows that only square or fat plants can be robustly regulated, and reduces the matrix coprimeness test of Theorem (2) to a scalar coprimeness criterion. Once one is sure that the robust tracking problem does indeed have a solution in the case at hand, the formula (3) describing *all* compensators C that achieve robust tracking has a very simple interpretation: In Figure 7.5, choose C_1 to be any compensator that stabilizes the feedback system. Then $C = C_1/\alpha_r$ achieves (R1)-(R3). Moreover, every such C is of the above form. The presence of the $1/\alpha_r$ term within the compensator is referred to elsewhere as the "internal model principle" [103].

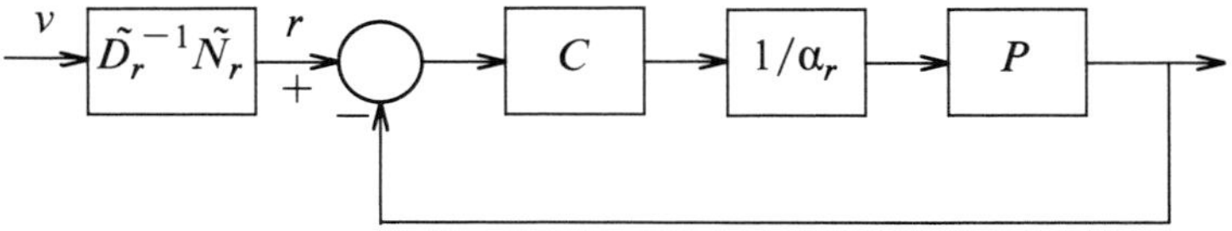

Figure 7.5. Solution to the Robust Tracking Problem.

The proof of Theorem (2) requires two preliminary results. The first of these is the same as Lemma (5.7.6), and is reproduced here for convenience.

5 *Lemma* Suppose $T,U,V \in \mathbf{M}(\mathbf{S})$ with $|U| \neq 0$ and U,V left-coprime. Then $TU^{-1}V \in \mathbf{M}(\mathbf{S})$ if and only if $TU^{-1} \in \mathbf{M}(\mathbf{S})$.

Because of Lemma (5), condition (R2) can be replaced by

(R2') $(I+PC)^{-1}\tilde{D}_r^{-1} \in \mathbf{M}(\mathbf{S})$.

The second result is actually a special case of a fact established in the next section (see Proposition (7.6.15)). It is stated here without proof, and the reader may wish to provide a specialized proof that applies to the ring $\mathbf{S}$ (see

Problem 1).

6 *Lemma* Suppose $A, B \in \mathbf{M}(\mathbf{S})$ are square, have the same dimensions, and suppose $|A| \neq 0$. Then for every $\varepsilon > 0$, the ball

7 $$\mathbf{B}(B,\varepsilon) = \{B_1 \in \mathbf{M}(\mathbf{S}): \| B_1 - B \| < \varepsilon\}$$

contains a matrix B_1 such that A, B_1 are right-coprime.

The proof of Theorem (2) is very long. The first step, which takes us most of the way, is to establish the next result.

8 *Lemma* Suppose (N_c, D_c) is an r.c.f. of $C \in S(P)$. Then C achieves (R1)-(R3) if and only if $\alpha_r^{-1} D_c \in \mathbf{M}(\mathbf{S})$ (i.e. if and only if α_r divides every component of D_c).

Proof Since the robustness condition (R3) involves perturbing the plant from its nominal description P, we can assume at the outset that $\tilde{D}$ and $\tilde{D}_r$ are right-coprime. If this is not so originally, it can be achieved by an arbitrarily small perturbation of $\tilde{D}$, by Lemma (6). Also, since the truth or falsity of the condition $\alpha_r^{-1} D_c$ is unaffected by multiplying D_c by a unimodular matrix, we can as well choose the particular r.c.f. (N_c, D_c) of C that satisfies $\tilde{D} D_c + \tilde{N} N_c = I$. Finally, note that since α_r is the largest invariant factor of $\tilde{D}_r$, we have that $\alpha_r 1 \tilde{D}_r^{-1} \in \mathbf{M}(\mathbf{S})$ (see Problem 2).

"if" Suppose $D_c = \alpha_r T$, where $T \in \mathbf{M}(\mathbf{S})$. Now (R1) holds by assumption, since $C \in S(P)$. Next,

9 $$(I + PC)^{-1} \tilde{D}_r^{-1} = D_c \tilde{D} \tilde{D}_r^{-1} = T \tilde{D} \alpha_r \tilde{D}_r^{-1} \in \mathbf{M}(\mathbf{S}),$$

so that (R2') holds. To prove (R3), suppose $P = \tilde{D}^{-1} \tilde{N}$ is replaced by $P_1 = \tilde{D}_1^{-1} \tilde{N}_1$, where $\tilde{D}_1 = \tilde{D} + \Delta \tilde{D}$, $\tilde{N}_1 = \tilde{N} + \Delta \tilde{N}$. Then the corresponding return difference matrix is

10 $$M = \tilde{D}_1 D_c + \tilde{N}_1 N_c = I + [\Delta \tilde{D} \;\; \Delta \tilde{N}] \begin{bmatrix} D_c \\ N_c \end{bmatrix}$$

which is unimodular provided

11 $$\| [\Delta\tilde{D} \;\; \Delta\tilde{N}] \| < \delta = \| \begin{bmatrix} D_c \\ N_c \end{bmatrix} \|^{-1}.$$

Hence (R1) holds robustly. To prove that (R2') also holds robustly, note that

12 $$(I+P_1C)^{-1}\tilde{D}_r^{-1} = D_c M^{-1}\tilde{D}_1\tilde{D}_r^{-1}$$
$$= TM^{-1}\tilde{D}_1\alpha_r\tilde{D}_r^{-1} \in \mathbf{M}(\mathbf{S})$$

since $M^{-1} \in \mathbf{M}(\mathbf{S})$ due to the unimodularity of M.

"only if" Suppose (R1), (R2'), (R3) hold; it is to be shown that $\alpha_r^{-1}D_c \in \mathbf{M}(\mathbf{S})$. This part of the proof is divided into several steps.

First, since $\tilde{D}$ and $\tilde{D}_r$ are right-coprime, let (A,B) be an l.c.f. of $\tilde{D}\tilde{D}_r^{-1}$. Then, by Theorem (4.1.49), A and $\tilde{D}_r$ have the same invariant factors. In particular, α_r is the largest invariant factor of A.

Step 1 It is shown that

13 $$D_c A^{-1} \in \mathbf{M}(\mathbf{S}).$$

By assumption, (R2') holds, i.e. $D_c\tilde{D}\tilde{D}_r^{-1} = D_c A^{-1}B \in \mathbf{M}(\mathbf{S})$. An application of Lemma (5) gives (13).

Step 2 It is shown that

14 $$D_c Q N_c A^{-1} \in \mathbf{M}(\mathbf{S}) \;\forall Q \in \mathbf{M}(\mathbf{S}).$$

Let $Q \in \mathbf{M}(\mathbf{S})$ be arbitrary, and choose ε in the interval $(0, 1/\| QN_c \|)$. Then $I - \varepsilon QN_c$ is unimodular. Define $M = (I - \varepsilon QN_c)^{-1}$. Then

15 $$M = I + \sum_{i=1}^{\infty} (\varepsilon QN_c)^i = I + R_\varepsilon N_c$$

for some *rational* matrix R_ε (note that M is rational). Moreover, $\| R_\varepsilon \| \to 0$ as $\varepsilon \to 0$. Hence $\| R_\varepsilon \| < \delta$ for sufficiently small ε, where δ is the number defined in (1). Now let $P_1 = \tilde{D}^{-1}(\tilde{N} + R_\varepsilon)$. By assumption, $(I+P_1C)^{-1}\tilde{D}_r^{-1} \in \mathbf{M}(\mathbf{S})$, since P_1 is a valid perturbation of P. But

16 $$(I+P_1C)^{-1}\tilde{D}_r^{-1} = D_c[\tilde{D}D_c + (N+R_\varepsilon)N_c]^{-1}\,\tilde{D}\tilde{D}_r^{-1}$$
$$= D_c(I+R_\varepsilon N_c)^{-1}A^{-1}B$$
$$= D_c(I-\varepsilon QN_c)A^{-1}B\in\mathbf{M}(\mathbf{S}).$$

Applying Lemma (5) gives

17 $$D_c(I-\varepsilon QN_c)A^{-1}\in\mathbf{M}(\mathbf{S}).$$

Subtracting (13) yields $\varepsilon D_cQN_cA^{-1}\in\mathbf{M}(\mathbf{S})$, and dividing by ε leads to (14).

Step 3 It is shown that

18 $$D_cQA^{-1}\in\mathbf{M}(\mathbf{S})\ \forall Q\in\mathbf{M}(\mathbf{S}).$$

Indeed, since $\tilde{D}D_c+\tilde{N}N_c = I$,

19 $$D_cQA^{-1} = D_cQ(\tilde{D}D_c+\tilde{N}N_c)A^{-1}$$
$$= (D_cQ\tilde{D})D_cA^{-1} + D_c(Q\tilde{N})N_cA^{-1}\in\mathbf{M}(\mathbf{S})$$

by (13) and (14) respectively.

Step 4 It is shown that

20 $$\alpha_r^{-1}D_c\in\mathbf{M}(\mathbf{S})$$

which is the desired conclusion. Select unimodular $U,V\in\mathbf{U}(\mathbf{S})$ such that

21 $$UAV = \begin{bmatrix} \alpha_1 & & & 0 \\ & \cdot & & \\ & & \cdot & \\ & & & \cdot \\ 0 & & & \alpha_r \end{bmatrix} = S$$

is the Smith form of A. By (13),

22 $$D_cQA^{-1}\in\mathbf{M}(\mathbf{S})\ \forall Q \iff D_cQVS^{-1}U\in\mathbf{M}(\mathbf{S})\ \forall Q$$

$$\Longleftrightarrow D_c Q V S^{-1} \in \mathbf{M}(\mathbf{S})\ \forall Q$$

$$\Longleftrightarrow D_c Q_1 S^{-1} \in \mathbf{M}(\mathbf{S})\ \forall Q_1$$

where Q_1 denotes QV, and Q, Q_1 range over $\mathbf{M}(\mathbf{S})$. Now let $Q_1 = I$. Then $D_c S^{-1} \in \mathbf{M}(\mathbf{S})$, which shows that α_r divides the last column of D_c. Next, successively let Q_1 equal the permutation matrix that interchanges column j with the last column, and let the index j vary over all possible values. Then $D_c Q_1 S^{-1} \in \mathbf{M}(\mathbf{S})$ implies that α_r divides every column of D_c. □

The argument used in the last step is sufficiently useful that it deserves display as a separate item.

23 *Proposition* Suppose $A, B \in \mathbf{M}(\mathbf{S})$, have the same number of columns, and in addition A is square and nonsingular. Then

24 $$BQA^{-1} \in \mathbf{M}(\mathbf{S})\ \forall Q \in \mathbf{M}(\mathbf{S})$$

if and only if the largest invariant factor of A divides every element of B.

Proof of Theorem (2) It is first shown that the robust regulation problem has a solution if and only if N and $\alpha_r I$ are left-coprime. Select $\tilde{X}, \tilde{Y} \in \mathbf{M}(\mathbf{S})$ such that $\tilde{N}\tilde{X} + \tilde{D}\tilde{Y} = I$. Then by Theorem (5.2.1),

25 $$S(P) = \{(\tilde{X}+DR)(\tilde{Y}-NR)^{-1} \colon R \in \mathbf{M}(\mathbf{S})\}.$$

Thus the set of all D_c as C varies over $S(P)$ is given by $\{\tilde{Y}-NR \colon R \in \mathbf{M}(\mathbf{S})\}$. By Lemma (8), the robust tracking problem has a solution if and only if $\alpha_r^{-1}(\tilde{Y}-NR) \in \mathbf{M}(\mathbf{S})$ for some $R \in \mathbf{M}(\mathbf{S})$, or equivalently, there exist $R, W \in \mathbf{M}(\mathbf{S})$ such that

26 $$\tilde{Y} - NR = \alpha_r W.$$

Now (26) can be rearranged as

27 $$NR + \alpha_r W = \tilde{Y}.$$

Thus the proof is complete if it can be shown that the left-coprimeness of N and $\alpha_r I$ is necessary and sufficient for (27) to have a solution. Accordingly, suppose first that N, $\alpha_r I$ are left-coprime, and select T, $U \in \mathbf{M}(\mathbf{S})$ such that $NT + \alpha_r U = I$; then $R = T\tilde{Y}$, $W = U\tilde{Y}$ is a solution of (27). Conversely, suppose N and $\alpha_r I$ are not left-coprime, and let V be a nontrivial (i.e. nonunimodular) common left-divisor of N and $\alpha_r I$. Then $NR + \alpha_r W$ is a left-multiple of V for all R, W. On the other hand, from Corollaries (4.1.67) and (4.1.26), N and $\tilde{Y}$ are left-coprime, so that $\tilde{Y}$ is *not* a left-multiple of V. Hence (27) does not have a solution.

To complete the proof, it is shown that the set of all $C \in S(P)$ such that $\alpha_r D_c^{-1} \in \mathbf{M}(\mathbf{S})$ is precisely $\{C_1/\alpha_r: C_1 \in S(P/\alpha_r)\}$, provided of course that N and $\alpha_r I$ are left-coprime. First, suppose $D_c = \alpha_r T$ for some $T \in \mathbf{M}(\mathbf{S})$. Then, since

28 $$I = \tilde{D}D_c + \tilde{N}N_c = \alpha_r \tilde{D}T + \tilde{N}N_c,$$

we see that T and N_c are right-coprime, and that $C_1 = N_c T^{-1}$ stabilizes $(\alpha_r \tilde{D})^{-1}\tilde{N} = P/\alpha_r$. Hence every solution to the robust tracking problem is of the form C_1/α_r for some $C_1 \in S(P/\alpha_r)$. Conversely, suppose $C_1 \in S(P/\alpha_r)$. Now $P/\alpha_r = (\alpha_r \tilde{D})^{-1}\tilde{N}$; moreover, it can be shown that $\alpha_r \tilde{D}$ and $\tilde{N}$ are left-coprime. Hence $(\alpha_r \tilde{D}, \tilde{N})$ is a l.c.f. of P/α_r. Since $C_1 \in S(P/\alpha_r)$, C_1 has an r.c.f. (N_{c_1}, D_{c_1}) satisfying

29 $$\alpha_r \tilde{D}D_{c_1} + \tilde{N}N_{c_1} = I.$$

Thus $C = C_1/\alpha_r = N_{c_1} DC1^{-1}$ stabilizes P, and the denominator matrix of C is a multiple of α_r. Hence C solves the robust tracking problem. □

30 *Corollary* Suppose N and $\alpha_r I$ are left-coprime, and let T, $U \in \mathbf{M}(\mathbf{S})$ be particular solutions of

31 $$NT + \alpha_r U = I.$$

Let (B, A) be an r.c.f. of N/α_r. Then the set of all $C \in S(P)$ that solve the robust tracking problem is

32 $$\{[\tilde{X}+D(T\tilde{Y}+AS)][\alpha_r(U\tilde{Y}-BS)]^{-1} : S \in \mathbf{M}(\mathbf{S})\}.$$

Proof From the proof of Theorem (2), we see that C solves the robust tracking problem if and only if $C = (\tilde{X}+DR)(\tilde{Y}-NR)^{-1}$ for some R satisfying (27). Now it is routine to verify that the set of all solutions of (27) is

33 $$R = T\tilde{Y}+AS,\ W = U\tilde{Y}-BS,\ S \in \mathbf{M}(\mathbf{S}).$$

Substituting for R from (33) and noting that $\tilde{Y}-NR = \alpha_r W = \alpha_r(U\tilde{Y}-BS)$ leads to (32). □

Proof of Lemma (4) Recall from Section 4.2 that N and $\alpha_r I$ are left-coprime if and only if the matrix $F = [N \;\; \alpha_r I]$ has full row rank at all $s \in C_{+e}$. If $\alpha_r(s) \neq 0$, the rank condition holds; therefore it is only necessary to verify the rank condition at the C_{+e}-zeros of α_r. If $\alpha_r(s)=0$, then $F(s)$ has full row rank if and only if $N(s)$ has full row rank. Hence we conclude that N and $\alpha_r I$ are left-coprime if and only if $N(s)$ has full row rank at all C_{+e}-zeros of α_r. Clearly this is the same as requiring the largest invariant factor of N and α_r to be coprime. □.

Lemma (4) also holds over any principal ideal domain, and not just over $\mathbf{S}$; but the proof in the general case is a little more complicated (see Problem 3).

Now let us consider the problem of robust tracking using a two-parameter compensation scheme. Recall from Section 5.6 that such a scheme can be depicted as shown in Figure 7.2. In this scheme, the tracking error between the reference input r and the plant output y is given by

34 $$e = [I - N(\tilde{N}_{c_2}N+\tilde{D}_c D)^{-1}\tilde{N}_{c_1}]\, r$$

Hence the transfer matrix from v to e is given by

35 $$F(N,D) = [I - NM^{-1}\tilde{N}_{c_1}]\tilde{D}_r^{-1}\tilde{N}_r,$$

where

36 $$M = M(N,D) = \tilde{N}_{c_2}N + \tilde{D}_cD.$$

As before, the objectives of the robust tracking problem are to determine the compensator parameters $\tilde{N}_{c_1}$, $\tilde{N}_{c_2}$, $\tilde{D}_c$ such that the closed-loop system is stable and $F(N,D) \in \mathbf{M}(\mathbf{S})$ for the nominal plant as well as for all plants in some neighborhood of the nominal plant.

A partial solution to the problem is given by the next theorem.

37 *Theorem* Consider the system of Figure 7.2, and suppose P has at least as many inputs as outputs. Let (N,D) be an r.c.f. of the nominal plant, and let (N_1,D_1) denote perturbations of (N,D). Let (N_{c_2},D_c) be an r.c.f. of the feedback compensator $\tilde{D}_c^{-1}\tilde{N}_{c_2}$. Then the triple $(\tilde{N}_{c_1}, \tilde{N}_{c_2}, \tilde{D}_c)$ solves the robust tracking problem if and only if

(i) $\alpha_r^{-1}D_c \in \mathbf{M}(\mathbf{S})$, and

(ii) $N_1[M(N_1,D)]^{-1}(\tilde{N}_{c_2}-\tilde{N}_{c_1})\tilde{D}_r^{-1} \in \mathbf{M}(\mathbf{S})\ \forall (N_1,D_1)$.

If P is square, (ii) is equivalent to

(ii') $(\tilde{N}_{c_2}-\tilde{N}_{c_1})\tilde{D}_r^{-1} \in \mathbf{M}(\mathbf{S})$.

If P is nonsquare, (ii') implies but is not implied by (ii).

Remarks If we use a one-parameter, or feedback compensator, then $\tilde{N}_{c_2} = \tilde{N}_{c_1}$, so that (ii) above is automatically satisfied. In this case, (i) is a necessary and sufficient condition for robust tracking. This is precisely the result contained in Lemma (8).

Proof "if" Suppose first that (i) and (ii) hold. It is desired to show that

38 $$(I - N_1M^{-1}\tilde{N}_{c_1})\tilde{D}_r^{-1} \in \mathbf{M}(\mathbf{S})\ \forall (N_1,D_1).$$

where $M = M(N_1,D_1)$ is as in (36). Define

39 $$S = \begin{bmatrix} D_1 & -N_{c_2} \\ N_1 & D_c \end{bmatrix}$$

By the robustness of closed-loop stability, S is unimodular for (N_1,D_1)

sufficiently close to (N,D). Let $T = S^{-1}$, and partition T as

40 $$T = \begin{bmatrix} A & B \\ K & L \end{bmatrix}$$

Now $[A \quad B]$ is equivalent to $[I \quad 0]$, since it is part of a unimodular matrix. This, plus the fact that

41 $$[A \quad B] \begin{bmatrix} -N_{c_2} \\ D_c \end{bmatrix} = 0$$

implies that $[A \quad B]$ is an l.c.f. of $\tilde{D}_c^{-1}\tilde{N}_{c_2}$, i.e. that

42 $$[A \quad B] = U\,[\tilde{D}_c \quad \tilde{N}_{c_2}] \text{ for some } U \in \mathbf{U}(\mathbf{S}).$$

Similarly,

43 $$[K \quad L] = [-\tilde{N}_1 \quad \tilde{D}_1],$$

where $(\tilde{D}_1,\tilde{N}_1)$ is an l.c.f. of the perturbed plant $N_1D_1^{-1}$. Now $TS = I$ implies that

44 $$U(\tilde{D}_cD_1+\tilde{N}_{c_2}N_1) = I,$$

or $U = M^{-1}$. Also, $ST = I$ implies in particular that

45 $$N_1B+D_cL = I,$$

or

46 $$NM^{-1}\tilde{N}_{c_2}+D_c\tilde{D}_1 = I.$$

Therefore

47 $$(I - N_1M^{-1}\tilde{N}_{c_1})\tilde{D}_r^{-1} = D_c\tilde{D}_1\tilde{D}_r^{-1} + N_1M^{-1}(\tilde{N}_{c_2} - \tilde{N}_{c_1})\tilde{D}_r^{-1}$$

which belongs to $\mathbf{M}(\mathbf{S})$ by (i) and (ii). Hence robust tracking is achieved.

"only if" Suppose the triple $(\tilde{N}_{c_1}, \tilde{N}_{c_2}, \tilde{D}_c)$ achieves robust tracking. It is desired to show that (i) and (ii) hold. The hypothesis is that

48 $$\{I - N_1[M(N_1,D_1)]^{-1}\tilde{N}_{c_1}\}\tilde{D}_r^{-1}\tilde{N}_r \in \mathbf{M}(\mathbf{S})\ \forall (N_1,D_1),$$

where (N_1,D_1) range over some neighborhood of (N,D). By Lemma (5), (48) is equivalent to

49 $$\{I - N_1[M(N_1,D_1)]^{-1}\tilde{N}_{c_1}\}\tilde{D}_r^{-1} \in \mathbf{M}(\mathbf{S})\ \forall (N_1,D_1).$$

As a prelude to the proof, observe first that it can be assumed without loss of generality that the numerator matrix N of the nominal plant and $\alpha_r I$ are left-coprime. If this is not so originally, it can be achieved by an arbitrarity small perturbation of the plant: It suffices to perturb N to a nearby matrix which has full row rank at all C_{+e}-zeros of α_r (see Lemma (4)). Also, we may as well assume that the r.c.f. (N,D) of the nominal plant and the triple $(\tilde{N}_{c_1}, \tilde{N}_{c_2}, \tilde{D}_c)$ satisfy

50 $$\tilde{N}_{c_2}N + \tilde{D}_c D = I.$$

Finally, let $(\tilde{D},\tilde{N})$ be an l.c.f. of the nominal plant defined by the equation

51 $$\begin{bmatrix} D & -N_{c_2} \\ N & D_c \end{bmatrix} = \begin{bmatrix} \tilde{D}_c & \tilde{N}_{c_2} \\ -\tilde{N} & \tilde{D} \end{bmatrix}^{-1}.$$

Then, since $\tilde{N}$ and N have the same invariant factors, it follows from Lemma (4) that $\tilde{N}$ and $\alpha_r I$ are also left-coprime.

The first step of the proof is to observe that the hypotheses imply that $\tilde{N}_{c_1}$ and $\tilde{D}_r$ are right-coprime. To see this, note that (49) implies that

52 $$W = (I - N\tilde{N}_{c_1})\tilde{D}_r^{-1} \in \mathbf{M}(\mathbf{S}).$$

Hence

53 $N\tilde{N}_{c_1} + W\tilde{D}_r = I,$

which shows that $\tilde{N}_{c_1}$ and $\tilde{D}_r$ are right-coprime. Let (A,B) be an l.c.f. of $\tilde{N}_{c_1}\tilde{D}_r^{-1}$, and note that α_r is also the largest invariant factor of A. Next, fix the plant numerator at N and vary its denominator. Then (49) states that

54 $\{I - N[M(N,D_1)]^{-1}\tilde{N}_{c_1}\}\tilde{D}_r^{-1} \in \mathbf{M}(\mathbf{S})\ \forall D_1.$

Subtracting (52) from (54) leads to

55 $$N\{[M(N,D_1)]^{-1} - I\}\tilde{N}_{c_1}\tilde{D}_r^{-1} = N\{[M(N,D_1)]^{-1} - I]A^{-1}B \in \mathbf{M}(\mathbf{S})\ \forall D_1.$$

Hence, from Lemma (5),

56 $N\{[M(N,D_1)]^{-1} - I\}A^{-1} \in \mathbf{M}(\mathbf{S})\ \forall D_1.$

Now $M = M(N,D_1) = I + \tilde{D}_c(D_1 - D)$. By an argument analogous to that in the proof of Lemma (8) (see especially Step 2), it can be shown that $M^{-1} - I$ can be made equal $\varepsilon\tilde{D}_c Q$ for an arbitrary matrix $Q \in \mathbf{M}(\mathbf{S})$, by suitable choice of D_1. Therefore (56) implies that

57 $N\tilde{D}_c Q A^{-1} \in \mathbf{M}(\mathbf{S})\ \forall Q \in \mathbf{M}(\mathbf{S}).$

Now Proposition (23) leads to the conclusion that α_r, the largest invariant factor of A, divides each element of $N\tilde{D}_c$. However, from (51), $N\tilde{D}_c = D_c\tilde{N}$. Hence

58 $D_c\alpha_r^{-1}\tilde{N} \in \mathbf{M}(\mathbf{S}).$

Finally, since $\alpha_r I$ and $\tilde{N}$ are left-coprime, an application of Lemma (5) to (58) gives

59 $D_c\alpha_r^{-1} \in \mathbf{M}(\mathbf{S}).$

This proves (i).

The proof of (ii) is now very easy. From previous arguments, robust tracking is equivalent to (47). However, since α_r divides every element of D_c, we have that

60 $$D_c \tilde{D}_1 \tilde{D}_r^{-1} \in \mathbf{M}(\mathbf{S}) \ \forall (N_1, D_1).$$

Subtracting (60) from (47) shows that

61 $$N_1 M^{-1} (\tilde{N}_{c_2} - \tilde{N}_{c_1}) \tilde{D}_r^{-1} \in \mathbf{M}(\mathbf{S}) \ \forall (N_1, D_1).$$

This proves (ii).

To complete the proof, it is necessary to show two things: 1) Condition (ii') implies (ii); 2) If the plant is square, (ii) implies (ii'). Of these, the first statement is obvious: If $\tilde{N}_{c_2} - \tilde{N}_{c_1}$ is a left-multiple of $\tilde{D}_r$, then (61) is clearly satisfied. To prove the second statement, define

62 $$Q = (\tilde{N}_{c_2} - \tilde{N}_{c_1}) \tilde{D}_r^{-1}.$$

Then (61) implies in particular that $NQ \in \mathbf{M}(\mathbf{S})$. Now select unimodular matrices $U, V \in \mathbf{U}(\mathbf{S})$ such that UQV is in Smith-McMillan form, and let

63 $$S = U^{-1} Q V^{-1} = \text{Diag } \{ s_1/t_1, \ldots, s_r/t_r \}.$$

Since $NQ \in \mathbf{M}(\mathbf{S})$, it follows that t_i divides the i-th column of NU, for all i. Now, if Q itself does not belong to $\mathbf{M}(\mathbf{S})$, then t_1 is not a unit. Note that t_1 is a divisor of α_r, because of (62). Now let s be any C_{+e}-zero of t_1. Then, since t_1 divides the first column of NU, $NU(s)$ is singular, which in turn implies, by the unimodularity of U, that $N(s)$ is singular. However, since t_1 is a divisor of α_r, this contradicts the left-coprimeness of N and $\alpha_r I$. Hence $Q \in \mathbf{M}(\mathbf{S})$. □

Theorem (37) shows that (ii) implies (ii') for square plants. It is now shown by example that there exist situations involving nonsquare plants where (ii) holds but (ii') is false.

64 *Example* Suppose the plant

$$P = [\frac{3s}{2s-1} \quad \frac{4s+1}{2s-1}]$$

is to be compensated so as to robustly track a step input. In this case, we may take $\tilde{D}_r = \alpha_r = s/(s+1)$. For convenience, let z denote $s/(s+1)$, and note that $z \in \mathbf{S}$. Then

$$P = [3z \quad 3z+1]\,(3z-1)^{-1},$$

and an r.c.f. for P is given by

$$N = [-z \quad 3z+1],$$

$$D = \begin{bmatrix} 1+z & -1-3z \\ -2z & 6z-1 \end{bmatrix}$$

Now choose

$$\tilde{D}_c = \begin{bmatrix} 1 & 0 \\ z & z \end{bmatrix}, \quad \tilde{N}_{c_2} = \begin{bmatrix} 1 \\ 1-z \end{bmatrix}.$$

Then it is claimed that robust tracking is achieved provided α_r divides the second component of $\tilde{N}_{c_2} - \tilde{N}_{c_1}$, even if α_r does not divide the first component of $\tilde{N}_{c_2} - \tilde{N}_{c_1}$. This is proved by establishing that condition (ii) of Theorem (37) holds. For this purpose, suppose N, D are perturbed to N_1, D_1, and partition these as

$$N_1 = [n_a \quad n_b], \; D_1 = [D_a \quad D_b].$$

Also, partition $\tilde{N}_{c_1}, \tilde{N}_{c_2}, \tilde{D}_c$ as

$$\tilde{N}_{c_1} = \begin{bmatrix} x_a \\ x_b \end{bmatrix}, \quad \tilde{N}_{c_2} = \begin{bmatrix} y_a \\ y_b \end{bmatrix}, \quad \tilde{D}_c = \begin{bmatrix} \tilde{D}_a \\ \tilde{D}_b \end{bmatrix}.$$

The key point to note here is that $\tilde{D}_b$ is a multiple of $z = \alpha_r$. Now partition $M = \tilde{N}_{c_2} N_1 + \tilde{D}_c D_1$ in the form

$$M = \begin{bmatrix} m_{aa} & m_{ab} \\ m_{ba} & m_{bb} \end{bmatrix}$$

Then

$$N_1 M^{-1}(\tilde{N}_{c_2} - \tilde{N}_{c_1})/\alpha_r =$$

$$| M |^{-1} [(n_a m_{bb} - n_b m_{ba})(y_a - x_a)]/\alpha_r +$$

$$| M |^{-1}[(-n_a m_{ab} - n_b m_{aa})(y_b - x_b)]/\alpha_r.$$

Since α_r divides $y_b - x_b$, the second term belongs to **S**, provided the plant perturbation is small enough to maintain the unimodularity of M, i.e. to maintain closed-loop stability. As for the first term,

$$m_{ba} = y_b n_a + \tilde{D}_b D_a, \; m_{bb} = y_b n_b + \tilde{D}_b D_b.$$

$$n_a m_{bb} - n_b m_{ba} = \tilde{D}_b (n_a D_b - n_b D_a)$$

is a multiple of α_r, since α_r divides $\tilde{D}_b$. Hence (ii) holds and robust tracking is achieved, whatever be x_a!

Now let us consider the problem of optimal regulation. The problems studied in Chapter 6 can be interpreted as minimizing the effect of a *stable* disturbance on the plant output, or minimizing the tracking error in response to a *stable* reference input. With the aid of the results derived thus far in this section, it is now possible to tackle the same problems in the presence of *unstable* disturbances or reference inputs. Specifically, suppose P is a given plant, $\tilde{D}_r^{-1}\tilde{N}_r$ is a given disturbance or reference input generator, and W_1, W_2 are given square weighting matrices. The objective is to minimize the objective function

65 $$J_p = \| W_1 (I + PC)^{-1} \tilde{D}_r^{-1} \tilde{N}_r W_2 \|_p$$

over all C that achieve robust regulation. This can be interpreted as the problem of optimal disturbance rejection, or optimal tracking using a one-parameter compensator. Using Theorem (2) and Corollary (30), this problem can be transformed into one of the type solved in Chapter 6. With all symbols as in Theorem (2) and Corollary (30), the set of all C over which J_p is to be minimized is given by (32). Substituting from (32) into (65) shows that

66 $$(I+PC)^{-1}\tilde{D}_r^{-1} = (U\tilde{Y}-BS)\tilde{D}L,$$

67 $$J_p = \| E - W_1 BSF \|_p,$$

where

68 $$L = \alpha_r \tilde{D}_r^{-1},$$

$$E = W_1 U\tilde{Y}F,$$

$$F = \tilde{D}L\tilde{N}_r W_2.$$

Note that the robust regulation problem has a solution if and only if the plant has at least as many inputs as outputs. Hence F is square or fat, and as a result the minimization of J_p can be carried out using the techniques presented in Section 6.5.

Consider now the problem of optimal robust tracking of square plants using a two-parameter compensation scheme. From Corollary (30) and Theorem (37) it follows that the set of all triples $(\tilde{N}_{c_1}, \tilde{N}_{c_2}, \tilde{D}_c)$ that achieve robust regulation is given by

69 $$\tilde{D}_c^{-1}\tilde{N}_{c_2} = [(\tilde{X}+D(T\tilde{Y}+AS)][\alpha_r(U\tilde{Y}-BS)]^{-1},$$

$$\tilde{N}_{c_1} = \tilde{N}_{c_2} - Q\tilde{D}_r,$$

$$S, Q \in \mathbf{M}(\mathbf{S}).$$

The above expression is not very convenient, as it does not explicitly specify the compensator parameters. Such an expression can be obtained using a doubly coprime factorization for the plant. Suppose

70 $$\begin{bmatrix} Y & X \\ -\tilde{N} & \tilde{D} \end{bmatrix} \begin{bmatrix} D & -\tilde{X} \\ N & \tilde{Y} \end{bmatrix} = I.$$

Then by Problem (5.2.2) it follows that

71 $$(\tilde{X}+DR)(\tilde{Y}-NR)^{-1} = (Y-R\tilde{N})^{-1}(X+R\tilde{D}) \quad \forall R \in \mathbf{M}(\mathbf{S}).$$

Hence the set of all compensators that achieve robust tracking can be described with the aid of Corollary (30) and (71) as

72 $$\tilde{D}_c = Y-(T\tilde{Y}+AS)\tilde{N},$$
$$\tilde{N}_{c_2} = X+(T\tilde{Y}+AS)\tilde{D},$$
$$\tilde{N}_{c_1} = X+(T\tilde{Y}+AS)\tilde{D}+Q\tilde{D}_r.$$

For the class of compensators described above, the tracking error transfer matrix can be computed from (47) as

73 $$(D_c\tilde{D}\tilde{D}_r^{-1} + NQ)\tilde{N}_r = ((U\tilde{Y}-BS)\tilde{D}L + NQ)\tilde{N}_r.$$

Comparing (73) with (66), we see that the extra flexibility arising from using a two-parameter compensation scheme is the presence of the term NQ in (73). The expression (73) can actually be simplified. Since the plant is square, N and $\alpha_r I$ are both left- as well as right-coprime. Hence $(N, \alpha_r I)$ is a r.c.f. for $\alpha_r^{-1}N$, or in other words, we may take $A = \alpha_r I$, $B = N$. Hence in (73),

74 $$(U\tilde{Y}-BS)\tilde{D}L+NQ = U\tilde{Y}\tilde{D}L+N(Q-S\tilde{D}L)$$
$$= U\tilde{Y}\tilde{D}L+NR$$

where $R = Q-S\tilde{D}$ is a free parameter. To reiterate, the set of stable transfer matrices achievable using a feedback compensator is

75 $$(U\tilde{Y}\tilde{D}L-NS\tilde{D}L)\tilde{N}_r, \; S\in\mathbf{M}(\mathbf{S}),$$

whereas with a two-parameter scheme, the set is

76 $$(U\tilde{Y}\tilde{D}L+NQ)\tilde{N}_r, \; Q\in\mathbf{M}(\mathbf{S}).$$

Now, the problem of minimizing

77 $$J_p = \| W_1(I-N\tilde{N}_{c_2})\tilde{D}_r^{-1}\tilde{N}_r W_2 \|_p$$

over all robustly stabilizing two-parameter compensators can be reformulated as that of minimizing

78 $J_p = \| E - W_1 N Q \tilde{N}_r W_2 \|_p$

as Q varies over $\mathbf{M}(\mathbf{S})$, where E is defined in (68). It is easy to see that the infimum in (78) is no larger than that in (67), since (78) contains (67) as a special case by choosing $Q = S\tilde{D}L$. Moreover, as in Section 6.6, there always exist infinitely many optimal compensators if a two-parameter scheme is used, and this flexibility can be exploited to achieve both good disturbance rejection and good tracking *simultaneously*.

79 *Example* For the sake of variety we consider discrete-time systems in the next two examples. Thus z denotes the unit delay and its z-transform. Note that the z-transform of a sequence $\{f_0, f_1, \ldots\}$ is given by

$$f(z) = \sum_{i=0}^{\infty} f_i z^i.$$

Consider the set-up shown in Figure 7.3, and suppose it is desired to robustly reject a step input disturbance. Since the z-transform of a sequence of one's is $1/(1-z)$, we may take

$$n_r = 1, \; d_r = \alpha_r = 1-z.$$

Suppose the plant has two inputs and one output, and has the transfer matrix

$$P = [z(z-0.5) \;\; z(z+0.5)]\,(z+0.25)^{-1}.$$

An l.c.f. of this plant is given by

$$\tilde{D} = z+0.25, \; \tilde{N} = [z(z-0.5) \;\; z(z+0.5)]$$

and a particular solution of the Bezout identity $\tilde{N}\tilde{X} + \tilde{D}\tilde{Y} = I$ is given by

$$\tilde{X} = [4 \;\; -4]', \; \tilde{Y} = 4.$$

An r.c.f. of P is given by

$$N = [z \;\; 0],$$

$$D = \begin{bmatrix} -1 & -z-0.5 \\ 1 & z-0.5 \end{bmatrix}.$$

Now N and α_r are left-coprime. In fact (31) is satisfied with

$$T = [1 \;\; 0]', U = 1$$

Hence the robust disturbance rejection problem has a solution. Further, all solutions of this problem can be parametrized using Corollary (30). An r.c.f. of N/α_r is

$$B = [z \;\; 0], A = \begin{bmatrix} z-1 & 0 \\ 0 & 1 \end{bmatrix}.$$

From (32), the set of all compensators that achieve robust disturbance rejection is given by

$$C = N_c D_c^{-1},$$

where

$$D_c = (4-zs_1),$$

$$N_c = (z+0.25) \begin{bmatrix} -4z+3+(1-z)s_1-(z+0.5)s_2 \\ 4z-3+(z-1)s_1+(z-0.5)s_2 \end{bmatrix},$$

$s_1, s_2 \in \mathbf{H}_\infty$ are rational.

Now consider the problem of optimal robust disturbance rejection as stated in (65), with both weighting functions equal to unity. From (66) and (67), the set of achievable transfer functions from v to y is

$$f(s_1,s_2) = (4-zs_1)(z+0.25).$$

Suppose it is desired to minimize the $\mathbf{H}_\infty$-norm of this transfer function. Using the methods described in Section 6.4, one can verify that the minimum is achieved with $s_1 = 0$, with a minimum cost of 5. The collection of compensators that achieve the minimum is

$$D_c = 4(1-z),\ N_c = \begin{bmatrix} -4z+3-(z+0.5)s_2 \\ 4z-3+(z-0.5)s_2 \end{bmatrix},\ s_2 \in \mathbf{H}_\infty.$$

80 *Example* Consider the set-up shown in Figure 7.2, where the input to be tracked is the sinusoidal sequence $\{0,1,0,-1,0,1,....\}$. Since the z-transform of this sequence is $z/(1+z^2)$, we may take

$$n_r = z,\ \alpha_r = 1+z^2.$$

Suppose the plant is scalar, with the transfer function

$$p = \frac{z}{z+0.5}.$$

Then, dropping the tildes because everything is a scalar, we hav

$$n = z,\ d = z+0.5,\ x = -2,\ y = 2.$$

$$a = \alpha_r = 1+z^2,\ b = n = z,\ t = -z,\ u = 1.$$

Hence the set of r, w that satisfy (27) is

$$r = ty+as = -2z+(1+z^2)s,\ w = uy-bs = 2-zs,\quad s \in \mathbf{H}_\infty.$$

From Corollary (30), the set of all feedback, or one-parameter compensators that achieve robust tracking is given by $c = n_c/d_c$, where

$$d_c = (2-zs)(1+z^2),\ n_c = -2+(z+0.5)(-2z+(1+z^2)s),$$

and $s \in \mathbf{H}_\infty$ is arbitrary. If it is desired to have optimal tracking by minimizing the objective function (65) with unity weighting, then one has to minimize

$$J_\infty = \| (2-zs)(z+0.5)z \|_\infty,\ s \in \mathbf{H}_\infty.$$

Using the techniques of Section 6.4, one can verify that the minimum is 3, corresponding to the choice $s = 0$. Hence the *unique* feedback compensator that achieves optimal robust tracking is

$$c = \frac{-(1+0.5z+z^2)}{1+z^2}.$$

Now let us see how much better we can do using a two-parameter compensation scheme. The set of transfer functions from v to the tracking error that can be achieved is

$$n_r((uy-bs)d+nq) = z[(2-zs)(z+0.5)+zq], \ s, q \in \mathbf{H}_\infty.$$
$$= z(1+zq_1),$$

where $q_1 = q-(z+0.5)s+2$. The $\mathbf{H}_\infty$-norm of the above is minimized when $q_1 = 0$, and equals 1. Hence not only does the two-parameter scheme result in a reduction in the optimal tracking error by a factor of 3, but there are *infinitely many* compensators that achieve this minimum. In this way it is possible to achieve both optimal disturbance rejection as well as good tracking. For example, a compensator that achieves optimal tracking corresponds to the choice $q = -2$, $s = 0$, which leads to the compensator

$$d_c = 2+2z^2,$$

$$n_{c_2} = -2(1+0.5+z^2),$$

$$n_{c_1} = n_{c_2}+2\alpha_r = -z.$$

Note that the feedback compensator is the same as the optimal compensator discussed above. But the use of a two-parameter scheme reduces the tracking error by a factor of 3.

Problems

1. Prove Lemma (6).

2. Suppose $A \in \mathbf{M}(\mathbf{S})$ is square and nonsingular and let a denote its largest invariant factor. Show that $aA^{-1} \in \mathbf{M}(\mathbf{S})$.

3. Prove Lemma (4) over a general principal ideal domain $\mathbf{R}$ (Hint: Partition all the full-size minors of the matrix $[N \ \ \alpha_r I]$ into several groups, namely: those that do not involve any columns of $\alpha_r I$, those that involve

one column, two columns, etc.).

4. Suppose $\tilde{D}$ and $\tilde{N}$ are left-coprime, as are $\tilde{N}$ and $\alpha_r I$. Show that $\alpha_r \tilde{D}$, $\tilde{N}$ are left-coprime.

5. Using the techniques of Section 6.7, construct a sequence of plants P_i and a sequence of reference inputs $\tilde{D}_{r_i}^{-1}\tilde{N}_{r_i}$ such that the ratio of the optimal tracking error achievable using a two-parameter compensation scheme to that achievable using a one-parameter scheme approaches zero as i approaches ∞.

7.6 GENERICITY

In this section, the focus is on the notion of genericity, which is a mathematically precise way of expressing the intuitive idea that some statement or property is almost always true. The concept of genericity is defined first, and then it is shown that various useful properties of control systems are generically true. These include coprimeness, strong stabilizability, and simultaneous stabilizability (subject of course to suitable restrictions).

We begin with a formal definition of genericity. Let $\mathbf{X}$ denote a topological space. Recall that a *binary relation* $\mathbf{R}$ on $\mathbf{X}$ is a subset of the product space $\mathbf{X} \times \mathbf{X}$. It is *symmetric* if $(x,y) \in \mathbf{R} \Leftrightarrow (y,x) \in \mathbf{R}$. More generally, an n-ary relation is a subset of the product space $\mathbf{X}^n$.

1 *Definition* A binary relation $\mathbf{R}$ on $\mathbf{X}$ is *generic* if it is an open dense subset of $\mathbf{X}$. It is *strongly generic* if the following is true: For each $x \in \mathbf{X}$, define

2 $$\mathbf{B}(x) = \{y \in \mathbf{X}: (x,y) \in \mathbf{R}\}.$$

Then the set of x for which $\mathbf{B}(x)$ is an open dense subset of $\mathbf{X}$ is itself an open dense subset of $\mathbf{X}$.

One can state the same definition in words: $\mathbf{R}$ is generic if two statements are true: (i) If a pair (x,y) belongs to $\mathbf{R}$, then there is a neighborhood of (x,y) such that every point in the neighborhood also belongs to $\mathbf{R}$. (ii) If a pair (x,y) does not belong to $\mathbf{R}$, then every neighborhood of this pair contains a point that belongs to **R**. In other words, **R** is generic if it has two features: (i) If a pair satisfies the relation, then sufficiently small perturbations will not destroy the relation. (ii) If a pair does not satisfy the relation, then arbitrarily small perturbations in the pair will make it satisfy the relation. A simple (and pertinent) example is that, in the polynomial ring $\mathbb{R}[s]$ topologized as in Problem C.2.3, the relation **R** defined by

3 $$\mathbf{R} = \{(x,y): x \text{ and } y \text{ have } \aleph_0 \text{ common zeros}\}$$

is generic. The exact relationship between genericity and strong genericity is brought out in Problem 1. But, roughly speaking, the symmetric relation **R** is strongly generic if, for almost all x, almost all y satisfy the relation $(x,y) \in \mathbf{R}$.

4 *Proposition* Let $\mathbf{A}$, $\mathbf{B}$ be open dense subsets of a topological space $\mathbf{X}$. Then $\mathbf{C} = \mathbf{A} \cap \mathbf{B}$ is also open and dense.

Proof Clearly $\mathbf{C}$ is open, being the intersection of two open sets. Hence it only remains to show that it is dense. For this, it is necessary to show the following: Let $x \in \mathbf{X}$ be arbitrary. Then every neighborhood of x contains a point of $\mathbf{A} \cap \mathbf{B}$. Accordingly, let $x \in \mathbf{X}$ be chosen arbitrarily, and let $\mathbf{N}$ be a neighborhood of x. Since $\mathbf{A}$ is dense, $\mathbf{N}$ contains an element $a \in \mathbf{A}$. Moreover, since $\mathbf{A}$ is open, there is a neighborhood $\mathbf{N}_1 \subseteq \mathbf{A}$ containing a. One can assume that $\mathbf{N}_1$ is also a subset of $\mathbf{N}$ by replacing $\mathbf{N}_1$ by $\mathbf{N} \cap \mathbf{N}_1$ if necessary. Next, since $\mathbf{B}$ is also dense, the neighborhood $\mathbf{N}_1$ contains an element $b \in \mathbf{B}$. Since b also belongs to $\mathbf{A}$, b belongs to $\mathbf{A} \cap \mathbf{B}$ and is contained in the neighborhood $\mathbf{N}$ of x. □

The topological space under study in this section is the ring **S** of proper stable rational functions. Recall that the topology on **S** is the natural one induced by its norm. Thus a base for the topology on **S** is provided by open balls of various radii centered at various points. Most of the arguments given below can be extended to an arbitrary topological ring whose set of units is open and where inversion of a unit is continuous, but at a considerable increase in the level of abstraction.

We begin by deriving some results concerning the genericity of coprimeness.

5 *Proposition* Define a symmetric binary relation $\mathbf{R}$ on $\mathbf{S}$ by

6 $$\mathbf{R} = \{(f,g): f \text{ and } g \text{ are coprime}\}.$$

Then $\mathbf{R}$ is strongly generic as well as generic.

Proof It is shown only that $\mathbf{R}$ is strongly generic. The proof of genericity is similar and is left as an exercise (see Problem 2).

Let $f \in \mathbf{S}$, and define

7 $$\mathbf{R}_f = \{g \in \mathbf{S}: f \text{ and } g \text{ are coprime}\}.$$

It is shown below that $\mathbf{R}_f$ is an open dense subset of $\mathbf{S}$ whenever $f \neq 0$. Since the set of nonzero elements is open and dense in $\mathbf{S}$, this would prove the strong genericity of $\mathbf{R}$.

To show that $\mathbf{R}_f$ is open, suppose that f and g are coprime, and choose $a, b \in \mathbf{S}$ such that $af + bg = 1$. If $\| h - g \|$ is less than $\| b \|^{-1}$, then

8 $$af + bh = 1 + b(h - g)$$

is a unit of $\mathbf{S}$, which shows that h and f are also coprime. Thus, if $g \in \mathbf{R}_f$, then $h \in \mathbf{R}_f$ whenever h belongs to the ball $\mathbf{B}(g, \| b \|^{-1})$. This shows that $\mathbf{R}_f$ is open.

To show that $\mathbf{R}_f$ is dense whenever $f \neq 0$, let g be an arbitrary element of $\mathbf{S}$. If f and g are coprime, then nothing further needs to be done. On the other hand, if f and g are not coprime, it must be shown that every neighborhood of g contains an element which is coprime with f. Towards this end, let S denote the set of C_{+e}-zeros of the function f, and note that S is a *finite* set since f is nonzero. Now partition the set S into two disjoint subsets S_1 and S_2, as follows:

9 $$S_1 = \{s \in S: g(s) = 0\}, S_2 = \{s \in S: g(s) \neq 0\}.$$

Define

10 $$r = \min_{s \in S_2} |g(s)|.$$

and note that $r > 0$. Now let $\varepsilon > 0$ be arbitrary; it will be shown that the ball $\mathbf{B}(g,\varepsilon)$ contains an element h which is coprime with f. Select a positive $\delta < \min(r,\varepsilon)$ and define $h = g + \delta \cdot 1$. Then, it is claimed, f and h are coprime, since they do not have any common C_{+e}-zeros. To establish the claim, note that the only possible common C_{+e}-zeros of f and h are the elements of the set S. Now, if $s \in S_1$, then $h(s) = \delta \neq 0$, whereas if $s \in S_2$, then $|h(s)| > r - \delta > 0$. Hence f and h are coprime. □

Thus Proposition (5) shows that, generically, any two functions in $\mathbf{S}$ are coprime. In order to extend this result to matrices, a preliminary lemma is needed.

11 *Proposition* Suppose $m > 1$ is some integer. Then the set

12 $$\mathbf{M}_m = \{D \in \mathbf{S}^{m \times m} : |D| \neq 0\}$$

is an open dense subset of $\mathbf{S}^{m \times m}$.

Proof The map $D \rightarrow |D|$ is continuous, and the set of nonzero elements in $\mathbf{S}$ is open. Hence $\mathbf{M}_m$, which is the preimage of an open set under a continuous map, is itself open. To show that it is also dense, suppose D is a matrix in $\mathbf{S}^{m \times m}$ with $|D| = 0$, and let ε be an arbitrary positive number. It is shown next that the ball $\mathbf{B}(D,\varepsilon)$ contains a matrix with a nonzero determinant. Select matrices U, $V \in \mathbf{U}(\mathbf{S})$ such that

13 $$UDV = Diag\{d_1, \cdots, d_r, 0, \cdots, 0\}$$

is the Smith form of D, and r is the rank of D. Observe now that d_i divides d_{i+1} for all i. Now define

14 $$S = Diag\{d_1, \cdots, d_r, \delta d_r, \cdots, \delta d_r\},$$

and let $D_1 = U^{-1} S V^{-1}$. Then $|D_1| \neq 0$, and D_1 can be made to lie in the ball $\mathbf{B}(D,\varepsilon)$ by choosing the constant δ sufficiently small. □

The next proposition generalizes Proposition (5) to the matrix case.

15 *Proposition* Suppose $m<n$ are positive integers, and define

16 $$\mathbf{R}(m,n) = \{A \in \mathbf{S}^{m \times n}: \text{The g.c.d. of all minors of } A \text{ is } 1\}.$$

Then $\mathbf{R}(m,n)$ is an open dense subset of $\mathbf{S}^{m \times n}$.

Remarks 1) If $\mathbf{R}(m,n)$ is viewed as mn-ary relation on $\mathbf{S}$, then the proposition states that this relation is generic.

2) Another way of stating the proposition is: Suppose $F \in \mathbf{S}^{m \times k}$, $G \in \mathbf{S}^{m \times l}$. Then F and G are generically coprime if $k+l>m$.

3) The definition (16) can be equivalently stated as:

17 $$\mathbf{R}(m,n) = \{A \in \mathbf{S}^{m \times n}: A \sim [I \;\; 0]\}.$$

Proof Clearly it is enough to prove the proposition for the case where $n = m+1$. The openness of the set $\mathbf{R}(m,n)$ is obvious, and it only remains to show that the set is dense. This is done by induction on m. Note that the proposition is true when $m = 1$, by Proposition (5). Suppose the proposition is true for matrices with $m-1$ rows and m (or more) columns. To establish the statement for matrices with m rows, suppose $A \in \mathbf{S}^{m \times n}$ but does not belong to $\mathbf{R}(m,n)$. To be shown is that every neighborhood of A contains a matrix belonging to $\mathbf{R}(m,n)$. Let $\varepsilon>0$ be any specified number, and partition A in the form

18 $$A = [A_1 \;\; a_2].$$

where A_1 and a_2 have dimensions $m \times m$ and $m \times 1$, respectively. Now let A_{11} denote the submatrix consisting of the first $m-1$ rows of A_1. By the inductive hypothesis and Proposition (11), there exists an $m \times m$ matrix

19 $$B_1 = \begin{bmatrix} B_{11} \\ B_{12} \end{bmatrix}$$

such that (i) $|B_1| \neq 0$, (ii) $[I \;\; 0]$ is a Smith form of B_{11}, and (iii) $\| B_1 - A_1 \| < \varepsilon/2$. Now let S denote the set of C_{+e}-zeros of $|B_1|$, and partition the set S into two disjoint sets S_1 and S_2, as follows:

20 $$S_1 = \{s \in S: \text{all minors of } [B_1 \;\; a_2] \text{ are zero}\},$$

$$S_2 = \{s \in S: \text{not all minors of } [B_1 \;\; a_2] \text{ are zero}\}.$$

The idea, as in the proof of Proposition (5), is to perturb a_2 in such a way as to destroy all the common zeros in the set S_1 while ensuring that none of the points in S_2 becomes a common zero of all the minors. For this purpose, note that all minors of the matrix $[B_1 \;\; a_2]$ belong to $\mathbf{S}$ and are thus bounded over the closed right half-plane. Let μ be an upper bound on the norms of all these minors. Now, at each point $s \in S_1$, find an $(m-1)\times(m-1)$ minor of B_1 which is nonzero at s; such a minor must exist, since $[I \;\; 0]$ is a Smith form of B_1. Suppose such a minor is obtained by omitting the j-th column of B_1. Let e_m denote the m-th elementary unit vector. Then the $m \times m$ minor of

21 $$B = [B_1 \;\; a_2 + \delta e_m]$$

obtained by omitting the j-th column is nonzero at s, whenever $\delta \neq 0$. Hence, at all $s \in S_1$, at least one of the $m \times m$ minors of the matrix B in (21) is nonzero. The next step is to show that, if δ is sufficiently small, then at least one of the $m \times m$ minors of the matrix B is nonzero at all points in the set S_2 as well. This can be shown as follows: In analogy with (10), choose a real number $r > 0$ such that, at each $s \in S_2$, at least one $m \times m$ minor of the matrix $[B_1 \;\; a_2]$ is nonzero. Now choose $\delta < r/\mu$. Then, at each $s \in S_2$, the corresponding $m \times m$ minor of B is also nonzero. Finally, since $S = S_1 \cup S_2$ is the zero set of $|B_1|$, these are the only possible points in the closed RHP where all $m \times m$ minors of the matrix B could vanish. Hence we conclude that, whenever $\delta < min(r/\mu, \varepsilon)$, B has the Smith form $[I \;\; 0]$. □

Proposition (15) shows that a *rectangular* matrix generically has the Smith form $[I \;\; 0]$. The next result is addressed to the Smith form of *square* matrices.

22 *Proposition* Suppose m is a positive integer, and define

23 $$\mathbf{T}_m = \{A \in \mathbf{S}^{m \times m}: A \sim \text{Diag}\ \{1, \cdots, 1, |A|\}\}.$$

Then $\mathbf{T}_m$ is an open dense subset of $\mathbf{S}^{m \times m}$.

Proof The openness of $\mathbf{T}_m$ is left to the reader to prove. To prove that $\mathbf{T}_m$ is dense, suppose A does not belong to $\mathbf{T}_m$, and let $\varepsilon > 0$ be a given number. To construct an element A_1 belonging to $\mathbf{T}_m$ lying in the ball $\mathbf{B}(A, \varepsilon)$, we proceed as follows: First, by Proposition (11), it can be assumed that $|A| \neq 0$, by perturbing the matrix A slightly if necessary. Now select Unimodular matrices U, V such that

24 $$S = UAV = \text{Diag}\ \{a_1, \cdots, a_m\}$$

is a Smith form of A. Then, for any $\delta > 0$, one can find functions $b_i \in \mathbf{B}(a_i, \delta)$ such that the functions $b_1, \cdots, b_m$ are all pairwise coprime (see Proposition (5)). Hence the matrix

25 $$B = U^{-1}\ \text{Diag}\ \{b_1, \cdots, b_m\} V^{-1}$$

has the Smith form Diag $\{1, \cdots, 1, |B|\}$. By adjusting δ, one can ensure that $B \in \mathbf{B}(A, \varepsilon)$. □

The first principal result of this section states that a collection of r multivariable plants is generically simultaneously stabilizable provided the plants have at least r inputs or r outputs. A useful corollary is that generically a multivariable plant can be strongly stabilized. These results make use of Proposition (15) concerning the generic Smith form of rectangular matrices.

26 *Theorem* Suppose l, m are given positive integers, and let $\mathbf{M}(\mathbb{R}(s))$ denote the set $[\mathbb{R}(s)]^{l \times m}$, equipped with the graph topology. Let r be a positive integer, and define the r-ary relation $\mathbf{R}$ on $\mathbf{M}(\mathbb{R}(s))$ by

27 $$\mathbf{R} = \{(P_1, \cdots, P_r): \bigcap_i S(P_i) \neq \emptyset\}.$$

Then $\mathbf{R}$ is an open dense subset of $\mathbf{M}(\mathbb{R}(s))$ whenever $r \leq \max(l, m)$.

The proof of the theorem is divided into two parts: First, it is shown that the result is true in the case where $min(l,m) = 1$, i.e. the case where all plants are either single-input or single-output. Then it is shown that the general case can be reduced to this special case.

28 *Lemma* Suppose $min(l,m) = 1$. Then **R** defined in (27) is an open dense subset of $\mathbf{M}(\mathbb{R}(s))$ if $max(l,m) \geq r$.

Proof The proof is given for the case where $l = 1$ and $m \geq r$. The other case can be proved by taking transposes.

Given the plants $P_1,\cdots,P_r$, each of dimension $1 \times m$, find l.c.f.'s $(d_p^{(i)}, [n_{p_1}^{(i)} \cdots n_{p_m}^{(i)}])$, for $i = 1,\cdots,r$. Define the matrix $Q \in \mathbf{M}(\mathbb{R}(s))^{r \times (m+1)}$ by

29 $$q_{ij} = d_p^{(i)} \text{ if } j = 1,$$

$$q_{ij} = n_{p(j-1)}^{(i)} \text{ if } j > 1.$$

Let $C \in \mathbf{M}(\mathbb{R}(s))^{m \times 1}$ be a compensator, and let $([n_{c_1} \cdots n_{c_m}]', d_c)$ be an r.c.f. of C. Then by Theorem (5.1.25), C stabilizes each of the plants P_i if and only if each of the functions u_i defined by

30 $$\begin{bmatrix} u_1 \\ \vdots \\ u_r \end{bmatrix} = Q \begin{bmatrix} d_c \\ n_{c_1} \\ \vdots \\ n_{c_m} \end{bmatrix}$$

is a unit. Turning this argument around, the set of plants $\{P_1,\cdots,P_r\}$ can be simultaneously stabilized if and only if there exist units u_i, $i = 1,\cdots,r$ such that (30) has a solution in $\mathbf{M}(\mathbb{R}(s))$ for the unknowns d_c and n_{c_i}, $i = 1,\cdots,m$, where the first component of the solution (corresponding to d_c) is nonzero. Now, if $m \geq r$, then the matrix Q has *more* columns than rows. Hence by Proposition (15), Q is generically right-invertible. That is, if Q is not right-invertible, it can be made so by an arbitrarily small perturbation in each of its elements, which is precisely the same as slightly perturbing each of the plants P_i in the graph topology. Thus by slightly perturbing each of the plants, we can ensure that (30) has a solution in $\mathbf{M}(\mathbb{R}(s))$, whatever be the left hand side. In particular, one can choose any arbitrary set of units $\{u_1,\cdots,u_r\}$, and (30) has a solution in $\mathbf{M}(\mathbb{R}(s))$. The only additional point to be worried about is that the first element of this solution (corresponding to d_c) should be nonzero. But this too can be guaranteed by perturbing the units slightly if necessary. The details of this last argument are routine and are left to the reader. □

The next step is to reduce the case of an $l \times m$ system where $l < m$ to the case of a single-output system (the case where $l > m$ is handled by transposition).

31 *Lemma* Suppose l, m are given positive integers. Define the relation $\mathbf{R}$ on the product space $\mathbf{M}(\mathbf{S})^{(m+l)\times m} \times \mathbb{R}^l$ as the set of all (D,N,v) such that vN, D are right-coprime. Then the relation $\mathbf{R}$ is generic.

Proof Let $n^{(1)},\cdots,n^{(l)}$ denote the rows of the matrix N. By Proposition (15), it is generically true that the Smith form of the matrix

32 $$M_i = \begin{bmatrix} D \\ n^{(i)} \end{bmatrix}$$

is $[I \ \ 0]'$ for all i. In other words, this means that the g.c.d. of all $m \times m$ minors of the matrix

33 $$M = \begin{bmatrix} D \\ N \end{bmatrix}$$

involving exactly one row of N is equal to 1. Now, if v is an $l \times 1$ vector, then the pair (D,vN) is right-coprime if and only if the g.c.d. of all $m \times m$ minors of the matrix

34 $$Q = \begin{bmatrix} D \\ vN \end{bmatrix}$$

is equal to 1. We show that generically this is the case for almost all $v \in \mathbb{R}^l$. Now the matrix Q has dimensions $(m+1)\times m$, as do the matrices M_i

defined in (32). Moreover, since the determinant function is multilinear, a minor of Q involving the last row can be expressed as a linear combination of the corresponding minors of the various M_i. More precisely, let q_i denote the minor of Q obtained by omitting the j-th row the matrix D, and define m_i^j analogously. Then

35 $$q^j = \sum_{i=1}^{l} v_i m_i^j.$$

By hypothesis, it is generically true that

36 $$\underset{i,j}{g.c.d.} \ \{m_i^j\} = 1.$$

Hence, if the elements m_i^j satisfy the rank condition of [V12], it follows that, for almost all vectors v,

37 $$\underset{j}{g.c.d.} \ \{q^j\} = 1.$$

It is left to the reader to show that the rank condition is generically satisfied. This shows that (D, vN) are right-coprime and completes the proof that the relation **R** is generic. □

The significance of Lemma (31) is in showing that, generically, a multivariable plant can be stabilized using a rank one compensator. Suppose P is a given multivariable plant with an r.c.f. (N, D), and suppose there exists a vector v such that vN, D are right-coprime. Then one can find a pair of elements $(a, B) \in \mathbf{M}(\mathbf{S})$ such that $avN + BD = I$. Note that a is a *column vector.* Now the compensator $C = B^{-1}av$ has rank one as an element of $\mathbf{M}(\mathbb{R}(s))$ and stabilizes P. Hence, if one can stabilize P using a rank one compensator if one can find a row vector v such that vN, D are right-coprime. Lemma (31) states that generically this is always possible.

With the aid of Lemma (31), the proof of Theorem (26) can be completed.

Proof of Theorem (26) If $min(l,m) = 1$, then the truth of the theorem is established by Lemma (28). If not, suppose without loss of generality that $r \leq m \geq l$. For each plant P_i, it is generically true that the set of row vectors v such that the m-input-scalar-output plant vP_i can be stabilized by an appropriate compensator is open and dense in $\mathbb{R}^l$. Since the intersection of a *finite* number of open and dense subsets of $\mathbb{R}^l$ is again open and dense (see Proposition (4)), generically the set of row vectors v such that *each* of the plants vP_i can be stabilized by an appropriate compensator is open and dense. Moreover, by Lemma (31), generically there exists a *common* stabilizing compensator C_1, since $r \leq m$. Hence $C = vC_1$ simultaneously stabilizes each of the plants P_i, $i = 1,\cdots,r$. □

In Section 5.4 it was shown that the problem of simultaneously stabilizing r plants is equivalent to that of simultaneously stabilizing $r-1$ plants using a stable compensator. Now Theorem (1) shows that r plants of dimension $l \times m$ can be generically simultaneously stabilized provided $r \leq max(l,m)$. Thus it is natural to conjecture that r plants of dimension $l \times m$ can generically be simultaneously stabilized using a stable compensator provided $r < max(l,m)$. In fact this is true, and the proof follows very nearly the reasoning above. However, a few minor adjustments are needed. The reason for this is the following: Given r plants P_i, $i = 1,\cdots,r$, they can be simultaneously stabilized using a stable compensator if and only if the plants $r+1$ plants $0, P_i$, $i = 1,\cdots,r$ can be simultaneously stabilized. Now, the reasoning used in Lemma (28) and Theorem (26) relies on being able to perturb *all* of the plants as needed. However, in the present instance, one can only perturb *all except one* plant (namely the zero plant). However, there is a saving feature: In choosing an l.c.f. of the zero plant, the numerator matrix has to be zero, but the *denominator matrix* is arbitrary. Hence the proof of the above result is complete if it can be shown that the Smith form of an $(r+1) \times (m+1)$ matrix of the form

38 $$M = \begin{bmatrix} d_1 & N_1 \\ d_2 & 0 \end{bmatrix}$$

is generically equal to $[I \;\; 0]$ whenever $r < m$. Comparing the matrix in (38) to that in Proposition (15), we see that the only difference is the presence of a row of zeros in the bottom of the matrix M in (38). Now, checking back to the proof of Proposition (15), the ingredients to the proof by induction were: (i) the fact that the $r \times m$ matrix N_1 generically had the Smith form $[I \;\; 0]$, and (ii) the ability to adjust the corner element d_2. As both of these features are present in (38), we can conclude that the matrix M is generically right-invertible if $r < m$. Hence the following corollary to Theorem (1) is proved:

39 *Corollary* Suppose the plants P_i, $i = 1, \cdots, r$ have dimension $l \times m$. Then they can generically be simultaneously stabilized using a stable compensator if $r < m$.

Problems

1. (a) Show that if $\mathbf{R}$ is open then $\mathbf{R}_x$ is open for all x.

   (b) Show that if $\mathbf{R}_x$ is a dense subset of $\mathbf{X}$ for all x belonging to a dense subset of $\mathbf{X}$, then $\mathbf{R}$ is a dense subset of $\mathbf{X} \times \mathbf{X}$.

2. Show that the relation $\mathbf{R}$ of (6) is generic.

3. Show that the relation $\mathbf{T}_m$ defined in (23) is open.

Notes and References

The graph topology of Section 7.2 and the graph metric of Section 7.3 were introduced in [92]. For another metric on the set of square rational plants, see [110]. The robustness results in Section 7.4 are due to several researchers, including Doyle and Stein [30], Chen and Desoer [13], Kimura [54], and Vidyasagar and Kimura [95]. For other results on the robustness of feedback stability, specifically the gain margin of systems, see [53, 63]. Some of the results on robust regulation are from [36]. The genericity of simultaneous stabilizability was first established for two plants by Vidyasagar and Viswanadham [98]. The general results given here were first obtained by Ghosh and Byrnes [40] using state-space methods. The present treatment follows [96].

Chapter 8. EXTENSIONS TO GENERAL SETTINGS

The previous six chapters concentrated on the problems of controller synthesis in the case where the underlying ring of "stable" transfer functions is the ring **S** defined in Section 2.1. As pointed out in Section 4.1, the only property of **S** that is used repeatedly in Chapters 2-7 is the fact that **S** is a principal ideal domain. Thus, most if not all of the contents of these chapters remain valid if **S** is replaced by any p.i.d. This is why the synthesis theory of Chapters 2-7 encompasses, within a single framework, lumped continuous-time as well as discrete-time systems, and situations where the stable region in the complex plant is something other than the left half-plane (or the closed unit disc in the discrete-time case).

The objective of the present chapter is to generalize the earlier synthesis theory by removing the assumption that the underlying set of "stable" systems is a p.i.d. In effect, the generalization extends the scope of the theory to include linear distributed systems, 2-D systems and the like.

In contrast with earlier chapters, this chapter, and in particular Section 8.1, presupposes a mathematical background on the part of the reader well beyond that contained in the appendices. Several references to the literature are given at appropriate places, but in general, [4,112] are good references for concepts from abstract algebra, while [39] is a handy source for results on Banach algebras.

8.1 COPRIME FACTORIZATIONS OVER A COMMUTATIVE RING

Throughout this section, **R** denotes a commutative domain with identity, and **F** denotes the field of fractions associated with **R**. As in the rest of the book, the symbols **M**(**R**), **M**(**F**), **U**(**R**) denote respectively the set of all matrices with elements in **R**, the set of matrices with elements in **F**, and the set of all unimodular matrices with elements in **R**. If **R** is a principal ideal domain, then the contents of Section 4.1 show that every matrix over $\mathbf{F}$ has both right- and left-coprime factorizations over $\mathbf{R}$. In fact, for this it is enough that $\mathbf{R}$ be a Bezout domain, i.e. that every *finitely generated* ideal in $\mathbf{R}$ be principal. However, if $\mathbf{R}$ is not Bezout, then the questions of the existence of r.c.f.'s and l.c.f.'s and their interrelationships need to be examined anew. It is the purpose of this section to do so.

The concepts of coprimeness and coprime factorizations introduced in Section 4.1 can be extended in a natural way to a general ring $\mathbf{R}$. Suppose $A, B \in \mathbf{M}(\mathbf{R})$ have the same number of columns. Then A and B are *right-coprime* if there exist $X, Y \in \mathbf{M}(\mathbf{R})$ such that

1 $$XN + YD = I.$$

Suppose $F \in \mathbf{M}(\mathbf{F})$. A pair (N, D) where $N, D \in \mathbf{M}(\mathbf{R})$ is a *right-coprime factorization (r.c.f.) of* F if (i) D is square and $|D| \neq 0$, (ii) $F = ND^{-1}$, and (iii) N, D are right-coprime. Left-coprime matrices and l.c.f.'s are defined analogously.

Given an arbitrary $F \in \mathbf{M}(\mathbf{F})$, the existence of an r.c.f. for F is not automatic unless $\mathbf{R}$ is a Bezout domain (see Corollary (8) below). However, if F does have an r.c.f., then it is easy enough to characterize *all* r.c.f.'s of F. In fact, the relevant part of Theorem (4.1.43) carries over *in toto*. For ease of reference, the result is repeated here, but the proof is omitted.

2 *Lemma* Suppose $F \in \mathbf{M}(\mathbf{F})$ has an r.c.f. (N, D). Then the set of all r.c.f.'s of F consists of the set of pairs (NU, DU) as U ranges over all matrices in $\mathbf{U}(\mathbf{R})$.

But when does an $F \in \mathbf{M}(\mathbf{F})$ have an r.c.f.? We begin with the scalar case.

3 *Lemma* Suppose $f \in F$, and express f as a/b where $a, b \in \mathbf{R}$. Then f has a coprime factorization if and only if the ideal generated by a and b in $\mathbf{R}$ is principal.

Proof "if" Suppose g is a generator of the ideal generated by a and b. Then $a = ng$, $b = dg$, $g = xa + yb$ for appropriate elements $n,d,x,y \in \mathbf{R}$. Since b is the denominator of a fraction, it is nonzero, which in turn implies that both d and g are nonzero. This, together with earlier relations, leads to $f = n/d$, $xn + yd = 1$. Hence (n,d) is a coprime factorization of f.

"only if" Suppose (n,d) is a coprime factorization of f, and choose $x,y \in \mathbf{R}$ such that $xn + yd = 1$. Let $r = b/d$. Then

4 $$r = bd^{-1} = b(xnd^{-1} + y)$$
$$= b(xab^{-1} + y) = xa + yb \in \mathbf{R}.$$

Hence d divides b in $\mathbf{R}$, and $dr = b$. This implies that $a = nr$, and that the ideal generated by a,b is the same as the principal ideal generated by r. □

The matrix case is essentially similar, if one observes that every matrix $F \in \mathbf{M}(\mathbf{F})$ can be expressed in the form AB^{-1} where $A,B \in \mathbf{M}(\mathbf{R})$ and $|B| \neq 0$. To see this, suppose $f_{ij} = p_{ij}/q_{ij}$ where $p_{ij}, q_{ij} \in \mathbf{R}$. Define $b = \prod_i \prod_j q_{ij}$, and let $a_{ij} = bp_{ij}/q_{ij} \in \mathbf{R}$. If A denotes the matrix of the a_{ij}'s, then $F = A(bI)^{-1}$. The result in the matrix case is stated without proof.

5 *Lemma* Suppose $F \in \mathbf{F}^{n \times m}$, and express F in the form AB^{-1} where $A,B \in \mathbf{M}(\mathbf{R})$ and $|B| \neq 0$. Define

6 $$\mathbf{I} = \{XA + YB : X \in \mathbf{R}^{m \times n}, Y \in \mathbf{R}^{m \times m}\},$$

and note that $\mathbf{I}$ is a left ideal in $\mathbf{R}^{m \times m}$. Then F has an r.c.f. if and only if $\mathbf{I}$ is a left principal ideal.

7 *Lemma* Every $f \in F$ has a coprime factorization if and only if $\mathbf{R}$ is a Bezout domain.

Proof The following statement is easy to prove: **R** is a Bezout domain if and only if every *pair* of elements in **R** generates a principal ideal. So, if **R** is not Bezout, then there exist $a,b \in \mathbf{R}$ such that the ideal generated by a and b is not principal. Obviously, both a and b are nonzero. Hence $f = a/b \in \mathbf{F}$, but does not have a coprime factorization, by Lemma (3). This proves the "only if" part. The "if" part follows readily from Lemma (3). □

8 *Corollary* Every $F \in \mathbf{F}^{n \times m}$ has an r.c.f. for all integers $n,m \geq 1$ if and only if **R** is a Bezout domain.

Proof If **R** is a Bezout domain, the proof of Theorem (4.1.43) applies to show that every $F \in \mathbf{F}^{n \times m}$ has an r.c.f. for all $n,m \geq 1$. □

The next several paragraphs are devoted to deriving conditions for a matrix in $\mathbf{M}(\mathbf{R})$ to have a left (or right) inverse in $\mathbf{M}(\mathbf{R})$, using the idea of reduction modulo maximal ideals. Since a finite set of matrices $A_1, \cdots ,A_n$ in $\mathbf{M}(\mathbf{R})$ are left-coprime if and only if the matrix $[A_1 \cdots A_n]$ has a right inverse, this problem arises naturally in the study of the coprimeness of two (or more) matrices. In the case where **R** is a Bezout domain, several equivalent criteria for left invertibility are given in Section 4.1. It turns out that Corollary (4.1.19) is the result that extends most naturally to the case of a general ring **R** (see Theorem (12) below).

An ideal **I** in **R** is *maximal* if $\mathbf{I} \neq \mathbf{R}$ and **I** is not properly contained in any ideal of **R** other than **R** itself. Equivalently (see [112, p.150]), **I** is maximal if and only if the factor ring $\mathbf{R}/\mathbf{I}$ is a field. Let $\max(\mathbf{R})$ denote the set of maximal ideals of **R**. Then $\mathbf{R}/\mathbf{I}$ is a field for each $\mathbf{I} \in \max(\mathbf{R})$, even though for different **I** this field could be different. However, if **R** is in addition a Banach algebra over the complex field, then $\mathbf{R}/\mathbf{I}$ is isomorphic to the complex field C for all $\mathbf{I} \in \max(\mathbf{R})$ [39, p.31]. For $x \in \mathbf{R}$ and $\mathbf{I} \in \max(\mathbf{R})$, let $\hat{x}_{\mathbf{I}}$ denote the coset $x + \mathbf{I}$ in the factor ring $\mathbf{R}/\mathbf{I}$.

9 *Lemma* Suppose $a_1, \cdots ,a_n \in \mathbf{R}$. Then there exist $x_1, \cdots ,x_n \in \mathbf{R}$ such that

10 $$\sum_{i=1}^{n} x_i a_i = 1$$

if and only if

11 $$\text{rank } [\hat{a}_{1\mathbf{I}} \cdots \hat{a}_{n\mathbf{I}}] = 1 \ \forall \mathbf{I} \in \max(\mathbf{R}).$$

Remarks Note that (10) is equivalent to the statement that $a_1, \cdots, a_n$ generate **R**, while (11) states that no maximal ideal of **R** contains all of the a_i's. Thus Lemma (9) is a natural restatement of a well-known result [39, p.24] that an element in **R** is a unit if and only if it is not contained in any maximal ideal of **R**.

Proof "if" Suppose (11) holds and let **J** denote the ideal generated by the a_i's. If $\mathbf{J} \neq \mathbf{R}$, then **J** is contained in a maximal ideal of **R** [112, p.151]. Suppose $\mathbf{J} \subseteq \mathbf{I} \in \max(\mathbf{R})$. Then **I** contains all the a_i's, in contradiction to (11). Hence $\mathbf{J} = \mathbf{R}$, i.e. (10) holds.

"only if" Suppose (10) holds. If (11) is false, then there exists an $\mathbf{I} \in \max(\mathbf{R})$ that contains all the a_i's. In view of (10), this implies that $1 \in \mathbf{I}$, which is absurd. □

12 *Theorem* Suppose $A \in \mathbf{R}^{n \times m}$ with $n \geq m$. Then A has a left inverse in $\mathbf{R}^{m \times n}$ if and only if[1]

13 $$\text{rank } \hat{A}_{\mathbf{I}} = m \ \forall \mathbf{I} \in \max(\mathbf{R}).$$

An equivalent, and extremely useful, way of stating this theorem is: $A \in \mathbf{R}^{n \times m}$ has a left inverse if and only if the set of $m \times m$ minors of A together generate **R**.

Proof "only if" Suppose $B \in \mathbf{R}^{m \times n}$ is a left inverse of A, so that $BA = I_m$. Since the map $x \rightarrow \hat{x}_{\mathbf{I}}$ from **R** into $\mathbf{R}/\mathbf{I}$ is a homomorphism for each $\mathbf{I} \in \max(\mathbf{R})$, it follows, upon taking cosets modulo **I**, that

[1] Note that $\hat{A}_{\mathbf{I}}$ is an $n \times m$ matrix, with elements in the field $\mathbf{R}/\mathbf{I}$, whose ij-th element is $\hat{a}_{ij\mathbf{I}}$.

14 $\hat{B}_{\mathbf{I}}\hat{A}_{\mathbf{I}} = I_m \ \forall \mathbf{I} \in \max(\mathbf{R})$.

Now (14) implies (13).

"if" This part of the proof is made more intelligible by using the symbol $S(m,n)$ defined in Section B.1. Recall that $S(m,n)$ denotes the set of all strictly increasing m-tuples taken from $\{1, \cdots, n\}$.

The case $m = n$ is very easy: In this case (13) merely states that $|A|$ is not contained in any maximal ideal of $\mathbf{R}$. Hence $|A|$ is a unit of $\mathbf{R}$, and $A^{-1} \in \mathbf{R}^{m \times m}$.

Now suppose $m < n$. For each m-tuple $J \in S(m,n)$, let $F_J \in \mathbf{R}^{m \times m}$ denote the $m \times m$ submatrix of A corresponding to the rows in $\mathbf{J}$. In other words, if $J = \{j_1, \cdots, j_m\}$, then

15 $$F_J = \begin{bmatrix} a_{j_1 1} & \cdots & a_{j_1 m} \\ \vdots & & \vdots \\ a_{j_m 1} & \cdots & a_{j_m m} \end{bmatrix}.$$

Let $f_J = |F_J|$ denote the $m \times m$ *minor* of A corresponding to the rows in J. Then (13) states that, for each $\mathbf{I} \in \max(\mathbf{R})$, we have $\hat{f}_{J\mathbf{I}} \neq 0$ for some $J \in S(m,n)$. Hence, by Lemma (9), the minors f_J, $J \in S(m,n)$ together generate $\mathbf{R}$; i.e. there exist elements x_J, $J \in S(m,n)$ such that

16 $$\sum_{J \in S(m,n)} x_J f_J = 1.$$

Next, for each $J \in S(m,n)$, define a matrix $B_J \in \mathbf{R}^{m \times n}$ as follows: Suppose $J = \{j_1, \cdots, j_m\}$, define F_J as in (15), and let $G_J \in \mathbf{R}^{m \times m}$ denote the adjoint matrix of F_J. Let $g_J^1, \cdots, g_J^m$ denote the rows of G_J. Then

17 $$B_J = \begin{bmatrix} 0 \\ g_J^1 \\ g_J^2 \\ 0 \\ \vdots \\ g_J^m \\ 0 \end{bmatrix}.$$

In other words, rows $j_1, \cdots, j_m$ of B_J equal $g_J^1, \cdots, g_J^m$, respectively, while all other rows of B_J are zero. By construction,

18 $$B_J A = G_J F_J = f_J I_m.$$

Now define

19 $$B = \sum_{J \in S(m,n)} x_J B_J.$$

Then $BA = I_m$, from (18) and (16).[2] □

20 *Corollary* Suppose $A, B \in \mathbf{M}(\mathbf{R})$ have the same number of columns. Then A and B are right-coprime if and only if the matrix $[\hat{A}_{\mathbf{I}}' \ \ \hat{B}_{\mathbf{I}}']'$ has full column rank for all $\mathbf{I} \in \max(\mathbf{R})$.

21 *Application* [106] Let k be an algebraically closed field (e.g. the complex numbers), and let $\mathbf{R}$ be the ring $k[x_1, \cdots, x_r]$ of polynomials in the r indeterminates $x_1, \cdots, x_r$ with coeficients in k. This ring arises in the study of multidimensional linear systems. For each r-tuple $p = (p_1, \cdots, p_r) \in k^r$, the set

22 $$\mathbf{I}_p = \{\phi \in k[x_1, \cdots, x_r]: \phi(p) = 0\}$$

of polynomials that vanish at p is a maximal ideal of $\mathbf{R}$. Moreover, since k is algebraically closed, *every* maximal ideal of $\mathbf{R}$ is of the form (22) for some $p \in k^r$ [4, p.82]. Now $\mathbf{R}/\mathbf{I}_p$ is isomorphic to k for all $p \in k^r$. Further, the canonical map $\phi \rightarrow \phi + \mathbf{I}_p$ takes the polynomial ϕ into the field element $\phi(p)$.

---

[2] This proof was communicated to the author by N.K. Bose.

In this case, Lemma (9) is the "weak" form of Hilbert's *nullstellensatz*, namely: A finite collection of polynomials $\phi_1, \cdots, \phi_n$ generates $\mathbf{R}$ if and only if, at each point $p \in k^r$, at least one of the $\phi_i(p)$ is nonzero. Corollary (20) states that two matrices $A, B \in \mathbf{M}(\mathbf{R})$ are right-coprime if and only if the matrix $[A'(p) \;\; B'(p)]'$ has full column rank for all $p \in k^r$. □

If $\mathbf{R}$ is a Banach algebra over the complex field, then Lemma (9) and Theorem (12) continue to apply, but some further refinements are possible because of the topological structure of the maximal ideal space of $\mathbf{R}$. As mentioned earlier, in this case $\mathbf{R}/\mathbf{I}$ is isomorphic to C for all $\mathbf{I} \in \max(\mathbf{R})$. Hence, in this case one can think of the canonical projection $x \longrightarrow \hat{x}_\mathbf{I}$ as a homomorphism from $\mathbf{R}$ into C. Thus, with every $\mathbf{I} \in \max(\mathbf{R})$ one can associate a nonzero homomorphism from $\mathbf{R}$ to C. The converse is also true: If $\phi : \mathbf{R} \longrightarrow C$ is a nonzero homomorphism, then the kernel of ϕ is a maximal ideal of $\mathbf{R}$. Let us now switch notation. Let Ω denote the maximal ideal space of $\mathbf{R}$, and for each $\omega \in \Omega$ and each $x \in \mathbf{R}$, let $\hat{x}_\omega$ denote the image (in C) of x under the homomorphism associated with ω. Let Ω be equipped with the weakest topology in which the map $\omega \longrightarrow \hat{x}_\omega$ is continuous for each $x \in \mathbf{R}$. Such a topology can be defined in terms of its base, consisting of sets of the form

23 $$\mathbf{B}(\omega; x_1, \cdots, x_n) = \{\theta \in \Omega : | \hat{x}_{i\theta} - \hat{x}_{i\omega} | < \varepsilon \; \forall i\}.$$

Then Ω becomes a compact topological space. Let $\mathbf{C}(\Omega)$ denote the Banach algebra of continuous (complex-valued) functions on Ω, with the supremum norm; i.e.

24 $$\| f \|_C = \max_{\omega \in \Omega} | f(\omega) | ,$$

and multiplication defined pointwise. Then, with each element $x \in \mathbf{R}$, one can associate an element $\hat{x} \in \mathbf{C}(\Omega)$ by the rule $\hat{x}(\omega) = \hat{x}_\omega \; \forall \omega \in \Omega$. The function $\hat{x} \in \mathbf{C}(\Omega)$ is called the *Gel'fand transform* of x. Let $\hat{\mathbf{R}}$ denote the subset of $\mathbf{C}(\Omega)$ consisting of the set of Gel'fand transforms of all $x \in \mathbf{R}$. Then $\hat{\mathbf{R}}$ is a subalgebra of $\mathbf{C}(\Omega)$, but does not in general equal $\mathbf{C}(\Omega)$; in fact, $\hat{\mathbf{R}}$ may not even be a *closed* subalgebra of $\mathbf{C}(\Omega)$.

The utility of the above theory arises from the fact that, for several Banach algebras of interest, the maximal ideal space can be identified explicitly.

25 *Examples* Let $\mathbf{R} = A$, the disc algebra defined in Section 2.4. One can think of A as the set of transfer functions of $\mathbf{l}_2$-stable shift-invariant digital filters whose frequency responses are continuous. The maximal ideal space Ω of A is the closed unit disc $\mathbf{D}$. For each $z \in \mathbf{D}$, the map $f \rightarrow f(z)$ (the "evalutation" map) defines a homomorphism from A into C; conversely, all nonzero homomorphisms from A into C are of this type. Hence, given $f \in A$, the function $z \rightarrow f(z)$ is the Gel'fand transform of f. Since the space $\mathbf{C}(\Omega)$ is just the set of continuous functions $\mathbf{C}(\mathbf{D})$, in this case the Gel'fand transform is just the injection map of A into $\mathbf{C}(\mathbf{D})$.

Let $\mathbf{R} = \mathbf{l}_1(\mathbf{Z}_+)$, the set of sequences $\{f_i\}_{i=0}^{\infty}$ that are absolutely summable, with convolution as the multiplication. As shown in [26,93], $\mathbf{l}_1(\mathbf{Z}_+)$ is precisely the set of unit-pulse responses of $\mathbf{l}_\infty$-stable shift-invariant causal digital filters. The maximal ideal space of $\mathbf{l}_1(\mathbf{Z}_+)$ is once again the closed unit disc $\mathbf{D}$. Corresponding to each $z \in \mathbf{D}$, one can associate the homomorphism

26 $$\hat{f}(z) = \sum_{i=0}^{\infty} f_i z^i,$$

which maps $\mathbf{l}_1(\mathbf{Z}_+)$ into C. The map $z \rightarrow \hat{f}(z)$ which is a function in $\mathbf{C}(\mathbf{D})$ is the Gel'fand transform of $f \in \mathbf{l}_1(\mathbf{Z}_+)$. In this case, it is just the usual z-transform of the sequence $\{f_i\}$.

More generally, let $\mathbf{R} = \mathbf{l}_1(\mathbf{Z}_+^n)$ denote the set of all absolutely summable n-indexed sequences $\{f_{i_1 \cdots i_n}\}$ where the each index ranges over the nonnegative integers. The maximal ideal space of $\mathbf{R}$ is $\mathbf{D}^n$, the cartesian product of the unit disc with itself n times, also referred to as the n-polydisc. Suppose $z = (z_1, \cdots, z_n) \in \mathbf{D}^n$. Then the function

27 $$\hat{f}(z) = \sum_{i_1=0}^{\infty} \cdots \sum_{i_n=0}^{\infty} f_{i_1 \cdots i_n} z_1^{i_1} \cdots z_n^{i_n}$$

defines a homomorphism from $\mathbf{l}_1(\mathbf{Z}_+^n)$ into C. The map $z \rightarrow \hat{f}(z)$ which belongs to $\mathbf{C}(\mathbf{D}^n)$ is the Gel'fand transform of f, which is again the usual z-transform.

Now suppose $\mathbf{R} = \mathbf{l}_1(\mathbf{Z})$, the set of absolutely summable *two-sided* sequences $\{f_i\}_{i=-\infty}^{\infty}$. In this case the maximal ideal space of $\mathbf{R}$ is the unit circle $\mathbf{T}$. For each $z \in \mathbf{T}$, the corresponding homomorphism from $\mathbf{l}_1(\mathbf{Z})$ to C is again given by (26), with the summation now running from $-\infty$ to ∞. If $\mathbf{R} = \mathbf{l}_1(\mathbf{Z}^n)$, then the maximal ideal space of $\mathbf{R}$ is $\mathbf{T}^n$, and (27) defines a homomorphism of $\mathbf{R}$ into C (with the obvious change of 0 to $-\infty$ everywhere). One can replace $\mathbf{Z}^n$ by any locally compact Abelian group and the theory is still complete; see [76].

In some applications, one does not know the maximal ideal space Ω of $\mathbf{R}$ explicitly, but does know a dense subset of Ω. Two such instances are discussed in Examples (38) and (40) below. Theorem (12) can be easily extended to this case.

28 *Lemma* Suppose $\mathbf{R}$ is a Banach algebra over C, with maximal ideal space Ω. Suppose Γ is a dense subset of Ω, and suppose $a_1, \cdots, a_n \in \mathbf{R}$. Then there exist $x_1, \cdots, x_n \in \mathbf{R}$ such that

29 $$\sum_{i=1}^{n} x_i a_i = 1$$

if and only if

30 $$\inf_{\omega \in \Gamma} \sum_{i=1}^{n} | \hat{a}_i(\omega) | > 0.$$

Proof "if" Suppose (30) holds; then it is claimed that

31 $$\sum_{i=1}^{n} | \hat{a}_i(\omega) | \neq 0 \ \forall \omega \in \Omega,$$

or equivalently, that at least one $\hat{a}_i(\omega)$ is nonzero at each $\omega \in \Omega$. The proof of (31) is by contradiction. Suppose that, for some $\omega_0 \in \Omega$, we have $\hat{a}_i(\omega_0) = 0 \ \forall i$. Since Γ is dense in Ω, there is a sequence[3] $\{\omega_j\}$ in Γ converging to ω_0, and

32 $$\sum_{i=1}^{n} | \hat{a}_i(\omega_j) | \to \sum_{i=1}^{n} | \hat{a}_i(\omega_0) | = 0.$$

But (32) contradicts (30). Hence (31) is true. The existence of the x_i's satisfying (29) now follows from Lemma (9).

"only if" Suppose (29) holds. Then, by Lemma (9), (31) is also true. Now, since Ω is a *compact* topological space, (31) implies that

33 $$\inf_{\omega \in \Omega} \sum_{i=1}^{n} | \hat{a}_i(\omega) | > 0.$$

Finally, (33) implies (30), since Γ is a subset of Ω. □

34 *Lemma* Let $\mathbf{R}, \Omega, \Gamma$ be as in Lemma (28), and suppose $A \in \mathbf{R}^{n \times m}$ with $n \geq m$. Then A has a left inverse in $\mathbf{R}^{m \times n}$ if and only if

35 $$\inf_{\omega \in \Gamma} \sigma_m(\hat{A}(\omega)) > 0,$$

where $\sigma_m(\cdot)$ denotes the smallest singular value of a matrix.

Remarks An equivalent way of stating (35) is as follows: There exists an $\alpha > 0$ such that $\hat{A}^*(\omega)\hat{A}(\omega) \geq \alpha I \ \forall \omega \in \Gamma$.

Proof "if" Suppose $\sigma_m(\hat{A}(\omega)) \geq \delta \ \forall \omega \in \Gamma$. Then a simple exercise in numerical analysis shows that

[3] If Ω is not a first countable topological space, this argument has to be modified by replacing all sequences by nets.

36 $$\sum_{J \in S(m,n)} |\hat{a}_J(\omega)| \geq \frac{n\delta^2}{c} > 0 \ \forall \omega \in \Gamma,$$

where $\hat{a}_J, J \in S(m,n)$ are the various $m \times m$ minors of $\hat{A}$, and $c = n!/[m!(n-m)!]$. From Lemma (28), (36) implies that the minors $\hat{a}_J, J \in S(m,n)$ generate $\mathbf{R}$. The existence of a left inverse for A now follows as in the proof of Theorem (12).

"only if" Suppose A has a left inverse. Then, by Theorem (12), the minors $a_J, J \in S(m,n)$ of A generate $\mathbf{R}$. Hence, by Lemma (28),

37 $$\inf_{\omega \in \Omega} \sum_J |\hat{a}_J(\omega)| > 0.$$

It is now routine to prove (35). □

38 *Application* Let $\mathbf{R} = \mathbf{H}_\infty$, which can be interpreted as the set of transfer functions of causal, $\mathbf{l}_2$-stable, shift-invariant digital filters, without the requirement that the frequency response be continuous. The maximal ideal space of $\mathbf{H}_\infty$ is not known explicitly. However, an extremely deep result due to Carleson [32, Chapter 12], called the corona theorem, states that the open unit disc is a dense subset of the maximal ideal space of $\mathbf{H}_\infty$. Hence Lemma (28) states, in this case, that two functions $f, g \in \mathbf{H}_\infty$ are coprime if and only if

39 $$\inf_{|z|<1} \{ |f(z)+g(z)| \} > 0.$$

A similar interpretation can be given for Lemma (34).

40 *Application* [8] Let $\mathbf{A}$ denote the set of distributions $f(\cdot)$ with support in $[0,\infty)$ of the form

41 $$f(t) = \sum_{i=0}^{\infty} f_i \delta(t-t_i) + f_a(t),$$

where $0 \leq t_1 < t_2 < \cdots$ are real constants, $f_a(\cdot)$ is absolutely continuous, and in addition

42 $$\|f\|_{\mathbf{A}} := \sum_{i=0}^{\infty} |f_i| + \int_0^{\infty} |f_a(t)|\, dt < \infty.$$

One can interpret **A** as the set of regular measures (without singular part) of bounded variation on the half-line $[0,\infty)$; see [44, 76]. It is shown in [26] that the set **A** is a good model for the set of impulse responses of BIBO-stable linear time-invariant continuous-time systems. Now the maximal ideal space of **A** is very messy to describe. However, it is shown in [44, pp.145 *et seq.*] that the closed right half of the complex plane is a dense subset of this maximal ideal space, in the following sense: For each $s \in C_+$, define a homomorphism $\phi_s : \mathbf{A} \rightarrow C$ by

43 $$\phi_s(f) = \hat{f}(s) = \sum_{i=0}^{\infty} f_i \exp(-st_i) + \int_0^{\infty} f_a(t) \exp(-st)\, dt.$$

It is recognized that $\hat{f}$ is just the Laplace transform of $f(\cdot)$. For each $s \in C_+$, the set of $f(\cdot) \in \mathbf{A}$ such that $\hat{f}(s) = 0$ is a maximal ideal of **A**. The result in [44] is that the collection of maximal ideals corresponding to all $s \in C_+$ is dense in the maximal ideal space of **A**. Hence[8], by Lemma (28), two functions $f, g \in \mathbf{A}$ are coprime if and only if

44 $$\inf_{s \in C_+} \{ |\hat{f}(s)| + |\hat{g}(s)| \} > 0.$$

Now we change direction somewhat. Lemma (7) shows that if **R** is not a Bezout domain, then not every matrix in **M(R)** has an r.c.f., and not every matrix has an l.c.f. At this stage it is natural to ask: If a matrix has one, does it have the other? The rest of this section is devoted to answering this question for various rings of interest.

Suppose $A \in \mathbf{R}^{m \times n}$ with $m < n$; then we say that A can be *complemented* if there exists a unimodular matrix $U \in \mathbf{R}^{n \times n}$ containing A as a submatrix. Suppose A can be complemented, and let $U = [A' \;\; B']' \in \mathbf{R}^{n \times n}$ be unimodular. By expanding $|U|$ along the first m rows using Laplace's expansion (Fact (B.1.16)), we see that the $m \times m$ minors of A must together generate **R**, or equivalently (by Theorem (12)), that A has a right inverse in $\mathbf{R}^{n \times m}$. Thus only right-invertible matrices can be complemented. But is right-invertibility sufficient?

We postpone answering this question in order to demonstrate the relevance of complementability to the issue at hand.

45 *Lemma* Suppose $F \in \mathbf{M}(\mathbf{F})$ has an l.c.f. $(\tilde{D}, \tilde{N})$. Then F has an r.c.f. if and only if the matrix $A = [\tilde{D} \;\; \tilde{N}]$ can be complemented.

Proof "if" Suppose A can be complemented. Since $B = [-\tilde{N} \;\; \tilde{D}]$ is obtained from A by elementary column operations, B can also be complemented.[4] Suppose

46 $$U = \begin{bmatrix} Y & X \\ -\tilde{N} & \tilde{D} \end{bmatrix}$$

is a unimodular matrix containing B as a submatrix, and partition $U^{-1} \in \mathbf{M}(\mathbf{R})$ as

47 $$U^{-1} = \begin{bmatrix} D & -\tilde{X} \\ N & \tilde{Y} \end{bmatrix}.$$

The claim is that (N, D) is an r.c.f. of F. From $UU^{-1} = I$, we get

48 $$YD + XN = I,$$

49 $$-\tilde{N}D + \tilde{D}N = 0.$$

Thus D and N are right-coprime from (48), and the proof is complete if it can be established that $|D| \neq 0$, because (49) would then imply that $ND^{-1} = \tilde{D}^{-1}\tilde{N} = F$.

To show that D is nonsingular, begin by observing that $\tilde{D}$ is nonsingular, since $(\tilde{D}, \tilde{N})$ is an l.c.f. of F. Therefore the matrix

---

[4] The change from $[\tilde{D} \;\; \tilde{N}]$ to $[-\tilde{N} \;\; \tilde{D}]$ is made simply to make U resemble the matrix in Theorem (4.1.60).

50 $$\begin{bmatrix} I & 0 \\ -\tilde{N} & \tilde{D} \end{bmatrix} U^{-1} = \begin{bmatrix} D & -\tilde{X} \\ 0 & I \end{bmatrix}$$

is also nonsingular. It now follows from (50) that $|D| \neq 0$.

"only if" This part exactly parallels the proof of Theorem (4.1.60). Suppose (N,D) is an r.c.f. of F, and select $X,Y,\tilde{X},\tilde{Y} \in \mathbf{M}(\mathbf{R})$ such that $XN + YD = I$, $\tilde{N}\tilde{X} + \tilde{D}\tilde{Y} = I$. Define

51 $$U = \begin{bmatrix} Y & X \\ -\tilde{N} & \tilde{D} \end{bmatrix}, W = \begin{bmatrix} D & -\tilde{X} \\ N & \tilde{Y} \end{bmatrix}.$$

Then

52 $$UW = \begin{bmatrix} I & -Y\tilde{X}+X\tilde{Y} \\ 0 & I \end{bmatrix}.$$

Since $|UW| = 1$, it follows that $|U|$ is a unit and that U is unimodular. This shows that $B = [-\tilde{N} \;\; \tilde{D}]$ can be complemented. The same is true of $A = [\tilde{D} \;\; \tilde{N}]$, since A can be transformed to B by elementary column operations. □

53 *Corollary* Suppose $F \in \mathbf{M}(\mathbf{F})$ has an r.c.f. (N,D). Then F has an l.c.f. if and only if the matrix $[D' \;\; N']'$ can be complemented.

The proof is easy and is left as an exercise.

Suppose $A \in \mathbf{R}^{m \times n}$ can be complemented. Then by definition there exists a matrix $B \in \mathbf{R}^{(n-m) \times m}$ such that $[A' \;\; B']'$ is unimodular. The next result parametrizes all such matrices B.

54 *Lemma* Suppose $A \in \mathbf{R}^{m \times n}$ can be complemented, and suppose $[A' \;\; B_0']'$ is unimodular. Then the set of $B \in \mathbf{R}^{(n-m) \times m}$ such that $[A' \;\; B']'$ is unimodular is given by

55 $$\{B : B = RA + VB_0, \text{ where } R \in \mathbf{M}(\mathbf{R}),\ V \in \mathbf{U}(\mathbf{R})\}.$$

Proof First, suppose $B = RA + VB_0$ where $V \in \mathbf{U}(\mathbf{R})$. Then

56 $$\begin{bmatrix} A \\ B \end{bmatrix} = \begin{bmatrix} A \\ RA+VB_0 \end{bmatrix} = \begin{bmatrix} I & 0 \\ R & V \end{bmatrix} \begin{bmatrix} A \\ B_0 \end{bmatrix},$$

which is unimodular since both matrices on the right side are in $\mathbf{U}(\mathbf{R})$. Conversely, suppose $[A' \; B']' \in \mathbf{U}(\mathbf{R})$. Partition $[A' \; B_0']'^{-1}$ as $[F \; G]$, and note that $AF = I, AG = 0$. Now, by assumption,

57 $$\begin{bmatrix} A \\ B \end{bmatrix} \begin{bmatrix} A \\ B_0 \end{bmatrix}^{-1} = \begin{bmatrix} A \\ B \end{bmatrix} [F \; G] = \begin{bmatrix} I & 0 \\ BF & BG \end{bmatrix}$$

is in $\mathbf{U}(\mathbf{R})$, which shows that $BG \in \mathbf{U}(\mathbf{R})$. Let $V = BG$, $R = BF$. Then, from (54),

58 $$\begin{bmatrix} A \\ B \end{bmatrix} = \begin{bmatrix} I & 0 \\ R & V \end{bmatrix} \begin{bmatrix} A \\ B_0 \end{bmatrix} = \begin{bmatrix} A \\ RA+VB_0 \end{bmatrix},$$

which implies that $B = RA + VB_0$. □

Now we return to the question: When can every right-invertible matrix over $\mathbf{R}$ be complemented? The answer is provided by the notion of a Hermite ring. A row $[a_1 \cdots a_n] \in \mathbf{R}^{1 \times n}$ is a *unimodular row* if $a_1, \cdots, a_n$ together generate $\mathbf{R}$. A ring $\mathbf{R}$ is *Hermite* if every unimodular row can be complemented. Lemma (B.2.1) shows that every Bezout domain is Hermite. But it is shown below that several other commonly encountered rings are Hermite, though they are not Bezout. Before this, a minor technicality is disposed of.

59 *Lemma* The following statements are equivalent:

(i) $\mathbf{R}$ is a Hermite ring.

(ii) If $A \in \mathbf{R}^{k \times n}$ has a right inverse, then A can be complemented.

Remarks (i) The objective of the lemma is to deal with "fat" matrices, not just row vectors; (ii) compare (ii) with Corollary (4.1.26).

Proof (ii) $\Longrightarrow$ (i) Obvious: just let $m = 1$.

(i) $\Longrightarrow$ (ii) Fix n. The proof is by induction on k. The definition of a Hermite ring shows that (ii) is true when $k = 1$. Suppose (ii) is true for matrices in $\mathbf{R}^{m \times n}$ whenever $m \leq k-1 < n-1$, and suppose $A \in \mathbf{R}^{k \times n}$ has a right inverse, say C. Partition A and C as shown below:

60 $$A = \begin{bmatrix} A^1 \\ a^2 \end{bmatrix}, C = [C_1 \; c_2],$$

where a^2 is the last row of A and c_2 is the last column of C. Then $A^1 C_1 = I_{k-1}$. Hence by the inductive hypothesis (and taking transposes), there exists a unimodular matrix $U \in \mathbf{R}^{n \times n}$ of the form $U = [C_1 \; D]$. Now

61 $$AU = \begin{bmatrix} A^1 \\ a^2 \end{bmatrix} [C_1 \; D] = \begin{bmatrix} I_{k-1} & Z \\ 0 & y \end{bmatrix} =: M.$$

Since A has a right inverse, so does M, and hence so does $y \in \mathbf{R}^{1 \times (n-k+1)}$. Let $b \in \mathbf{R}^{(n-k+1) \times 1}$ be a right inverse of y. Since $\mathbf{R}$ is Hermite, there exists a unimodular matrix $B \in \mathbf{R}^{(n-k+1) \times (n-k+1)}$ containing b as its first column. Thus

62 $$AU \begin{bmatrix} I_{k-1} & 0 \\ 0 & B \end{bmatrix} = \begin{bmatrix} I_{k-1} & ZB \\ 0 & yB \end{bmatrix} = \begin{bmatrix} I_{k-1} & \times & \times \\ 0 & 1 & \times \end{bmatrix} =: N,$$

since the first element of $yB \in \mathbf{R}^{1 \times (n-k+1)}$ is a one. (In (62), $\times$ is a generic symbol denoting an element or a matrix whose value is unimportant.) Now the matrix N in (62) can be complemented, since the determinant of

63 $$\begin{bmatrix} I_{k-1} & \times & \times \\ 0 & 1 & \times \\ 0 & 0 & I_{n-k} \end{bmatrix}$$

is one. Since N is a right associate of A, it follows that A can also be complemented. □

64 *Corollary* Suppose $\mathbf{R}$ is a Hermite ring and that $A \in \mathbf{R}^{m \times n}$ with $m < n$. Then A has a right inverse if and only if there exist unimodular $V \in \mathbf{R}^{m \times m}$, $U \in \mathbf{R}^{n \times n}$ such that

65 $$VAU = [I_m \ \ 0].$$

Corollary (64) shows that in a Hermite ring, even though there is no Smith form in general for arbitrary matrices, there is one for right-invertible matrices (and of course for left-invertible matrices).

66 *Theorem* The following three statements are equivalent:

(i) $\mathbf{R}$ is a Hermite ring.

(ii) If $F \in \mathbf{M}(\mathbf{F})$ has an l.c.f., then it has an r.c.f.

(iii) If $F \in \mathbf{M}(\mathbf{F})$ has an r.c.f., then it has an l.c.f.

Proof The equivalence of (ii) and (iii) follows readily by taking transposes.

(i) $\Rightarrow$ (ii) Suppose $\mathbf{R}$ is Hermite, and suppose $F \in \mathbf{M}(\mathbf{F})$ has an l.c.f., say $(\tilde{D}, \tilde{N})$. Then $A = [\tilde{D} \ \ \tilde{N}]$ has a right inverse. From Lemma (59), A can be complemented, and from Lemma (45), it follows that F has an r.c.f.

(ii) $\Rightarrow$ (i) Suppose $[a_1 \cdots a_n] \in \mathbf{R}^{1 \times n}$ is an arbitrary unimodular row; it must be shown that this row can be complemented. Since $a_1, \cdots, a_n$ generate $\mathbf{R}$, at least one of the a_i must be nonzero. Suppose $a_1 \neq 0$ by renumbering if necessary, and define $F \in \mathbf{F}^{1 \times (n-1)}$ by

67 $$F = \left[\frac{a_2}{a_1} \cdots \frac{a_n}{a_1} \right].$$

Now the unimodularity of the row $[a_1 \cdots a_n]$ shows that F has an l.c.f., namely $(a_1, [a_2 \cdots a_n])$. Since (ii) is assumed to hold, F also has an r.c.f. By Lemma (45), this implies that the row $[a_1 \cdots a_n]$ can be complemented. Hence $\mathbf{R}$ is Hermite. □

Theorem (66) shows that Hermite rings are the next best thing to Bezout domains, at least as far as controller synthesis theory goes. If $\mathbf{R}$ is Bezout, then every matrix in $\mathbf{M}(\mathbf{F})$ has both an r.c.f. and an l.c.f., and all of the synthesis theory of Chapter 5 goes through very nicely. If **R** is not Bezout, then not every matrix in **M**(**F**) has an r.c.f., and not every matrix has an l.c.f. However, if **R** is at least Hermite, then the matrices in **M**(**F**) that have an r.c.f. are precisely those that have an l.c.f. Moreover, this property characterizes Hermite rings.

It is therefore worthwhile to explore whether various commonly encountered rings are Hermite. If **R** is a complex Banach algebra, then there is a very general result available. The proof of this result is well beyond the scope of this book and can be found in [58]; see [58, Theorem 3].

68 *Theorem* Suppose **R** is a complex Banach algebra with maximal ideal space Ω. Then a unimodular row $[a_1 \cdots a_n]$ in **R** can be complemented if and only if the row $[\hat{a}_1 \cdots \hat{a}_n]$ can be complemented in $\mathbf{C}(\Omega)$. In particular, if Ω is contractible, then **R** is Hermite.

Recall that a topological space Ω is *contractible* if the identity map on Ω is homotopic to a constant map. In other words, Ω is contractible if there exists an $\omega_0 \in \Omega$ and a continuous function $f:[0,1] \times \Omega \rightarrow \Omega$ such that

69 $$f(0,\omega) = \omega, f(1,\omega) = \omega_0, \ \ \forall \omega \in \Omega.$$

Theorem (68) is similar in spirit to one of the basic results in the Gel'fand theory of Banach algebras, which states that, given any complex Banach algebra **R** and any $f \in \mathbf{R}$, f can be inverted in **R** if and only if $\hat{f}$ can be inverted in $\mathbf{C}(\Omega)$.

Note that a convex set in C^n is contractible: If $\Omega \subseteq C^n$ is convex, let $\omega_0 \in \Omega$ be arbitrary and define

70 $$f(\lambda,\omega) = \lambda\omega_0 + (1-\lambda)\omega.$$

This observation leads to a very useful criterion.

71 *Corollary* Suppose **R** is a complex Banach algebra whose maximal ideal space Ω is a (compact) convex subset of C^n for some integer n. Then **R** is a Hermite ring.

72 *Examples* [6, 50] The disc algebra, $\mathbf{l}_1(\mathbf{Z}_+)$, $\mathbf{l}_1(\mathbf{Z}_+^n)$ are all Hermite.

Theorem (68) and Corollary (71) are not valid in general if $\mathbf{R}$ is a *real* Banach algebra. However, in [6, 50] it is shown that the following real algebras are indeed Hermite rings:

(i) The set A_s of functions in the disc algebra A whose Fourier coefficients are all real (i.e. the set of all symmetric functions in A).

(ii) The set $l_{1r}(\mathbf{Z}_+^n)$ consisting of all real sequences in $\mathbf{l}_1(\mathbf{Z}_+^n)$.

Theorem (68) suggests a simple example of a ring that is *not* Hermite.

73 *Example*[5] Let S^2 denote the unit sphere in $\mathbb{R}^3$, and let $\mathbf{R} = \mathbf{C}(S^2)$, the set of all continuous complex-valued functions on S^2. Then $\Omega = S^2$, which is not contractible. It is claimed that $\mathbf{R}$ is not a Hermite ring. To show this, it is enough to demonstrate a unimodular row in $\mathbf{R}^{1\times 3}$ that cannot be complemented. Define $\mathbf{a} = [a_1\ a_2\ a_3]$ by setting $\mathbf{a}(x)$ equal to the unit outward normal vector at x, for each $x \in S^2$. Then $\mathbf{a}(\cdot)$ is continuous and therefore belongs to $\mathbf{R}^{1\times 3}$. Moreover, since $\mathbf{a}(x) \neq 0\ \forall x \in S^2$, it follows from Lemma (9) that $\mathbf{a}$ is a unimodular row. Now suppose $\mathbf{a}$ could be complemented, and suppose $[\mathbf{a}'\ \mathbf{b}'\ \mathbf{c}'] \in \mathbf{R}^{3\times 3}$ is unimodular. Then for each $x \in S^2$, the three vectors $\mathbf{a}(x)$, $\mathbf{b}(x)$, $\mathbf{c}(x)$ must span the vector space C^3. Let $T(x)$ denote the (two-dimensional) tangent space to S^2 at x. Since span $\{\mathbf{a}(x), \mathbf{b}(x), \mathbf{c}(x)\} = C^3$ contains $T(x)$, and since $\mathbf{a}(x)$ is orthogonal to $T(x)$, we see that the projections of $\mathbf{b}(x)$, $\mathbf{c}(x)$ onto $T(x)$ must both be nonzero and must together span $T(x)$. Let $\mathbf{f}(x)$, $\mathbf{g}(x)$ denote the projections of $\mathbf{b}(x)$, $\mathbf{c}(x)$ onto $T(x)$, respectively. Then $\mathbf{f}(\cdot)$ is a vector field on S^2 that does not vanish anywhere. However, this contradicts the Borsuk-Ulam theorem [42, p.116], which states that every vector field on S^2 must have a zero. Hence $\mathbf{a}(\cdot)$ cannot be complemented, and $\mathbf{R}$ is not a Hermite ring.

If we define $F = [a_2/a_1\ \ a_3/a_1]$, then F has an l.c.f. but not an r.c.f.

We conclude this section with a theorem that is very important in several branches of control theory and indeed in mathematics.

[5] This example was suggested to the author by Ken Davidson.

74 *Theorem* (Quillen-Suslin) Let $\mathbf{k}$ be a field, and let $\mathbf{R} = \mathbf{k}[x_1, \cdots, x_l]$, the ring of polynomials in l indeterminates with coefficients in $\mathbf{k}$. Then $\mathbf{R}$ is a Hermite ring.

Remarks (i) Until it was proven, the above result was known as Serre's conjecture. See [57] for an excellent historical introduction to this result, its origins and its significance.

(ii) Quillen [72] shows that the result remains valid if $\mathbf{k}$ is replaced by any p.i.d. However, the proof given below does not apply to this case.

(iii) The ring $\mathbf{k}[[x_1, \cdots, x_l]]$ of formal power series in l indeterminates with coefficients in $\mathbf{k}$ is also a Hermite ring; see [59].

(iv) The proof given below is taken from [75, pp.138-145]. See [108] for another proof that applies to the special case when $\mathbf{k} = C$.

The elementary proof of Theorem (74) is rather long, and makes use of a few concepts not previously introduced. A ring $\mathbf{B}$ is *local* if it contains a unique maximal ideal [4, p.38]. Suppose $\mathbf{A}$ is a domain, and let $\mathbf{P}$ be a prime ideal in $\mathbf{A}$. The set consisting of all fractions of the form a/b where $a \in \mathbf{A}$, $b \in \mathbf{A} - \mathbf{P}$, is a local ring, called the *localization* of $\mathbf{A}$ at $\mathbf{P}$, and is denoted by $\mathbf{A_P}$. The unique maximal ideal of $\mathbf{A_P}$ is the set of fractions a/b where $a \in \mathbf{P}$, $b \in \mathbf{A} - \mathbf{P}$. The pertinent property of local rings is the following: If $\mathbf{B}$ is a local ring and $\mathbf{J}$ is its unique maximal ideal, then every $b \notin \mathbf{J}$ is a unit. This follows from Lemma (9).

If $\mathbf{B}$ is a ring and $b_1, \cdots, b_n \in \mathbf{B}$, the symbol $(b_1, \cdots, b_n)$ denotes the ideal generated by the b_i's. If $f \in \mathbf{B}[y]$, then $\delta(f)$ denotes the degree of the polynomial f. More generally, if $f \in \mathbf{B}[y_1, \cdots, y_m]$, then $\delta(f)$ denotes the total degree of f. That is, if f is of the form

$$75 \qquad f(y_1, \cdots, y_l) = \sum_{\mathbf{i}} f_{\mathbf{i}} y_1^{i_1} \cdots y_l^{i_l},$$

where $\mathbf{i}$ is a shorthand for the l-tuple $(i_1, \cdots, i_l)$, then $\delta(f)$ is the largest of the sums $i_1 + \cdots + i_l$ for which the coefficient $f_{\mathbf{i}}$ is nonzero.

The proof of Theorem (74) is basically inductive. Clearly $\mathbf{k}[x_1]$ is Hermite, since it is a p.i.d. (see Lemma (B.2.1)). The proof consists of showing that if $\mathbf{k}[x_1, \cdots, x_{l-1}]$ is Hermite, then so is $\mathbf{k}[x_1, \cdots, x_l]$. This proof consists of two independent steps. The first step (which is actually proved last here, since it is more complicated than the other) is to show that if $\mathbf{B}$ is a Hermite ring and $[a_1 \cdots a_n]$ is a unimodular row in $\mathbf{B}^{1 \times n}$, then it can be complemented *provided one of the a_i's is a monic polynomial.* The second step (which is proved first) is to show that in the ring $\mathbf{k}[x_1, \cdots, x_l]$, it is always possible to effect a change of variables in such a way that every unimodular row contains at least one monic polynomial in the new set of variables.

76 *Lemma* (Nagata) Suppose $a \in \mathbf{R}$, and let $m = \delta(a)+1$. Define a new set of indeterminates

77 $$y_l = x_l, y_i = x_i - x_l^{m^{l-i}}.$$

Then $a(x_1, \cdots, x_l) = rb(y_1, \cdots, y_l)$, where $r \neq 0$ belongs to $\mathbf{k}$, and b is a monic polynomial in y_l (with coefficients in $\mathbf{k}[y_1, \cdots, y_{l-1}]$).

Proof Denote an l-tuple $(j_1, \cdots, j_l)$ where each of the j_i is nonnegative by the symbol $\mathbf{j}$. Then the polynomial a can be written as

78 $$a = \sum_{\mathbf{j}} r_{\mathbf{j}} x_1^{j_1} \cdots x_l^{j_l},$$

where $r_{\mathbf{j}} \in \mathbf{k}$. Note that (77) has the inverse transformation

79 $$x_l = y_l, x_i = y_i + y_l^{m^{l-i}},$$

Substituting from (79) into (78) gives

80 $$b = \sum_{\mathbf{j}} r_{\mathbf{j}} (y_1 + y_l^{m^{l-1}})^{j_1} \cdots (y_{l-1} + y_l^{m})^{j_{l-1}} y_l^{j_l}.$$

Now let $\mathbf{m}$ denote the l-tuple $(m^{l-1}, \cdots, m, 1)$. The first point to note is that if $\mathbf{i} = (i_1, \cdots, i_l)$ and $\mathbf{j} = (j_1, \cdots, j_l)$ are two *distinct* l-tuples with $0 \leq i_k, j_k \leq m-1 \; \forall k$, then $\mathbf{i} \cdot \mathbf{m}$ and $\mathbf{j} \cdot \mathbf{m}$ are distinct integers, where $\cdot$ denotes the usual vector dot product. This follows upon noting that

$\mathbf{i}\cdot\mathbf{m} < m^l$, and that the l-tuple $\mathbf{i}$ is just the representation of the integer $\mathbf{i}\cdot\mathbf{m}$ with respect to the base m; similarly for the integer $\mathbf{j}\cdot\mathbf{m}$. Returning to (80), let $\mathbf{J} = \{\mathbf{j}: r_{\mathbf{j}} \neq 0\}$. Then there is a *unique* l-tuple $\mathbf{j}_0 \in \mathbf{J}$ such that $\mathbf{j}_0 \cdot \mathbf{m} > \mathbf{j}\cdot\mathbf{m}\ \forall \mathbf{j} \in \mathbf{J}$. Hence, when expanded, (80) is of the form

81 $$b = r_{\mathbf{j}_0} y_l^{\mathbf{j}_0 \cdot \mathbf{m}} + \text{terms involving lower powers of } y_l.$$

Hence b equals $r_{\mathbf{j}_0}$ times a monic polynomial in y_l of degree $\mathbf{j}_0 \cdot \mathbf{m}$. □

Note that if $\mathbf{k}$ is an infinite field, then it is possible to achieve the same result as in Lemma (76) by using only a *linear* transformation of variables [108]. However, the more complicated transformation is needed in the case of an arbitrary (and possibly finite) field.

To prove that $\mathbf{R} = \mathbf{k}[x_1, \cdots, x_l]$ is a Hermite ring, suppose $[a_1 \cdots a_n]$ is an arbitrary unimodular row in $\mathbf{R}^{1\times n}$; it must be shown that this row can be complemented. In order to do this, we can first transform the indeterminates $x_1, \cdots, x_l$ into the set $y_1, \cdots, y_l$ using Lemma (76) such that one of the b_i's is monic in y_l, where $b_i(y_1, \cdots, y_l) = a(x_1, \cdots, x_l)\ \forall i$. Since the rings $\mathbf{k}[x_1, \cdots, x_l]$ and $\mathbf{k}[y_1, \cdots, y_l]$ are clearly isomorphic, it is enough to show that the row $[b_1 \cdots b_n]$ can be complemented. Thus the proof of Theorem (74) is complete once the following lemma is established. Note that this lemma is of independent interest.

82 *Lemma* Suppose $\mathbf{B}$ is a Hermite ring, and let $\mathbf{R} = \mathbf{B}[x]$. Suppose $[a_1 \cdots a_n] \in \mathbf{R}^{1\times n}$ is a unimodular row, and that at least one of the a_i's is monic. Then this row can be complemented.

The proof of Lemma (82) requires two preliminary results.

83 *Lemma* (Suslin) Suppose $\mathbf{B}$ is a ring, $\mathbf{R} = \mathbf{B}[x]$, and suppose $f, g \in \mathbf{R}$ are of the form

84 $$f(x) = x^r + a_1 x^{r-1} + \cdots + a_r,$$

$$g(x) = b_1 x^{r-1} + \cdots + b_r.$$

Then, for each $j \in \{1, \cdots, r\}$, the ideal (f, g) contains a polynomial h_j such that $\delta(h_j) \leq r-1$ and the leading coefficient of h_j is b_j.

Proof Define $\mathbf{I} \subseteq \mathbf{B}$ to be the set of leading coefficients of all h in the ideal (f,g) such that $\delta(h) \leq r-1$. Then $\mathbf{I}$ is an ideal in $\mathbf{B}$. To see this, suppose $p_1, q_1 \in \mathbf{I}$, and let $p, q \in (f,g)$ be the corresponding polynomials with leading coefficients p_1, q_1 and degrees less than or equal to $r-1$. Assume without loss of generality that $\delta(p) \leq \delta(q)$. Then the polynomial

85 $$h(x) = q(x) - x^{\delta(q)-\delta(p)} p(x)$$

has leading coefficient $q_1 - p_1$ and degree no larger than $r-1$. Hence $\mathbf{I}$ is closed under subtraction. Finally, it is clear that if $p_1 \in \mathbf{I}$ and $s \in \mathbf{B}$, then $p_1 s \in \mathbf{I}$; just take $h(x) = sp(x)$.

To prove the lemma, it is enough to establish the following statement: If $h \in (f,g)$ and $\delta(h) \leq r-1$, then all coefficients of h belong to $\mathbf{I}$. This is established by induction, by proving that: for each $j \in \{1, \cdots, r\}$, if $h \in (f,g)$ and $\delta(h) \leq r-1$, then the coefficients of $x^{r-1}, \cdots, x^{r-j}$ belong to $\mathbf{I}$. Now the statement is true for $j = 1$, from the definition of $\mathbf{I}$. Suppose the statement is true up to $j-1$, and suppose h is an arbitrary element of the ideal (f,g) with $\delta(h) \leq r-1$. Let $\delta = \delta(h)$, and write

86 $$h(x) = h_1 x^{r-1} + \cdots + h_r,$$

where some of the h_i may be zero. By the inductive hypothesis, $h_1, \cdots, h_{j-1}$ belong to $\mathbf{I}$. Now, $xh(x) - h_1 f(x) \in (f,g)$ has a degree no larger than $r-1$. Moreover,

87 $$xh(x) - h_1 f(x) = \sum_{i=1}^{r} (h_{i+1} - h_1 a_i) x^{r-i}.$$

The $(j-1)$-st coefficient of this polynomial is $h_j - h_1 a_{j-1}$, which belongs to $\mathbf{I}$ by the inductive hypothesis. Since $h_1 a_{j-1} \in \mathbf{I}$, it follows that $h_j \in \mathbf{I}$. This completes the proof of the inductive step. □

The last preliminary result is the same as Lemma (82), except that the ring $\mathbf{B}$ is restricted to be a local ring.

88 *Lemma* Let $\mathbf{B}$ be a local ring, let $\mathbf{R} = \mathbf{B}[x]$, and suppose $[a_1 \cdots a_n] \in \mathbf{R}^{1 \times n}$ is a unimodular row with at least one of the a_i being monic. Then this row can be complemented.

Proof (Suslin) Assume without loss of generality that $n \geq 3$,[6] and that a_1 is monic. The proof is by induction on the degree of a_1. If $\delta(a_1) = 0$, then $a_1 = 1$ and there is nothing further to prove. If $\delta(a_1) =: r > 0$, we will show that $[a_1 \cdots a_n]$ can be reduced, through elementary column operations, to a row $[b_1 \cdots b_n]$ where b_1 is monic and $\delta(b_1) < r$.

Accordingly, suppose $r = \delta(a_1)$. Since a_1 is monic, by performing column operations if necessary we can assume that $\delta(a_i) < r$ for $i = 2, \cdots, n$. Let $\mathbf{J}$ be the unique maximal ideal of $\mathbf{B}$, and for $b \in \mathbf{B}$ let $\hat{b} \in \mathbf{B}/\mathbf{J}$ denote the coset $b + \mathbf{J}$. For $f \in \mathbf{R} = \mathbf{B}[x]$, define $\hat{f} \in [\mathbf{B}/\mathbf{J}][x]$ in the obvious way, namely, if $f = \sum f_i x^i$, then $\hat{f} = \sum \hat{f}_i x^i$. Now the row $[\hat{a}_1 \cdots \hat{a}_n]$ is unimodular in $[\mathbf{B}/\mathbf{J}][x]$. Since $\delta(\hat{a}_1) = \delta(a_1) > 0$, not all of $a_2, \cdots, a_n$ can be zero. By renumbering if necessary, suppose $a_2 \neq 0$. Then some coefficient of a_2 does not belong to $\mathbf{J}$. Suppose

89 $$a_2(x) = \sum_{i=1}^{r} b_{2i} x^{r-i},$$

and suppose $b_{2j} \notin \mathbf{J}$. Since $\mathbf{B}$ is a local ring, this implies that b_{2j} is a unit. Let us now apply Lemma (82) with $f = a_1$, $g = a_2$, and define $\mathbf{I}$ as in Lemma (82). Then Lemma (82) implies that $b_{2j} \in \mathbf{I}$, i.e. that the ideal (a_1, a_2) contains a monic polynomial of degree less than r. So by elementary column operations, a_3 can be replaced by a monic polynomial of degree less than r. Now apply the inductive hypothesis. □

Proof of Lemma (82) The lemma is proved by establishing the following fact: There exists a unimodular matrix $U \in \mathbf{U}(\mathbf{R})$ such that $aU = b$, where $b = [b_1 \cdots b_n] \in \mathbf{B}^{1 \times n}$ is a unimodular row. In fact, such a transformation is possible (as the following proof will reveal) for *any* ring $\mathbf{B}$. Since $\mathbf{B}$ is

[6] Note that a unimodular row consisting of exactly two elements can always be complemented, in *any* ring.

Hermite, b can be complemented, whence a can also be complemented.

Suppose $f,g \in \mathbf{R} = \mathbf{B}[x]$; then $f \cdot g \in \mathbf{B}[x]$ can be defined in the obvious way, namely, if $f = \sum f_i x^i$, then $(f \cdot g)(x) = \sum f_i (g(x))^i$. Given the row a, define $\mathbf{I}$ to be the subset of $\mathbf{B}$ consisting of all $b \in \mathbf{B}$ which have the property that, for each $f,g \in \mathbf{R}$ there exists a unimodular matrix $U = U(f,g)$ such that $a(f+bg) = a(f)U$, where $a(f+bg)$, $a(f)$ are row vectors in $\mathbf{R}^{1 \times n}$ defined by

90 $$(a(f+bg))_i = a_i \cdot (f+bg),$$
$$(a(f))_i = a_i \cdot f.$$

Then $\mathbf{I}$ is an ideal in $\mathbf{B}$: Clearly $0 \in \mathbf{I}$ (take $U = I \ \forall f,g$). If $b,c \in \mathbf{I}$, then for each $f,g \in \mathbf{R}$ there exist unimodular matrices $U(f,g)$, $V(f,g)$ such that $a(f+bg) = a(f)U$, $a(f+cg) = a(f)V$. Hence we have $a(f+(b+c)g)$ $= a(f+bg+cg) = a(f+bg)V(f+bg,g) = a(f)U(f,g)V(f+bg,g)$. Since the matrix $U(f,g)V(f+bg,g)$ is unimodular for all $f,g \in \mathbf{R}$, it follows that $b+c \in \mathbf{I}$. Finally, if $b \in \mathbf{I}$ and $c \in \mathbf{B}$, then for any $f,g \in \mathbf{R}$ we have $a(f+bcg) = a(f)U(f,cg)$, which shows that $bc \in \mathbf{I}$.

If it can be shown that $\mathbf{I} = \mathbf{B}$, then the proof is complete. If $1 \in \mathbf{I}$, then for each $f,g \in \mathbf{R}$ there exists a $U(f,g) \in \mathbf{U}(\mathbf{R})$ such that $a(f+g) = a(f)$. In particular, let $f = x$, $g = -x$; then $a(0) = a(x)U$ for some $U \in \mathbf{U}(\mathbf{R})$. If we now define $b = a(0)$, then the desired statement follows upon observing that $a(0)$ is a unimodular row in $\mathbf{B}$.

To show that $\mathbf{I} = \mathbf{B}$, suppose by way of contradiction that $\mathbf{I} \neq \mathbf{B}$. Then $\mathbf{I}$ is contained in some maximal ideal of $\mathbf{B}$, say $\mathbf{J}$. Let $\mathbf{L}$ denote the localization $\mathbf{B}_{\mathbf{J}}$. Then $\mathbf{B} \subseteq \mathbf{L}$, and a is a (monic) unimodular row in $\mathbf{L}[x]$. By Lemma (88), a can be complemented in $\mathbf{L}[x]$. In other words, there exists a unimodular matrix U over $\mathbf{L}[x]$ such that

91 $$a(x) = [1 \cdots 0\ 0]U(x).$$

Now adjoin a new indeterminate y to $\mathbf{L}[x]$, and define

92 $$V(x,y) = [U(x+y)]^{-1}U(x).$$

Then V is also a unimodular matrix over $\mathbf{L}[x,y]$, and $V(x,0) = I_n$. From (91), it follows that

93 $$a(x+y) = [1 \cdots 0\ 0]U(x+y),$$

so that, from (93) and (92),

94 $$a(x+y)V(x,y) = [1 \cdots 0\ 0]U(x) = a(x).$$

Next, note that $V(x,y)$ is a matrix over $\mathbf{L}[x,y]$, so that each element of $V(x,y)$ is a polynomial in x and y with coefficients in $\mathbf{L}$. Moreover, since $V(x,0) = I_n$, it follows that each element of V is of the form

95 $$v_{ij}(x,y) = c_{ij} + \sum_{k=1}^{r} d_{ijk}(x)y^k,$$

where $c_{ij} = 0$ or 1. Loosely speaking, each term in the polynomial expansion of v_{ij} is either a "constant" 0 or 1, or else involves a positive power of y. Recall that $\mathbf{L}$ consists of fractions p/q where $p \in \mathbf{B}$, $q \in \mathbf{B}-\mathbf{J}$. Let b denote the product of the denominators of all the coefficients occuring in the matrix $V(x,y)$; thus b is the product of the denominators of the coefficients of $d_{ijk}(x)$, for all i,j,k. Then $V(x,by)$ is actually a unimodular matrix over $\mathbf{B}[x,y]$, since replacing y by by wipes out all the denominators. Now (94) implies that

96 $$a(x+by)V(x,by) = a(x).$$

So for any $f,g \in \mathbf{B}[x]$, (96) implies that

97 $$a(f+bg)V(f,bg) = a(f).$$

Therefore $b \in \mathbf{I}$, by definition. On the other hand, b is the product of elements in $\mathbf{B}-\mathbf{J}$, and hence itself belongs to $\mathbf{B}-\mathbf{J}$. In particular, $b \notin \mathbf{I} \subseteq \mathbf{J}$. This contradiction shows that $\mathbf{I} = \mathbf{B}$. □

8.2 COPRIME FACTORIZATIONS FOR DISTRIBUTED SYSTEMS

Suppose **R** is a commutative domain with identity, and let **F** denote the associated field of fractions. The previous section was concerned with the question of when a matrix in $\mathbf{M}(\mathbf{F})$ has a right or left-coprime factorization, and when the existence of one implies the existence of the other. It was shown that the notion of a Hermite ring plays an important role in answering the above questions. In particular, if **R** is a Hermite ring, then a matrix in $\mathbf{M}(\mathbf{F})$ has an r.c.f if and only if it has an l.c.f. However, aside from some general and abstract results, very little was said about the existence of coprime factorizations. The objective of this section is to study this issue in more detail for the case of linear distributed systems, both continuous-time as well as discrete-time. This is done by first proving a general theorem about the existence of coprime factorizations, and then specializing this theorem to the two special cases mentioned above. As a consequence, we shall obtain two very large classes of transfer matrices whose elements all have both r.c.f's as well as l.c.f.'s.

The main result of this section is given next.

1 *Theorem* Suppose **I** is a subset of **R** satisfying two conditions:

(I1) **I** is a saturated multiplicative system;

(I2) If $a \in \mathbf{I}$ and $b \in \mathbf{R}$, then the ideal in **R** generated by a and b is principal.

Under these conditions, every matrix $G \in \mathbf{M}(\mathbf{I}^{-1}\mathbf{R})$ has both an r.c.f. as well as an l.c.f.

2 *Remarks* 1) Recall from Section A.2 that a subset **I** of **R** is said to be *multiplicative* if $a, b \in \mathbf{I}$ implies that $ab \in \mathbf{I}$. It is *saturated* if $a \in \mathbf{I}$ and $b \in \mathbf{R}$ divides a together imply that $b \in \mathbf{I}$.

2) By induction, one can show that (I2) is equivalent to the following statement: (I2') If $a \in \mathbf{I}$ and $b_1,\ldots,b_n \in \mathbf{R}$, then the ideal in **R** generated by $a, b_1,\ldots,b_n$ is principal. Let d be a generator of this principal ideal. Then

one can show quite easily that d is a greatest common divisor (g.c.d.) of the set $\{a, b_1, \ldots, b_n\}$.

3) Recall from Section A.2 that $\mathbf{I}^{-1}\mathbf{R}$ denotes the ring consisting of all fractions of the form a/b, where $a \in \mathbf{R}$, and $b \in \mathbf{I}$. One may assume without loss of generality that $1 \in \mathbf{I}$. By identifying $a \in \mathbf{R}$ with $a/1$, one can imbed $\mathbf{R}$ as a subring of $\mathbf{I}^{-1}\mathbf{R}$.

The proof of Theorem (1) is facilitated by some preliminary results.

3 *Lemma* Suppose $a \in \mathbf{I}$, $b_1, \ldots, b_n \in \mathbf{R}$, and let d be a g.c.d. of the set $\{a, b_1, \ldots, b_n\}$. Then there exists a matrix $U \in \mathbf{U}(\mathbf{R})$ such that $|U| = d$ and $[a \; b_1 \cdots b_n]'$ is the first column of U.

The proof of this lemma exactly follows that of Lemma (B.2.1), if one notes that every subset $\{a \; b_1 \cdots b_i\}$ has a g.c.d. by condition (I2).

4 *Corollary* Suppose $a \in \mathbf{I}$, $b_1, \cdots, b_n$ have a g.c.d. of 1. Then there exists a $U \in \mathbf{U}(\mathbf{R})$ with $[a \; b_1 \cdots b_n]'$ as its first column.

5 *Lemma* Suppose $a \in \mathbf{I}$, $b_1, \cdots, b_n \in \mathbf{R}$ have a g.c.d. of d. Then there exists a $V \in \mathbf{U}(\mathbf{R})$ such that

6 $$V \begin{bmatrix} a \\ b_1 \\ \cdot \\ \cdot \\ \cdot \\ b_n \end{bmatrix} = \begin{bmatrix} d \\ 0 \\ \cdot \\ \cdot \\ \cdot \\ 0 \end{bmatrix}.$$

Proof Let $c_0 = a/d$, $c_i = b_i/d$, $i \geq 1$. Then $c_0 \in \mathbf{I}$ since $\mathbf{I}$ is saturated, and 1 is a g.c.d. of all the c_i's. By Corollary (4) above, there exists a $U \in \mathbf{U}(\mathbf{R})$ such that $[c_0 \cdots c_n]'$ is the first column of U. Now let $V = U^{-1}$. □

7 *Lemma* Suppose $G \in \mathbf{M}(\mathbf{I}^{-1}\mathbf{R})$ is of order $n \times m$, and $G = BA^{-1}$, where $B \in \mathbf{R}^{n \times m}$, $A \in \mathbf{R}^{m \times m}$. Then G has an r.c.f. provided there exists a $U \in \mathbf{U}(\mathbf{R})$ such that

8 $$U\begin{bmatrix} A \\ B \end{bmatrix} = \begin{bmatrix} D \\ 0_{n\times m} \end{bmatrix}.$$

The proof of this lemma is the same as that of Theorem (4.1.7) with a few inessential modifications, and is left to the reader.

Proof of Theorem (1) It is shown that every $G\in\mathbf{M}(\mathbf{I}^{-1}\mathbf{R})$ has an r.c.f. The existence of an l.c.f. for every such G will then follow by taking transposes.

Suppose $G\in\mathbf{M}(\mathbf{I})$ is of order $n\times m$. Then a typical element g_{ij} of G is of the form c_{ij}/a_{ij}, where $c_{ij}\in\mathbf{R}$, $a_{ij}\in\mathbf{I}$ $\forall i,j$. Define

9 $$a = \prod_{i,j} a_{ij},\ b_{ij} = g_{ij}a.$$

and let B be the matrix whose elements are the b_{ij}'s, $A = aI$. Then $G = BA^{-1}$, and the theorem is proved if it can be shown that there exists a $U\in\mathbf{U}(\mathbf{R})$ such that (8) holds. Such a U is constructed iteratively. As the first step, let d_m be a g.c.d. of the set $\{a, b_{1m}, \cdots, b_{n\times m}\}$. Such a d_m exists, since $a\in\mathbf{I}$ (by the multiplicativity of the set $\mathbf{I}$). Select a unimodular matrix $V_m\in\mathbf{R}^{(n+1)\times(n+1)}$ such that

10 $$V_m \begin{bmatrix} a \\ b_{1m} \\ \cdot \\ \cdot \\ \cdot \\ b_{n\times m} \end{bmatrix} = \begin{bmatrix} d_m \\ 0 \\ \cdot \\ \cdot \\ \cdot \\ 0 \end{bmatrix}.$$

Such a V_m exists, by Lemma (5). Now define

11 $$U_m = \begin{bmatrix} I_{m-1} & 0 \\ 0 & V_m \end{bmatrix}.$$

Then routine computations show that

12 $$U_m \begin{bmatrix} A \\ B \end{bmatrix} = \begin{bmatrix} aI_{m-1} & e_m \\ X & 0 \end{bmatrix},$$

where

13 $$e_m = [0 \cdots 0 \; d_m]',$$

and $X \in \mathbf{M}(\mathbf{R})$. It is now possible to repeat the procedure by working on the next-to-last column of the matrix in (12). After m repetitions of the reduction, one arrives at a matrix of the form shown in (8). □

As an application of the above theory, consider the ring **A** described in 8.1.40, which is recalled here for convenience. The ring **A** consists of all distributions $f(\cdot)$ with support in $[0,\infty)$ of the form

14 $$f(t) = \sum_{i=0}^{\infty} f_i \delta(t-t_i) + f_a(t),$$

where $0 \le t_1 < t_2 < \ldots$. are real numbers, f_i are real, $f(\cdot)$ is measurable, and in addition

15 $$\| f \|_{\mathbf{A}} = \sum_{i=0}^{\infty} |f_i| + \int_0^{\infty} |f_a(t)| \, dt < \infty.$$

It can be shown that **A** is precisely the set of impulse responses of BIBO-stable linear time-invariant sytems, and is a ring with pointwise addition and convolution as the product. Now, it can be shown [97] that **A** is *not* a Bezout domain. Hence not every element in the field of fractions of **A** has a coprime factorization. Moreover, it is not known whether **A** is a Hermite ring or not. Hence it is of interest to identify a subset **B** of the fraction field of **A** such that every matrix in $\mathbf{M}(\mathbf{B})$ has both an r.c.f. and an l.c.f. This is done in [8] and can be described in the current framework as follows: Let $\hat{\mathbf{A}}$ denote the set of Laplace transforms of distributions in **A**. Then $\hat{\mathbf{A}}$ is isomorphic to **A**, and has pointwise addition and multiplication as the ring operations. It is routine to verify that each Laplace transform in $\hat{\mathbf{A}}$ converges over the closed right half-plane C_{+e}, and is analytic over the open right half-plane. Now let $\hat{\mathbf{A}}_0$ denote the subset of $\hat{\mathbf{A}}$ defined as follows: $\hat{\mathbf{A}}_0$ consists of those functions

$f \in \hat{\mathbf{A}}$ which are analytic over the *closed* right half-plane. More precisely, $f \in \hat{\mathbf{A}}_0$ if there exists an $\varepsilon > 0$ (which may depend on f) such that $f(\cdot)$ is analytic over the half-plane $\{s: \text{Re } s > -\varepsilon\}$. Then $\hat{\mathbf{A}}_0$ is a subring of $\hat{\mathbf{A}}$. We take $\hat{\mathbf{A}}_0$ as the ring $\mathbf{R}$. The subset $\mathbf{I}$ of $\mathbf{R}$ is defined as follows: A function f in $\hat{\mathbf{A}}_0$ is *bounded away from zero at infinity* if, whenever $\{s_i\}$ is a sequence in C_{+e} such that $|s_i| \to \infty$, we have that $\liminf |f(s_i)| > 0$. Now $\mathbf{I}$ consists precisely of those f in $\hat{\mathbf{A}}_0$ that are bounded away from zero at infinity. It is easy to verify that $\mathbf{I}$ is a saturated multiplicative system in $\mathbf{R}$ (see Problem 1), so that condition (I1) is satisfied. The proof of (I2) is a little more complicated, but can be found in [97]. Hence, on the basis of Theorem (1), we conclude that every function in $\mathbf{B} = \mathbf{I}^{-1}\mathbf{R}$ has a coprime factorization, and more generally, every $G \in \mathbf{M}(\mathbf{B})$ has both an r.c.f. as well as an l.c.f.

In view of the above very useful property, it is worthwhile to understand a little better the class of functions $\mathbf{B}$. Since the zeros of an analytic function are always isolated, if $f \in \hat{\mathbf{A}}$ then any zero of f in the open RHP is isolated. Now, if $f \in \hat{\mathbf{A}}_0$, then every zero of f in the *closed* RHP is isolated. Finally, if $f \in \mathbf{I}$, then f is also bounded away from zero at infinity, which implies that all zeros of f in C_{+e} must in fact be contained in a compact subset thereof. Since any such zero is isolated and of finite multiplicity, we conclude that every $f \in \mathbf{I}$ has only a finite number of isolated zeros, each of finite order, in the closed RHP. Hence every $g \in \mathbf{B}$ has only a finite number of *singularities* in the closed RHP, and each of these is a *pole* of finite order.

Now let us consider discrete-time systems. In this case, a good choice for the set of transfer functions of stable systems is the disc algebra A, consisting of functions that are analytic on the open unit disc and continuous on the closed unit disc. The argument given in [97] shows that A is not a Bezout domain, but A *is* a Hermite ring as shown in Example (8.1.72). Hence every matrix in $\mathbf{M}(A)$ that has an r.c.f. also has an l.c.f. and vice versa. But it is still desirable to have simple sufficient conditions for the existence of an r.c.f. (or an l.c.f.). For this purpose, let $\mathbf{R}$ be the disc algebra A, and let $\mathbf{I}$ consist of those functions $f \in A$ that have only a finite number of zeros in the closed unit disc and none on the unit circle (hence they are all in the *open* unit disc).

With this definition, it is once again routine to verify condition (I1). To check (I2), suppose $f \in A$, $g \in \mathbf{I}$. Then f and g can have only a finite number of common zeros in the closed unit disc, and these must all be in the open unit disc. Suppose these common zeros are $\alpha_1, \ldots, \alpha_n$, with multiplicities $m_1, \ldots, m_n$. Let $\phi(z)$ be the polynomial

16 $$\phi(z) = \prod_{i=1}^{n} (z - \alpha_i)^{m_i}.$$

Then it can be shown that the ideal in A generated by f and g is the same as the principal ideal generated by ϕ (see Problem 2). Hence (I2) holds, and every matrix whose elements all belong to $\mathbf{I}^{-1}A$ have both an r.c.f. and an l.c.f.

Problems

1. Show that the set $\mathbf{I}$ consisting of the functions in $\hat{\mathbf{A}}_0$ that are bounded away from zero infinity satisfies condition (I1).

2. Show that the subset $\mathbf{I}$ of the disc algebra defined above satisfies conditions (I1) and (I2).

8.3 CONTROLLER SYNTHESIS OVER A GENERAL RING

The objective of this section is to extend the results contained in Chapter 5, and specifically Section 5.2, to the case where the set $\mathbf{S}$ is replaced by a general ring $\mathbf{R}$. The main obstacle to doing so is that the plant and compensator can not be assumed *a priori* to have right- or left-coprime factorizations. It is shown below that this difficulty can be mostly circumvented.

Suppose $\mathbf{R}$ is a commutative domain with identity, and denote its field of fractions by $\mathbf{F}$. In the present set-up, both the plant P and the compensator C are assumed to belong to $\mathbf{M}(\mathbf{F})$. The pair (P,C) is *stable* if

1 $$H(P,C) = \begin{bmatrix} (I+PC)^{-1} & -P(I+CP)^{-1} \\ C(I+PC)^{-1} & (I+CP)^{-1} \end{bmatrix}.$$

is well-defined and belongs to $\mathbf{M}(\mathbf{R})$. The symbol $S(P)$ denotes the set of $C \in \mathbf{M}(\mathbf{F})$ such that (P,C) is a stable pair. The first objective is to parametrize the set $S(P)$ corresponding to a given plant P. This is made easier by the next result.

2 *Lemma* Suppose P has an r.c.f (resp. an l.c.f.) and that $C \in S(P)$. Then C has an l.c.f. (resp. an r.c.f.).

Proof Consider the case where P has an r.c.f.; in case P has an l.c.f. the arguments are entirely similar. Suppose $P = ND^{-1}$ where $N,D \in \mathbf{M}(\mathbf{R})$ are right-coprime, and select $X,Y \in \mathbf{M}(\mathbf{R})$ such that $XN + YD = I$. Now

3 $$H(P,C) = H(ND^{-1},C)$$

$$= \begin{bmatrix} I - N(D+CN)^{-1}C & -N(D+CN)^{-1} \\ D(D+CN)^{-1}C & D(D+CN)^{-1} \end{bmatrix}$$

belongs to $\mathbf{M}(\mathbf{R})$, since $C \in S(P)$. Define $S,T \in \mathbf{M}(\mathbf{R})$ by

4 $$S = (D+CN)^{-1},\ T = (D+CN)^{-1}C.$$

Then

5 $$H(P,C) = \begin{bmatrix} I-NT & -NS \\ DT & DS \end{bmatrix},$$

which shows that NS, NT, DS, $DT \in \mathbf{M}(\mathbf{R})$. This in turn implies that

6 $$S = X(NS) + Y(DS) \in \mathbf{M}(\mathbf{R});$$

$$T = X(NT) + Y(DT) \in \mathbf{M}(\mathbf{R}).$$

Also, from (4), $C = S^{-1}T$. Thus, in order to prove that (S,T) is an l.c.f. of C, it only remains to establish the left-coprimeness of S and T. But this is immediate, since

7 $$SD + TN = (D+CN)^{-1}(D+CN) = I.$$

Hence (S,T) is an l.c.f. of C. □

8 *Corollary* Suppose P has both an r.c.f. and an l.c.f. and that $C \in S(P)$. Then C has both an r.c.f. and an l.c.f.

9 *Corollary* Suppose $P, C \in \mathbf{M}(\mathbf{F})$ and that the pair (P,C) is stable. Then the matrix

10 $$G = \begin{bmatrix} C & 0 \\ 0 & P \end{bmatrix}$$

has both an r.c.f. and an l.c.f.

Remarks It is *not* claimed that P and C *individually* have both an r.c.f. and an l.c.f. - only that G has both.

Proof Define

11 $$F = \begin{bmatrix} 0 & I \\ -I & 0 \end{bmatrix}.$$

Then (P,C) is stable if and only if (G,F) is stable. Since F is a constant matrix, it has both an r.c.f. as well as an l.c.f. The conclusion now follows from Corollary (8). □

Thus, if one is interested in designing stabilizing compensators for a plant that has both an r.c.f. and an l.c.f., the basic result of Chapter 5, namely Theorem (5.2.1), carries over completely. It is repeated here for emphasis.

12 *Theorem* Suppose $P \in \mathbf{M}(\mathbf{F})$ has an r.c.f. (N,D) and an l.c.f. $(\tilde{D},\tilde{N})$. Select $X, Y, \tilde{X}, \tilde{Y} \in \mathbf{M}(\mathbf{R})$ such that $XN + YD = I$, $\tilde{N}\tilde{X} + \tilde{D}\tilde{Y} = I$. Then

13 $$S(P) = \{(Y-R\tilde{N})^{-1}(X+R\tilde{D}): R \in \mathbf{M}(\mathbf{R}) \text{ and } |Y-R\tilde{N}| \neq 0\}$$

$$= \{(\tilde{X}+DR)(\tilde{Y}-NR)^{-1}: R \in \mathbf{M}(\mathbf{R}) \text{ and } |\tilde{Y}-NR| \neq 0\}.$$

Once a parametrization of $S(P)$ is available, Theorem (5.4.2) concerning the simultaneous stabilizability of two plants P_0 and P_1 carries over to a general ring $\mathbf{R}$.

14 *Theorem* Suppose P_0 and $P_1 \in \mathbf{M}(\mathbf{F})$ have the same dimensions, and that each has both an r.c.f. and an l.c.f. Define

15 $$A_0 = X_0 N_1 + Y_0 D_1, \; B_0 = -\tilde{N}_0 D_1 + \tilde{D}_0 N_1,$$

where the symbols are as in (5.4.1). Then there exists a $C \in \mathbf{M}(\mathbf{F})$ that stabililzes both P_0 and P_1 if and only if there exists an $M \in \mathbf{M}(\mathbf{R})$ such that $A_0 + MB_0 \in \mathbf{U}(\mathbf{R})$.

Thus, as before, the simultaneous stabilizability question for two plants can be reduced to a strong stabilizability question for a single plant. Unfortunately, this latter question is very difficult to answer, especially for multivariable plants.

The strong stabilizability question for scalar plants is the following: Given coprime elements $n, d \in \mathbf{R}$, when does there exist a $c \in \mathbf{R}$ such that $d + cn$ is a unit of $\mathbf{R}$? The answer to this question depends strongly on the specifics of the ring $\mathbf{R}$. In case $\mathbf{R} = A$, the disc algebra, and n has only a finite number of zeros inside the closed unit disc, the existence or otherwise of a suitable c can be decided using the parity interlacing test, as in Section 2.4. However, if n has a sequence of zeros clustering on the unit circle, the methods of Section 2.4 are no longer applicable. For multivariable plants, the strong stabilization question is: Given a right-coprime pair $N, D \in \mathbf{M}(\mathbf{R})$, does there exist a $C \in \mathbf{M}(\mathbf{R})$ such that $D + CN \in \mathbf{U}(\mathbf{R})$? If $\mathbf{R}$ is a proper Euclildean domain, the results of Section 4.4 can be used to transform this matrix problem into a scalar problem. Specifically, such a C exists if and only if the pair (n, d) can be strongly stabilized, where n is the g.c.d. of all elements of N and $d = |D|$. However, such a reduction is not possible over a general ring. As a result, the strong stabilization problem for multivariable plants over a general ring is still open.

The other contents of Chapter 5, such as the two-parameter compensation scheme for example, can also be extended to a general ring without undue difficulty. The details are left to the reader.

The results of Chapters 6 and 7 can be extended almost without modification to distributed systems. The reason is that these results are mainly based on the properties of the function spaces $\mathbf{H}_2$ and $\mathbf{H}_\infty$, and the rationality of the various functions involved does not play a central role.

This section is concluded with a discussion of the possibility of designing lumped compensators (i.e. those with rational transfer function matrices) for distributed systems. Such compensators are desirable because of ease of implementation. Consider first the discrete-time case, and suppose the ring $\mathbf{R}$ is the disc algebra A. Then the polynomials are dense in A. As a consequence, the set of rational functions (i.e. *ratios* of polynomials) are dense in the fraction field of A in the graph topology. Thus, if c is any element (compensator) in the fraction field of A, then every neighborhood of c contains a rational function. This fact can be exploited to design lumped stabilizing compensators, as follows: First, use Theorem (12) together with any other mathematical apparatus to design a good, or even optimal, compensator C in the form $N_c D_c^{-1}$ or $\tilde{D}_c^{-1}\tilde{N}_c$, where the matrices N_c, D_c or $\tilde{N}_c, \tilde{D}_c$ have their elements in the ring A. By approximating the elements of these matrices by polynomials or rational $\mathbf{H}_\infty$-functions to a sufficiently high degree of accuracy, one can come up with *lumped* compensators that stabilize the plant under study, or achieve near-optimal performance.

In the continuous-time case, the situation is a little more complicated, because the elements of $\mathbf{S}$ are **not** dense in the ring $\hat{\mathbf{A}}$. For example, the function $f(s) = exp(-s)$ belongs to $\hat{\mathbf{A}}$, since it is the Laplace transform of a delayed impulse distribution. But it cannot be uniformly approximated by any stable rational function. Hence it is not always possible to approximate compensators in the fraction field of $\hat{\mathbf{A}}$ by rational compensators. However, it is not too difficult to verify that any function which is the Laplace transform of a distribution of the form

16 $f(t) = f_0\delta(t) + f_a(t),$

where the function f_a is absolutely integrable (i.e. belongs to the space L_1), can be approximated arbitrarily closely by a function in $\mathbf{S}$. Thus, if $\mathbf{R}$ denotes the set of Laplace transforms of such functions, then $\mathbf{S}$ is a dense subset of $\mathbf{R}$. Hence any compensator of the form $\tilde{D}_c^{-1}\tilde{N}_c$, or $N_cD_c^{-1}$, where $D_c,N_c,\tilde{N}_c,\tilde{D}_c \in \mathbf{M}(\mathbf{R})$ can be approximated by a lumped compensator.

The situation is more satisfactory in the case of the ring $\mathbf{B}$ discussed in the previous section. Recall that $\mathbf{B}$ is the ring of fractions of the form $\mathbf{I}^{-1}\mathbf{A}$, where $\mathbf{I}$ and $\mathbf{A}$ are defined in Section 8.2. It can be shown that every function in $\mathbf{B}$ is the sum of a rational function and the Laplace transform of a function belonging to L_1 (see [8]). Thus, if $F \in \mathbf{M}(\mathbf{B})$, then F can be written as $F = G+H$, where $G \in \mathbf{M}(\mathbb{R}(s))$ and H is the Laplace transform of a matrix consisting of functions from L_1. Now, the problem of designing a stabilizing compensator for F is essentially that of designing one for the "unstable part" G of F. Suppose (N_g,D_g) is an r.c.f. of G over the ring $\mathbf{M}(\mathbf{S})$, and select matrices $X,Y \in \mathbf{M}(\mathbf{S})$ that satisfy the identity $XN_g + YD_g = I$. Then, as shown in Problem (4.1.12), the pair (N_g+HD_g,D_g) is an r.c.f. of F, this time over the ring $\hat{\mathbf{A}}$ (which contains $\mathbf{S}$ as a subring), and moreover,

17 $X(N_g+HD_g) + (Y-XH)D_g = I.$

Thus the compensator $(Y-XH)^{-1}X$ stabilizes the plant F. Next, note that $Y-XH \in \mathbf{M}(\mathbf{R})$, where $\mathbf{R}$ is the set of Laplace transforms of distributions of the form (16). Thus there exists a sequence of *rational* matrices $\{A_i\}$ in $\mathbf{M}(\mathbf{S})$ that converges to $Y-XH$. By choosing the index i sufficiently large, we can ensure that the *lumped* compensator $A_i^{-1}X$ stabilizes the plant F. If $|Y-XH| = 0$, this presents no problem, since one can always choose the sequence $\{A_i\}$ to consist of nonsingular matrices. Thus we have shown that every *distributed* plant whose transfer matrix belongs to the set $\mathbf{B}$ can be stabilized by a *lumped* compensator.

Notes and References

The references for the material in Section 8.1 are given at the appropriate places. The design of lumped compensators for distributed systems is discussed in [66, 67], and for the special case of time-delay systems in [51]. See [47] for results relating the internal and external stability of distributed systems, and [6, 50, 60, 73] for controller synthesis theory in very general settings.

The factorization approach carries over with some limitations to nonlinear systems as well. In this connection, see [1, 23, 24].

Appendix A. ALGEBRAIC PRELIMINARIES

This appendix contains a brief introduction to some basic properties of rings. It is important to note that the choice of results from ring theory that are presented here is highly selective; the reader is referred to [48] or [112] for a more comprehensive treatment.

A.1 RINGS, FIELDS AND IDEALS

1 *Definition* A *ring* is a nonempty set $\mathbf{R}$ together with two binary operations $+$ (addition) and $\cdot$ (multiplication) such that the following axioms are satisfied:

(R1) $(\mathbf{R},+)$ is a commutative group. This means that

$$a+(b+c) = (a+b)+c, \ \forall a,b,c \in \mathbf{R}.$$

$$a+b = b+a, \ \forall a,b \in \mathbf{R}.$$

There exists an element $0 \in \mathbf{R}$ such that

$$a+0 = 0+a = a, \ \forall a \in \mathbf{R}.$$

For every element $a \in \mathbf{R}$, there exists a corresponding element $-a \in \mathbf{R}$ such that $a+(-a) = 0$.

(R2) $(\mathbf{R},\cdot)$ is a semigroup. This means that

$$a\cdot(b\cdot c) = (a\cdot b)\cdot c, \ \forall a,b,c \in \mathbf{R}.$$

(R3) Multiplication is distributive over addition. This means that

$$a\cdot(b+c) = a\cdot b + a\cdot c,$$

$$(a+b)\cdot c = a\cdot c + b\cdot c, \ \forall a,b,c \in \mathbf{R}.$$

As is customary, $a\cdot b$ is denoted by ab, and $a+(-b)$ is denoted by $a-b$.

A ring $\mathbf{R}$ is said to be *commutative* if $ab = ba \ \forall a,b \in \mathbf{R}$, and is said to *have an identity* if there exists an element $1 \in \mathbf{R}$ such that $1 \cdot a = a \cdot 1 = a \ \forall a \in \mathbf{R}$.

2 *Examples* A classic example of a ring is the set of integers, denoted by $\mathbf{Z}$, with the usual notions of addition and multiplication. Clearly, $\mathbf{Z}$ is a commutative ring with identity. The set $\mathbf{E}$ of even integers, together with the usual addition and multiplication is an example of a ring without identity; but $\mathbf{E}$ is commutative. The set $\mathbf{Z}^{2\times 2}$ of 2×2 matrices with integer elements is a noncommutative ring with identity.

A ring $\mathbf{R}$ is said to be a *domain* (or an integral domain) if $a \in \mathbf{R}$, $b \in \mathbf{R}$, $ab = 0$ implies that either $a = 0$ or $b = 0$. In other words, $\mathbf{R}$ is a domain if the product of every pair of nonzero elements is nonzero.

3 *Example* In Example (2) above, both $\mathbf{Z}$ and $\mathbf{E}$ are domains, but $\mathbf{Z}^{2\times 2}$ is not.

4 *Example* Consider the ring $\mathbf{Z}/(6)$ consisting of the six elements $\{0,1,2,3,4,5\}$, with addition and multiplication defined modulo 6. For example, $3+5 = 8 \bmod 6 = 2$, and $2.5 = 10 \bmod 6 = 4$. This ring is commutative and has an identity. However, since $2 \cdot 3 = 0$, it is not a domain.

Suppose $\mathbf{R}$ is a ring with identity. An element $x \in \mathbf{R}$ is called a *unit* of $\mathbf{R}$ if there is a $y \in \mathbf{R}$ such that $xy = yx = 1$. It can be easily shown that such a y is unique; y is called the *inverse* of x and is denoted by x^{-1}.

5 *Example* Let $\mathbf{R} = \mathbf{C}[0,1]$, the set of continuous real-valued functions defined over the interval $[0,1]$. If addition and multiplication on $\mathbf{R}$ are defined pointwise, i.e.

$$(x+y)(t) = x(t) + y(t),$$

$$(xy)(t) = x(t)y(t), \ \forall t \in [0,1], \ \forall x,y \in \mathbf{R},$$

then $\mathbf{R}$ is a commutative ring with identity. However, it is not a domain; for example, let

$$x(t) = \begin{cases} 1-2t & 0 \le t \le 0.5 \\ 0 & 0.5 < t \le 1 \end{cases}$$

$$y(t) = \begin{cases} 0 & 0 \le t \le 0.5 \\ 2t-1 & 0.5 < t \le 1 \end{cases}$$

Then $xy = 0$, even though $x \neq 0$, $y \neq 0$. A function $x \in \mathbf{R}$ is a unit of $\mathbf{R}$ if and only if $x(t) \neq 0 \ \forall t \in [0,1]$. In other words, the units in $\mathbf{R}$ are the functions that do not change sign over [0,1].

6 *Definition* A *field* is a commutative ring $\mathbf{F}$ with an identity, satisfying two additional assumptions:

(F1) $\mathbf{F}$ contains at least two elements.

(F2) Every nonzero element of $\mathbf{F}$ is a unit.

7 *Examples* The rational numbers, real numbers and complex numbers are all well-known examples of fields. Another example is $\mathbf{Z}/(p)$ where p is a prime integer, consisting of the elements $\{0,1, \cdots ,p-1\}$. Addition and multiplication are defined modulo p. This set is clearly a commutative ring with identity, and it contains at least two elements. To show that it satisfies axiom (F2), suppose a is any nonzero element of $\mathbf{Z}/(p)$. Since the greatest common divisor of a and p (in the usual sense) is 1, there exist integers x and y such that

$$ax + py = 1.$$

Moreover, for *any* integer q, we have that

$$a(x-qp) + p(y+qa) = 1.$$

So by a suitable choice of q, the integer $x-qp$ can be made to lie between 0 and $p-1$. As a result, it follows that there exist integers b and c, with b lying between 0 and $p-1$, such that

$$ab + pc = 1, \text{ or } ab = 1-pc \equiv 1 \bmod \text{p}.$$

Hence a is a unit and b is its inverse. For example, in the ring $\mathbf{Z}/(13)$, the inverse of 5 is 8, and the inverse of 2 is 7.

The field $\mathbf{Z}/(2)$, consisting of the two elements $\{0,1\}$, is called the *binary field*.

A subset $\mathbf{S}$ of a ring $\mathbf{R}$ is a *subring* of $\mathbf{R}$ if it is a ring in its own right, i.e. if $0 \in \mathbf{S}$, and the sum, difference and product of two elements of $\mathbf{S}$ again belong to $\mathbf{S}$.

8 *Definition* A subset $\mathbf{I}$ in a ring $\mathbf{R}$ is said to be a *left ideal* if (i) $\mathbf{I}$ is a subgroup of the additive group of $\mathbf{R}$, and (iiL) $a \in \mathbf{I}$, $x \in \mathbf{R}$ imply that $xa \in \mathbf{I}$. $\mathbf{I}$ is a *right ideal* if (i) $\mathbf{I}$ is a subgroup of the additive group of $\mathbf{R}$ and (iiR) $a \in \mathbf{I}$, $x \in \mathbf{R}$ imply that $ax \in \mathbf{I}$. $\mathbf{I}$ is an *ideal* if it is both a left ideal and a right ideal.

In the above definition, condition (i) means the following: First, $0 \in \mathbf{I}$, and second, if $x, y \in \mathbf{I}$, then $x \pm y \in \mathbf{I}$. Condition (iiL) (resp. (iiR)) means that if any element of $\mathbf{I}$ is multiplied on the left (resp. right) by any element of $\mathbf{R}$, the product is once again in $\mathbf{I}$. Clearly, if $\mathbf{R}$ is a commutative ring, the concepts of a left ideal, right ideal, and ideal all coincide.

9 *Example* Let $\mathbf{R}$ be the ring $\mathbf{Z}^{2\times 2}$ of 2×2 matrices with integer elements. Let

$$\mathbf{I}_1 = \{M \in \mathbf{Z}^{2\times 2}: m_{11} = 0, m_{21} = 0\}.$$

Then it is easy to verify that $\mathbf{I}_1$ is a left ideal in $\mathbf{R}$. Similarly,

$$\mathbf{I}^2 = \{M \in \mathbf{Z}^{2\times 2}: m_{21} = 0, m_{22} = 0\}$$

is a right ideal in $\mathbf{R}$. The set of diagonal matrices

$$\mathbf{D} = \{M \in \mathbf{Z}^{2\times 2}: m_{21} = 0, m_{12} = 0\}$$

is a subring of $\mathbf{R}$ but is neither a left ideal nor a right ideal.

Let a be an integer. Then

$$\mathbf{M}(a) = \{M \in \mathbf{Z}^{2 \times 2}: m_{ij} \text{ is divisible by } a \ \ \forall i,j\}$$

is an ideal in $\mathbf{R}$.

10 *Example* Consider the ring $C[0,1]$ of Example (5), and let $t_0 \in [0,1]$. Then

$$\mathbf{I}_{t_0} = \{x(\cdot) \in C[0,1]: x(t_0) = 0\}$$

is an ideal. More generally, let S be any subset of [0,1]. Then

$$\mathbf{I}_S = \{x(\cdot) \in C[0,1]: x(t) = 0 \ \forall t \in S\}$$

is an ideal.

Suppose a is some element of a ring $\mathbf{R}$. Then the set of all elements of the form xa where $x \in \mathbf{R}$, i.e. the set of all left multiples of a, is a left ideal. It is called the *left principal ideal* generated by a. Similarly, the *right principal ideal* generated by a is the set of all elements ax where $x \in \mathbf{R}$.

An ideal $\mathbf{I}$ in a commutative ring $\mathbf{R}$ is a *prime ideal* if $a \in \mathbf{R}$, $b \in \mathbf{R}$, $ab \in \mathbf{I}$ implies that either $a \in \mathbf{I}$ or $b \in \mathbf{I}$. Equivalently, $\mathbf{I}$ is a prime ideal if and only if the set $\mathbf{R} - \mathbf{I}$ is closed under multiplication (i.e. the product of two elements not belonging to $\mathbf{I}$ cannot itself belong to $\mathbf{I}$).

11 *Example* Consider the ring of integers $\mathbf{Z}$, and let n be any integer. Then the set of multiples of n, denoted by (n), is the principal ideal generated by n. It is a prime ideal if and only if n divides ab implies that n divides a or n divides b, which is true if and only if n is a prime number.

12 *Example* Consider the ideal $\mathbf{I}_{t_0}$ of Example (10). If a product xy belongs to this ideal, then $(xy)(t_0) = 0$, which means that either $x(t_0) = 0$ or $y(t_0) = 0$. Hence $\mathbf{I}_{t_0}$ is a prime ideal. However, if S contains more than one point, then $\mathbf{I}_S$ is not a prime ideal.

By convention, the entire ring $\mathbf{R}$ is *not* considered to be a prime ideal. This is in conformity with the convention that 1 is not considered a prime number (see Example (11)). Thus "$\mathbf{I}$ is a prime ideal in $\mathbf{R}$" automatically implies, by convention, that $\mathbf{I}$ is a *proper* subset of $\mathbf{R}$.

Problems

1. (i) Show that the zero element of a ring is unique (hint: if 0_1 and 0_2 are two additive identities, then $0_1 = 0_1 + 0_2 = 0_2$).
(ii) Show that the additive inverse of an element is unique.

2. (i) Show that $a \cdot 0 = 0$ for all a in a ring (hint: $ab = a(b+0)$).
(ii) Show that, if a ring **R** contains an identity and has at least one nonzero element, then $1 \neq 0$ (hint: let $a \neq 0$; then $a0 \neq a = a1$).

3. Show that, if **R** is a domain, then the cancellation law holds, i.e. $ab = ac$, $a \neq 0$ implies that $b = c$.

4. Let **R** be a ring with identity and at least one nonzero element.
(i) Show that the set of units of **R** is a group under multiplication; i.e. show that (a) if u is a unit, then so is u^{-1}, and (b) if u,v are units, so are uv and vu.
(ii) Show that 0 can never be a unit (hint: see Problem 2).
(iii) Show that if x has an inverse, then it is unique.
(iv) Show that if x has a left inverse y and a right inverse z, then $y = z$ (i.e. show that $yx = 1$, $xz = 1$ implies $y = z$).

5. Let **R** be the set of functions mapping the interval [0,1] into the set of integers **Z**.
(i) Show that **R** is a ring under pointwise addition and multiplication.
(ii) What are the units of this ring?

6. Consider the ring $\mathbf{Z}/(p)$ of Example (7). Show that this ring is a field if and only if p is a prime number.

7. (i) Consider the ring $\mathbf{Z}/(9)$, consisting of $\{0,1, \cdots ,8\}$, with addition and multiplication modulo 9. Determine the units of this ring. (Answer: 1,2,4,5,7,8)
(ii) Consider the ring $\mathbf{Z}/(n)$. Show that m is a unit of this ring if and only if the greatest common divisor of m and n is 1.

8. Consider the ring $\mathbf{Z}^{n \times n}$ of $n \times n$ matrices with integer elements. Show that a matrix M in this ring is a unit if and only if its determinant is ± 1.

9. Show that a commutative ring $\mathbf{R}$ is a domain if and only if $\{0\}$ is a prime ideal.

A.2 RINGS AND FIELDS OF FRACTIONS

Throughout this section "ring" means a commutative ring with identity.

Suppose $\mathbf{R}$ is a ring. An element $a \in \mathbf{R}$ is an *absolute nondivisor of zero* if $b \in \mathbf{R}$, $ab = 0$ implies that $b = 0$. In every ring, there are absolute nondivisors of zero; for example, all the units are of this type. But there may be others as well.

1 *Examples* In the ring $\mathbf{Z}^{n \times n}$, a matrix M is an absolute nondivisor of zero if and only if the determinant of M is nonzero. Thus a nonunit matrix can be an absolute nondivisor of zero (cf. Problem A.1.8).

Now consider the ring $C[0,1]$ defined in Example (A.1.5). Suppose a function $x(\cdot)$ belonging to this ring vanishes at only a finite number of points. If $y \in C[0,1]$ and $xy = 0$, then $x(t)y(t) \equiv 0$, which means that $y(t) = 0$ for all except a finite number of values of t. However, since $y(\cdot)$ is continuous, this implies that $y(t) \equiv 0$, or that y is the zero element of the ring. Hence x is an absolute nondivisor of zero.

If a is an absolute nondivisor of zero and $ab = ac$, then $b = c$; in other words, absolute nondivisors of zero can be "cancelled."

Of course, if $\mathbf{R}$ is a domain, then *every* nonzero element of $\mathbf{R}$ is an absolute nondivisor of zero.

A set $\mathbf{M}$ in a ring $\mathbf{R}$ is said to be a *multiplicative system* if $a,b \in \mathbf{M}$ implies that $ab \in \mathbf{M}$. It is *saturated* if $a \in \mathbf{R}$, $b \in \mathbf{R}$, $ab \in \mathbf{M}$ implies that $a \in \mathbf{M}$, $b \in \mathbf{M}$.

2 *Fact* The set $\mathbf{N}$ of absolute nondivisors of zero in a ring $\mathbf{R}$ ia a multiplicative system.

Proof Suppose $a,b \in \mathbf{N}$, $y \in \mathbf{R}$, and $aby = 0$. Then, since $a,b \in \mathbf{N}$, it follows successively that $aby = 0 \Rightarrow by = 0 \Rightarrow y = 0$. Hence $ab \in \mathbf{N}$. □

Suppose **R** is a ring, **M** is a multiplicative system in **R** containing 1, and **M** is a subset of **N** (the set of absolute nondivisors of zero in **R**). We now begin a construction which will ultimately result in a ring **L** which contains **R** as a subring, and in which every element of **M** is a unit.

Consider the set $\mathbf{R} \times \mathbf{M}$, and define a binary relation $\sim$ on $\mathbf{R} \times \mathbf{M}$ as follows: $(a,b) \sim (c,d) \Longleftrightarrow ad = bc$. The relation $\sim$ is an equivalence relation: Clearly $\sim$ is reflexive and symmetric. To show that it is transitive, suppose $(a,b) \sim (c,d)$ and $(c,d) \sim (e,f)$. Then $ad = bc$ and $cf = de$. Multiplying the first equation by f and the second one by b gives $adf = bcf = bde$. Now $d \in \mathbf{N}$ since $d \in \mathbf{M}$ and $\mathbf{M} \subseteq \mathbf{N}$. Hence d can be cancelled in the above equation to give $af = be$, i.e. $(a,b) \sim (e,f)$.

Since $\sim$ is an equivalence relation, the set $\mathbf{R} \times \mathbf{M}$ can be partitioned into disjoint equivalence classes under $\sim$. The set of equivalence classes $\mathbf{R} \times \mathbf{M} / \sim$ is denoted by **L**. The set **L** therefore consists of *fractions* (a,b), or a/b in more familiar terms, where we agree to treat two fractions a/b and c/d as equivalent if $ad = bc$. Addition and multiplication of fractions in **L** are defined in the familiar way, namely

3 $$\frac{a}{b} + \frac{c}{d} = \frac{ad+bc}{bd}.$$

4 $$\frac{a}{b} \cdot \frac{c}{d} = \frac{ac}{bd}.$$

Note that, if $b,d \in \mathbf{M}$, then so does bd. Hence the right sides of (3) and (4) are valid fractions. Actually, (3) and (4) represent operations on equivalence classes; but the reader can verify that the sum and product of two fractions (i.e. two equivalence classes) do not depend on which representatives of the equivalence classes are used.

With addition and multiplication defined by (3) and (4), $\mathbf{L}$ is a ring. Moreover, if every $a \in \mathbf{R}$ is identified with the fraction $a/1$, then $\mathbf{R}$ is isomorphic to a subring of $\mathbf{L}$. The element $1/1$ serves as an identity for $\mathbf{L}$. Finally, if $d \in \mathbf{M}$, then d corresponds to $d/1 \in \mathbf{L}$; moreover, $d/1$ is a unit in $\mathbf{L}$ with the inverse $1/d$. Thus every element of (the isomorphic image of) $\mathbf{M}$ is a unit in $\mathbf{L}$. The ring $\mathbf{L}$ is called the *ring of fractions* of $\mathbf{R}$ with respect to $\mathbf{M}$, and is denoted by $\mathbf{M}^{-1}\mathbf{R}$.

Since $\mathbf{M}$ is a subset of $\mathbf{N}$ (the set of absolute nondivisors of zero), the ring $\mathbf{N}^{-1}\mathbf{R}$ contains the largest number of units. If in particular $\mathbf{R}$ is a domain, then $\mathbf{N} = \mathbf{R} - 0$, i.e. $\mathbf{N}$ consists of all nonzero elements of $\mathbf{R}$. In this case, the ring $\mathbf{F} = (\mathbf{R} - 0)^{-1}\mathbf{R}$ has the property that *every* nonzero element of $\mathbf{R}$ is a unit in $\mathbf{F}$. Moreover, every nonzero element of $\mathbf{F}$ is also a unit: If $a/b \in \mathbf{F}$ and $a \neq 0$, then b/a is the inverse of a/b. Hence $\mathbf{F}$ is a field; it is referred to as the *field of fractions* or *quotient field* associated with the domain $\mathbf{R}$.

There is a particularly important class of fraction rings that is frequently encountered. Suppose $\mathbf{R}$ is a ring, and that $\mathbf{I}$ is a prime ideal in $\mathbf{R}$. If $\mathbf{M}$ denotes the complement of $\mathbf{I}$, then $\mathbf{M}$ is a multiplicative system (see Problem 4). The corresponding fraction ring $\mathbf{M}^{-1}\mathbf{R}$ is called the *localization* of $\mathbf{R}$ with respect to $\mathbf{I}$.

5 *Examples* The set of integers $\mathbf{Z}$ is a domain. The field of fractions associated with the integers is the set of rational numbers.

In the ring $\mathbf{Z}$, let $\mathbf{M}$ denote the set of numbers that are *not* divisible by 3. Then $\mathbf{M}$ is a multiplicative system (since 3 is a prime number), and $1 \in \mathbf{M}$. The ring of fractions $\mathbf{M}^{-1}\mathbf{Z}$ is the set of rational numbers whose denominators (when expressed in reduced form) are not divisible by 3. This is a *subring* of the *field* of rational numbers.

Problems

1. Suppose $\mathbf{R}$ is a ring, $a,b,c \in \mathbf{R}$, and a is an absolute nondivisor of zero. Show that if $ab = ac$, then $b = c$.

2. Show that the addition and multiplication rules (3) and (4) are unambiguous, in the sense that the final answers do not depend on which representatives of the equivalence classes are used in the computation.

3. Consider the ring $\mathbf{Z}/(10)$, consisting of the integers $0, \cdots, 9$ with addition and multiplication defined modulo 10.

   (i) Determine the units of this ring.

   (ii) Show that the set $\mathbf{N}$ of absolute nondivisors of zero consists of just the units.

   (iii) Using (ii), show that the ring of fractions with numerators in $\mathbf{Z}/(10)$ and denominators in $\mathbf{N}$ is again just $\mathbf{Z}/(10)$.

4. Suppose $\mathbf{R}$ is a ring and that $\mathbf{I}$ is a prime ideal in $\mathbf{R}$. Let $\mathbf{M}$ denote the complement of $\mathbf{I}$. Show that $\mathbf{M}$ is a multiplicative system.

A.3 PRINCIPAL IDEAL DOMAINS

Throughout this section, "ring" means a commutative ring with identity.

1 *Definition* A ring $\mathbf{R}$ is said to be a *principal ideal ring* if every ideal in $\mathbf{R}$ is principal. $\mathbf{R}$ is a *principal ideal domain (p.i.d.)* if it is a domain as well as a principal ideal ring.

Recall that a principal ideal $\mathbf{I}$ consists of *all* multiples of some element a, i.e. $\mathbf{I} = \{xa : x \in \mathbf{R}\}$. Thus, in a principal ideal ring, every ideal is generated by a single element.

If x and y are elements of a ring $\mathbf{R}$ with $x \neq 0$, we say that x *divides* y, and y *is a multiple of* x, if there is an element $z \in \mathbf{R}$ such that $y = xz$; this is denoted by $x \mid y$. If x and y are elements of a ring $\mathbf{R}$ such that not both are zero, a *greatest common divisor (g.c.d.)* of x, y is any element $d \in \mathbf{R}$ such that

(GCD1) $d \mid x$ and $d \mid y$.

(GCD2) $c \mid x, c \mid y \Rightarrow c \mid d$.

In the above definition, it is implicit that $d \neq 0$, since d divides a nonzero element by virtue of (GCD1). We say *a* greatest common divisor because a g.c.d. is not unique. Certainly, if d is a g.c.d. of x and y, then so is $-d$. Actually, the following stronger result holds:

2 *Fact* Suppose d is a g.c.d. of x and y; then so is du whenever u is a unit. Suppose in addition that **R** is a domain; then every g.c.d. d_1 of x and y is of the form $d_1 = du$ for some unit u.

Proof The first sentence is obvious. To prove the second sentence, observe that if d_1 is another g.c.d. of x and y, then by (GCD2) $d_1 \mid d$ and $d \mid d_1$. Hence $d_1 = du$ for some u. Moreover, $du \mid d$ since $d_1 \mid d$. Since **R** is a domain and $d \neq 0$, this implies that $u \mid 1$, i.e. that u is a unit. □

Fact (2) states that if **R** is a domain, then once we have found *one* g.c.d. of a given pair of elements, we can quickly find them *all*. However, a question that is not answered by Fact (2) is: Does every pair of elements have a g.c.d.? The answer to this question is provided next.

3 *Theorem* Suppose **R** is a principal ideal ring. Then every pair of elements $x, y \in \mathbf{R}$, not both of which are zero, has a g.c.d. d which can be expressed in the form

4 $$d = px + qy$$

for appropriate elements $p, q \in \mathbf{R}$. Moreover, if **R** is a domain, then *every* g.c.d. of x and y can be expressed in the form (4).

Proof Given $x, y \in \mathbf{R}$, consider the set

5 $$\mathbf{I} = \{ax + by,\ a, b \in \mathbf{R}\}.$$

In other words, **I** is the set of all "linear combinations" of x and y. It is easy to verify that **I** is an ideal in **R**.[1] Since **R** is a principal ideal ring, **I** must be a principal ideal; that is, there exists an element $d \in \mathbf{I}$ such that **I** is the set of multiples of d. Since $x = 1 \cdot x + 0 \cdot y \in \mathbf{I}$, it follows that x is a multiple of d,

[1] **I** is referred to as the *ideal generated by* x and y.

i.e. $d \mid x$; similarly $d \mid y$. Thus d satisfies the axiom (GCD1). Next, since $d \in \mathbf{I}$ and $\mathbf{I}$ is the ideal generated by x, y, there exist $p, q \in \mathbf{R}$ such that (4) holds. Now suppose $c \mid x$, $c \mid y$. Then $c \mid (px+qy)$, i.e. $c \mid d$. Hence d also satisfies (GCD2) and is thus a g.c.d. of x and y.

Up to now, we have shown that every pair of elements x and y has *a* g.c.d. which can be written in the form (4) for a suitable choice of p and q. Now, if in addition $\mathbf{R}$ is a domain, then *every* g.c.d. d_1 of x, y is of the form $d_1 = du$ for some unit u, by Fact (2). Thus every g.c.d. d_1 can be written in the form $d_1 = du = upx + upy$. □

Two elements $x, y \in \mathbf{R}$ are *relatively prime* or simply *coprime* if every g.c.d. of x, y is a unit. In view of Fact (2), if $\mathbf{R}$ is a domain, this is equivalent to saying that x and y are coprime if and only if 1 is a g.c.d. of x and y. Now suppose $\mathbf{R}$ is a principal ideal domain. Then, by Theorem (3), $x, y \in \mathbf{R}$ are coprime if and only if the ideal $\mathbf{I}$ in (5) is the same as the ideal generated by 1. But the latter is clearly the entire ring $\mathbf{R}$. This can be summarized as follows:

6 *Fact* Let $\mathbf{R}$ be a principal ideal domain. Then $x, y \in \mathbf{R}$ are coprime if and only if there exist $p, q \in \mathbf{R}$ such that $px + qy = 1$.

One can also define a g.c.d. of an n-tuple of elements $(x_1, \cdots, x_n)$, not all of which are zero. An element $d \in \mathbf{R}$ is a g.c.d. of the collection $(x_1, \cdots, x_n)$ if

7 (i) $d \mid x_i \ \forall i$,

8 (ii) $c \mid x_i \ \forall i \implies c \mid d$.

As in Fact (2), it follows that a g.c.d. of a given n-tuple is unique to within a unit provided the ring is a domain. Moreover, Theorem (3) can be generalized as follows:

9 *Theorem* Suppose $\mathbf{R}$ is a principal ideal ring, and suppose $x_1, \cdots, x_n \in \mathbf{R}$, with at least one element not equal to zero. Then the n-tuple $(x_1, \cdots, x_n)$ has a g.c.d. d which can be expressed in the form

10 $$d = \sum_{i=1}^{n} p_i x_i$$

for appropriate elements $p_1, \cdots, p_n \in \mathbf{R}$. Moreover, if $\mathbf{R}$ is a domain, then every g.c.d. of this n-tuple can be expressed in the form (10).

The proof is entirely analogous to that of Theorem (3) and is left as an exercise.

In the field of rational numbers, every ratio a/b of integers has an equivalent "reduced form" f/g where f and g are coprime. The next result shows that such a statement is also true in the field of fractions associated with *any* principal ideal domain.

11 *Fact* Let $\mathbf{R}$ be a principal ideal domain, and let $\mathbf{F}$ be the field of fractions associated with $\mathbf{R}$. Given any fraction a/b in $\mathbf{F}$, there exists an equivalent fraction f/g such that f and g are coprime.

Proof If a and b are already coprime, then there is nothing to be done. Otherwise, let d be a g.c.d. of a,b, and define $f = a/d$, $g = b/d$. Then clearly $f/g = a/b$, and it only remains to show that f and g are coprime. >From Theorem (3), there exist $p,q \in \mathbf{R}$ such that

12 $$d = pa + qb = pfd + qgd.$$

Cancelling d from both sides of (12) gives $1 = pf + qg$, which shows that f,g are coprime. □

Thus the procedure for "reducing" a fraction is the natural one: namely, we extract a greatest common divisor from the numerator and the denominator.

Two elements $a,b \in \mathbf{R}$ are *associates* (denoted by $a \sim b$) if there is a unit u such that $a = bu$. One can readily verify that $\sim$ is an equivalence relation on $\mathbf{R}$. A nonunit, nonzero element $p \in \mathbf{R}$ is a *prime* if the only divisors of p are either units or associates of p. An equivalent definition is the following: p is a prime if $p = ab$, $a,b \in \mathbf{R}$ implies that either a or b is a unit. In the ring of integers, the primes are precisely the prime numbers.

A useful property of principal ideal domains is stated next, without proof (see [112]).

13 *Fact* Every nonunit, nonzero element of a principal ideal domain can be expressed as a product of primes. Moreover, this factorization is unique in the following sense: If

14 $$x = \prod_{i=1}^{n} p_i = \prod_{i=1}^{m} q_i,$$

where p_i, q_i are all primes, then $n = m$, and the q_i's can be renumbered such that $p_i \sim q_i \; \forall i$.

Thus the only nonuniqueness in the prime factorization of an element arises from the possibility of multiplying some of the prime factors by a unit (for example, $6 = 2 \cdot 3 = (-2) \cdot (-3)$).

Using prime factorizations, one can give a simple expression for a g.c.d. of a set of elements. The proof is left as an exercise (see Problem 7).

15 *Fact* Suppose $x_1, \cdots, x_n$ is a set of elements such that none is a unit nor zero. Express each element x_i in terms of its prime factors in the form

16 $$x_i = \prod_{i=1}^{n} p_j^{\alpha_{ij}},$$

where the p_j's are distinct (i.e. nonassociate) primes, $\alpha_{ij} \geq 0$, and $\alpha_{ij} = 0$ if p_j is not a divisor of x_i. Then a g.c.d. d of $(x_1, \cdots, x_n)$ is given by

17 $$d = \prod_{i=1}^{n} p_j^{\beta_j},$$

where

18 $$\beta_j = \min_i \alpha_{ij}.$$

19 *Corollary* Two elements x,y are coprime if and only if their prime divisors are distinct.

Up to now, we have talked about the greatest common divisor of a set of elements. A parallel concept is that of the least common multiple. Suppose $(x_1, \cdots, x_n)$ is a set of elements, none of which is zero. We say that y is a *least common multiple* of this set of elements if

(LCM1) $x_i \mid y \ \forall i$,

(LCM2) $x_i \mid z \ \forall i$ implies that $y \mid z$.

Using prime factorizations, one can give a simple expression for a l.c.m. that parallels Fact (15).

20 *Fact* Suppose $(x_1, \cdots, x_n)$ is a collection of nonzero, nonunit elements. Express each element x_i in terms of its prime factors as in (16). Then a l.c.m. of this collection of elements is given by

21 $$y = \prod_{i=1}^{n} p_j^{\gamma_j},$$

where

22 $$\gamma_j = \max_i \alpha_{ij}.$$

The proof is left as an exercise (see Problem 8).

Problems

1. Prove that two elements x, y in a p.i.d. $\mathbf{R}$ are coprime if and only if there exist $p, q \in \mathbf{R}$ such that $px + qy$ is a unit.

2. Let $\mathbf{R}$ be a domain, and let $<x,y>$ denote a g.c.d. of x, y which is unique to within a unit factor. Show that if u, v are units, then $<x,y> = <ux,vy>$ (in other words, every g.c.d. of x, y is also a g.c.d. of ux, vy and vice versa).

3. Let $\mathbf{R}$ be a p.i.d., and let $x, y, z \in \mathbf{R}$. Show that $<x,y,z> = <<x,y>,z>$. More generally, show that if $x_1, \cdots, x_n \in \mathbf{R}$, then $<x_1, \cdots, x_n> = <<x_1, \cdots, x_m>, <x_{m+1}, \cdots, x_n>>$ for $1 \leq m < n$.

4. Let **R** be a p.i.d., and let $x_1, \cdots, x_n \in \mathbf{R}$. Let d be a g.c.d. of this set of elements and select $p_1, \cdots, p_n \in \mathbf{R}$ such that $\sum p_i x_i = d$. Show that 1 is a g.c.d. of the set of elements $p_1, \cdots, p_n$.

5. Suppose x,y are coprime and z divides y, in a p.i.d. **R**. Show that x and z are coprime.

6. Suppose u is a unit and $x_1, \cdots, x_n$ are nonzero elements. Show that $<u, x_1, \cdots, x_n> = 1$.

7. Prove Fact (15).

8. Prove Fact (20).

9. (i) Find the prime factorizations of 8, 24, 42 in the ring of integers.
   (ii) Find a g.c.d. of the above three numbers using Fact (15)
   (iii) Find a l.c.m. of the above three numbers using Fact (20).

10. Show that the following three statements are equivalent:
    (i) x divides y.
    (ii) x is a g.c.d. of x,y.
    (iii) y is a l.c.m.of x,y.

11. Suppose **R** is a p.i.d., that $x,y,z \in \mathbf{R}$, x and y are coprime, and that x divides yz. Show that x divides z. (Hint: Use Corollary (19).)

A.4 EUCLIDEAN DOMAINS

In this section, we study a special type of ring that finds a lot of application in this book. Throughout this section "domain" means a commutative domain with identity.

1 *Definition* A domain **R** is a *Euclidean domain* if there is a degree function $\delta: \mathbf{R} - 0 \rightarrow \mathbf{Z}_+$ satisfying the following axioms:[2]

[2] Note that $\mathbf{Z}_+$ denotes the set of nonnegative integers.

(ED1) For every $x, y \in \mathbf{R}$ with $y \neq 0$, there exists a $q \in \mathbf{R}$ such that either $r := x - qy$ is zero, or else $\delta(r) < \delta(y)$.

(ED2) If $x \mid y$ then $\delta(x) \leq \delta(y)$.

One can think of q as a quotient, and r as a remainder, after "dividing" x by y. The axiom (ED1) states that we can always get a remainder that is either zero or else has a smaller degree than the divisor y. We speak of *a* quotient and remainder because q and r are not necessarily unique unless additional conditions are imposed on the degree function $\delta(\cdot)$.

The axiom (ED2) implies that $\delta(1) \leq \delta(x)\ \forall x \neq 0$, since 1 divides every nonzero element. Hence it can be assumed without loss of generality that $\delta(1) = 0$. The same axiom also implies that if x and y are associates, then they have the same degree (because if x and y are associates, then $x \mid y$ and $y \mid x$). In particular, $\delta(u) = 0$ whenever u is a unit.

2 *Fact* Let $\mathbf{R}$ be a Euclidean domain with degree function $\delta(\cdot)$, and suppose

3 $$\delta(x+y) \leq \max\{\delta(x), \delta(y)\},$$

4 $$\delta(xy) = \delta(x) + \delta(y).$$

Then, for every $x, y \in \mathbf{R}$ with $y \neq 0$, there exists a *unique* $q \in \mathbf{R}$ such that $\delta(x - yq) < \delta(y)$, where the degree of zero is taken as $-\infty$.

Proof By (ED1), there exists at least one such q. Now suppose $\delta(x-ay) < \delta(y)$, $\delta(x-by) < \delta(y)$, and define $r = x - ay$, $s = x - by$. Then $x = ay + r = by + s$. Rearranging gives $(a-b)y = s - r$. If $a \neq b$, then $\delta((a-b)y) = \delta(a-b) + \delta(y) \geq \delta(y)$, by (4). On the other hand, $\delta(s-r) \leq \max\{\delta(r), \delta(s)\} < \delta(y)$. This contradiction shows that $a = b$ and also $r = s$. □

5 *Definition* A Euclidean domain $\mathbf{R}$ with degree function $\delta(\cdot)$ is called a *proper Euclidean domain* if $\mathbf{R}$ is not a field and $\delta(\cdot)$ satisfies (4).[3]

[3] This is slightly different from the definition in [61, p.30].

Note that, in a proper Euclidean domain, the division process might still produce nonunique quotients and remainders, because (3) is not assumed to hold. This is the case, for example, in the ring of proper stable rational functions, which are studied in Chapter 2.

6 *Fact* Every Euclidean domain is a principal ideal domain.

Proof Let **R** be a Euclidean domain, and let **I** be an ideal in **R**. If **I** $= \{0\}$, then **I** is principal with 0 as the generator. So suppose **I** contains some nonzero elements, and let x be an element of **I** such that $\delta(x)$ is minimum over all nonzero elements of **I**. We claim that **I** is the ideal generated by x and is hence principal. To prove this, let $y \in \mathbf{I}$ be chosen arbitrarily; it is shown that x divides y. By axiom (ED1), there exists a $q \in \mathbf{R}$ such that either $r := y - qx$ is zero or else $\delta(r) < \delta(x)$. If $r \neq 0$, then $\delta(r) < \delta(x)$ contradicts the manner in which x was chosen. Hence $r = 0$, i.e. x divides y. □

We now present a very important example of a Euclidean domain, namely the ring of polynomials in one indeterminate with coefficients in a field. To lead up to this example, an abstract definition of a polynomial is given.

Let **R** be a ring. Then a *polynomial* over **R** is an infinite sequence $\{a_0, a_1, \cdots\}$ such that only finitely many terms are nonzero. The sum and product of two polynomials $a = \{a_i\}$ and $b = \{b_i\}$ are defined by

7 $$(a+b)_i = a_i + b_i,$$

8 $$(ab)_i = \sum_{j=0}^{i} a_{i-j}\, b_j = \sum_{j=0}^{i} a_j\, b_{i-j}.$$

For notational convenience, a polynomial $a = \{a_i\}$ can be represented by $a_0 + a_1 s + a_2 s^2 + \cdots$ where s is called the "indeterminate." The highest value of the index i such that $a_i \neq 0$ is called the *degree* of a polynomial $a = \{a_0, a_1, \cdots\}$.[4] Thus, if a is a polynomial of degree m, we can write

[4] The degree of the zero polynomial is taken as $-\infty$.

9 $$a(s) = a_0 + a_1 s + \cdots + a_m s^m = \sum_{i=0}^{m} a_i s^i .$$

The set of polynomials over **R** is denoted by **R**$[s]$, and is a commutative ring with identity. Moreover, if **R** is a domain, so is **R**$[s]$.

10 *Fact* Suppose **F** is a field. Then **F**$[s]$ is a Euclidean domain if the degree of a polynomial in **F**$[s]$ is defined as above.

Proof To prove axiom (ED1), suppose

11 $$f(s) = \sum_{i=0}^{n} f_i s^i ,\ g(s) = \sum_{i=0}^{m} g_i s^i ,\ g_m \neq 0.$$

It is necessary to show the existence of a $q \in \mathbf{F}[s]$ such that $\delta(f - gq) < \delta(g)$, where $\delta(0) = -\infty$. If $n < m$, take $q = 0$. If $n \geq m$, define $q_1(s) = (f_n / g_m) s^{n-m}$; then $\delta(f - gq_1) \leq n - 1$. By repeating this process if necessary on the polynomial $f - gq_1$, we can ultimately find a $q \in \mathbf{F}[s]$ such that $\delta(f - gq) < \delta(g)$. Thus (ED1) is satisfied. The proof of (ED2) is straight-forward. □

Since the degree function $\delta(\cdot)$ satisfies both (3) and (4), the Euclidean division process yields a *unique* remainder and quotient r,q corresponding to each pair f,g with $g \neq 0$. Moreover, it is clear that **F**$[s]$ is not a field (the polynomial s has no inverse). Hence **F**$[s]$ is a proper Euclidean domain.

The field of fractions associated with **F**$[s]$ is denoted by **F**(s), and is called the set of *rational functions* over **F**. Note that every element of **F**(s) is a ratio of two polynomials (hence the name rational function).

Problems

1. Show that the set of integers **Z** is a Euclidean domain if we define the degree of an integer to be its absolute value. Does the division process result in unique remainders?

2. Consider the ring $\mathbb{R}[s]$, consisting of polynomials with real coefficients. What are the primes of this ring?

3. Suppose **R** is a proper Euclidean domain and $x \in \mathbf{R}$. Show that if $\delta(x) = 1$, then x is a prime. Is the converse true? (Hint: See Problem 2)

4. Let **R** be a Euclidean domain. Show that $x \in \mathbf{R}$ is a unit if and only if $\delta(x) = 0$. (Hint: Use axiom (ED1))

Appendix B. PRELIMINARIES ON MATRIX RINGS

The objective of this appendix is to gather some well-known facts on matrices whose elements belong to a ring or a field, and to state them in as much generality as possible.

B.1 MATRICES AND DETERMINANTS

Let $\mathbf{R}$ be a ring, and let $\mathbf{R}^{n \times n}$ denote the set of *square* matrices of order $n \times n$ whose elements belong to $\mathbf{R}$. If the sum and product of two matrices in $\mathbf{R}^{n \times n}$ are defined in the familiar way, namely

1 $$(A+B)_{ij} = a_{ij} + b_{ij},$$

2 $$(AB)_{ij} = \sum_{k=1}^{n} a_{ik}\, b_{kj},$$

then $\mathbf{R}^{n \times n}$ becomes a ring, usually referred to as a *ring of matrices over* $\mathbf{R}$. If $\mathbf{R}$ contains an identity and $n \geq 2$, then $\mathbf{R}^{n \times n}$ is not commutative. For instance, if $n = 2$, we have

3 $$\begin{bmatrix} 1 & 0 \\ 0 & 0 \end{bmatrix} \begin{bmatrix} 0 & 0 \\ 1 & 0 \end{bmatrix} \neq \begin{bmatrix} 0 & 0 \\ 1 & 0 \end{bmatrix} \begin{bmatrix} 1 & 0 \\ 0 & 0 \end{bmatrix}.$$

Similar examples can be constructed if $n > 2$. Also, if $n \geq 2$, then $\mathbf{R}^{n \times n}$ is not a domain because

4 $$\text{Diag}\,\{1,0,\cdots,0\}\ \text{Diag}\,\{0,1,0,\cdots,0\} = 0_{n \times n}.$$

The *determinant* of a matrix $A \in \mathbf{R}^{n \times n}$ is denoted by $|A|$ and is defined in the familiar way, namely

5 $$|A| = \sum_{\phi \in \Pi_n} \text{sign } \phi \prod_{i=1}^{n} a_{i\phi(i)},$$

where Π_n denotes the set of permutations of the set $\mathbf{N} = \{1, \cdots, n\}$ into itself, and sign $\phi = \pm 1$ depending on whether ϕ is an even or odd permutation.[1] Most of the usual properties of determinants hold in the present abstract setting. The required results are stated without proof, and the reader is referred to [61] for further details.

Define a function $\Delta: \mathbf{R}^n \times \cdots \times \mathbf{R}^n \to \mathbf{R}$ as follows: For every $v_1, \cdots, v_n \in \mathbf{R}^n$, define $\Delta(v_1, \cdots, v_n)$ to be the determinant of the matrix $V \in \mathbf{R}^{n \times n}$ whose columns are $v_1, \cdots, v_n$ in that order. Thus Δ is just the determinant function viewed as a function of the columns of a matrix. Similarly, define $\bar{\Delta}(v_1, \cdots, v_n)$ to be the determinant of the matrix $V' \in \mathbf{R}^{n \times n}$ whose *rows* are $v_1, \cdots, v_n$ in that order.

6 *Fact* We have

7 $$\Delta(v_1, \cdots, v_n) = \bar{\Delta}(v_1, \cdots, v_n) \; \forall \; v_1, \cdots, v_n \in \mathbf{R}^n.$$

The function Δ (and hence $\bar{\Delta}$) is alternating and multilinear. That is, if two arguments of Δ are interchanged, then Δ changes sign, and if two arguments of Δ are equal, then Δ equals zero; finally,

8 $$\Delta(\alpha v_1 + \beta w_1, v_2, \cdots, v_n) = \alpha \Delta(v_1, v_2, \cdots, v_n) + \beta \Delta(w_1, v_2, \cdots, v_n)$$

for all possible choices of the arguments.

9 *Fact* Let $A \in \mathbf{R}^{n \times n}$, $n \geq 2$. Then, for any $i, j \in \{1, \cdots, n\}$,

10 $$|A| = \sum_{j=1}^{n} (-1)^{i+j} a_{ij} m_{ij}(A),$$

[1] Throughout this section, the symbol "1" is used to denote both the integer as well as the identity element of the ring **R**. It is usually clear from the context which is meant.

11 $$|A| = \sum_{i=1}^{n} (-1)^{i+j} a_{ij} m_{ij}(A),$$

where $m_{ij}(A)$ is the *ij-th minor* of A, defined as the determinant of the $(n-1)\times(n-1)$ matrix obtained from A by deleting its i-th row and j-th column.[2]

12 *Fact* Let $A \in \mathbf{R}^{n\times n}$, $n \geq 2$. Then

13 $$\sum_{j=1}^{n} a_{ij} (-1)^{k+j} m_{kj}(A) = \begin{cases} |A| & \text{if } i = k \\ 0 & \text{if } i \neq k \end{cases}$$

14 $$\sum_{i=1}^{n} a_{ik} (-1)^{i+j} m_{ij}(A) = \begin{cases} |A| & \text{if } j = k \\ 0 & \text{if } j \neq k \end{cases}$$

Now a bit of notation is introduced to make subsequent theorem statements more compact. Suppose m and n are positive integers, with $m \leq n$. Then $S(m,n)$ denotes the collection of all strictly increasing m-tuples $\{i_1, \cdots, i_m\}$, where $1 \leq i_1 < i_2 < \cdots < i_m \leq n$. For example,

15 $$S(3,5) = \{(1,2,3), (1,2,4), (1,2,5), (1,3,4), (1,3,5), (1,4,5), (2,3,4), (2,3,5), (2,4,5), (3,4,5)\}.$$

If $m = n$, then $S(m,n)$ is the singleton set $\{(1,2, \cdots ,n)\}$, while if $m = 0$ then $S(m,n)$ is just the empty set.

Suppose $A \in \mathbf{R}^{m\times n}$, and let $I \in S(l,m)$, $J \in S(l,n)$. Then a_{IJ} denotes the $l \times l$ minor of A consisting of the rows from I and the columns from J. In particular, if I and J are singleton sets of the form $I = \{i\}$, $J = \{j\}$, then a_{IJ} is just the element a_{ij}. Note that $a_{IJ} \in \mathbf{R}$. We use A_{IJ} to denote the $l \times l$ *matrix* consisting of the elements from the rows in I and the columns in J. Thus $a_{IJ} = A_{IJ}$. In some situations it is of interest to examine a submatrix of A consisting of the rows in I and *all* columns of A; such a submatrix is denoted by $A_{I.}$. The notation $A_{.J}$ is similarly defined.

[2] To be consistent with subsequent notation, one should write $a_{\mathrm{N}-i,\mathrm{N}-j}$ instead of $m_{ij}(A)$. But the latter notation is more convenient.

16 *Fact* (Laplace's Expansion of a Determinant) Suppose $A \in \mathbf{R}^{n \times n}$, and suppose $I \in S(m,n)$. Then

17 $$|A| = \sum_{J \in S(m,n)} (-1)^{\nu(I,J)} a_{IJ} a_{\mathbf{N}-I,\mathbf{N}-J},$$

where

18 $$\nu(I,J) = \sum_{i \in I} i + \sum_{j \in J} j.$$

Equation (17) generalizes (10). The corresponding generalization of (11) is similar and is left to the reader.

19 *Fact* (Binet-Cauchy Formula) Suppose $A \in \mathbf{R}^{n \times m}$, $B \in \mathbf{R}^{m \times l}$, and let $C = AB \in \mathbf{R}^{n \times l}$. Let $I \in S(p,n)$, $J \in S(p,l)$. Then

20 $$c_{IJ} = \sum_{K \in S(p,m)} a_{IK} b_{KJ}.$$

In particular, if $n = m = l$, then $|C| = |A| \cdot |B|$.

Note that (20) is a natural generalization of (2).

Using the multilinearity of the determinant function, one can obtain an expression for the determinant of the sum of two matrices. For example, if $A, B \in \mathbf{R}^{2 \times 2}$ and a_1, a_2, b_1, b_2 are the columns of the two matrices, then

21 $$\begin{aligned} |A+B| &= \Delta(a_1+b_1, a_2+b_2) \\ &= \Delta(a_1,a_2) + \Delta(a_1,b_2) + \Delta(b_1,a_2) + \Delta(b_1,b_2). \end{aligned}$$

If $A, B \in \mathbf{R}^{n \times n}$ then the formula for $|A+B|$ will involve the sum of 2^n terms. In case one of the matrices is diagonal, each term in this expansion can be neatly expressed as a product of principal minors of A and B.

22 *Fact* Suppose $A, B \in \mathbf{R}^{n \times n}$ and that A is diagonal (i.e. $a_{ij} = 0$ for $i \neq j$). Then

23 $$|A+B| = \sum_{l=0}^{n} \sum_{I \in S(l,n)} a_{II} b_{\mathbf{N}-I,\mathbf{N}-I},$$

where a_{II} is interpreted as 1 when I is empty.

If $A \in \mathbf{R}^{n \times n}$, its *adjoint matrix*, denoted by A^{adj}, is defined by

24 $$(A^{adj})_{ij} = (-1)^{i+j}\, m_{ji}(A).$$

In view of (13) and (14), it is seen that, for any $A \in \mathbf{R}^{n \times n}$,

25 $$A \cdot A^{adj} = A^{adj} \cdot A = |A|\, I_n,$$

where I_n denotes the $n \times n$ identity marix.

A matrix $A \in \mathbf{R}^{n \times n}$ is *unimodular* if it has an inverse in $\mathbf{R}^{n \times n}$, i.e. it is a unit in the ring $\mathbf{R}^{n \times n}$.

26 *Fact* $A \in \mathbf{R}^{n \times n}$ is unimodular if and only if $|A|$ is a unit in $\mathbf{R}$.

Proof "if" Suppose $|A|$ is a unit in $\mathbf{R}$ and let $b = |A|^{-1} \in \mathbf{R}$. Then $bA^{adj} \in \mathbf{R}^{n \times n}$ and $A \cdot bA^{adj} = bA^{adj} \cdot A = I_n$. Hence A is unimodular.

"only if" Suppose A is unimodular, and let $B \in \mathbf{R}^{n \times n}$ be the inverse of A. Then $1 = |I_n| = |A| \cdot |B|$, which shows that $|A|$ is a unit in $\mathbf{R}$.

□

Now consider the set $\mathbf{F}^{n \times n}$ of matrices with elements in a *field* $\mathbf{F}$. Since a field is also a ring, all of the preceding discussion applies. In addition, since every nonzero element of $\mathbf{F}$ is a unit, we see that every $A \in \mathbf{F}^{n \times n}$ such that $|A| \neq 0$ has an inverse in $\mathbf{F}^{n \times n}$. It is customary to call a matrix A *nonsingular* if $|A| \neq 0$. >From Fact (12), we see that if A is nonsingular, then A^{-1} is given by

27 $$(A^{-1})_{ij} = (-1)^{i+j}\, m_{ji}(A) / |A|.$$

The relation (27) is a special case of the following result, which gives the relationship between the minors of A and A^{-1}. The proof can be found in [38, pp.21-22].

28 *Fact* Suppose $A \in \mathbf{F}^{n \times n}$ is nonsingular and let $B = A^{-1}$. Let $\mathbf{N} = \{1, \cdots, n\}$ and suppose $I, J \in S(l, n)$ for some l. Then

29 $$b_{IJ} = (-1)^{\nu(\mathbf{N}-I, \mathbf{N}-J)} a_{\mathbf{N}-I, \mathbf{N}-J} \, | A |^{-1},$$

where the function ν is defined in (18).

If I and J are singleton sets then (29) reduces to (27).

Suppose $F \in \mathbf{F}^{r \times s}$, and suppose F can be partitioned as

30 $$F = \begin{bmatrix} A & B \\ C & D \end{bmatrix},$$

where A is a nonsingular matrix of order $n \times n$. Then the matrix

31 $$G := D - CA^{-1}B \in \mathbf{F}^{r-n \times s-m}$$

is called the *Schur complement* of F with respect to A, and is sometimes denoted by F/A. The next result relates the minors of G to those of F.

32 *Fact* Suppose $J \in S(t, m)$, $K \in S(t, l)$. Then

33 $$g_{JK} = | A |^{-1} f_{\mathbf{N} \cup (\{n\}+J), \mathbf{N} \cup (\{n\}+K)},$$

where $\{n\}+J$ denotes the set sum (thus if $J = \{j_1, \cdots, j_t\}$, then $\{n\}+J = \{n+j_1, \cdots, n+j_t\}$).

Proof Observe that

34 $$\begin{bmatrix} I & 0 \\ -CA^{-1} & I \end{bmatrix} F = \begin{bmatrix} A & B \\ 0 & G \end{bmatrix} =: E.$$

Now suppose $P \in S(n+t, n+m)$, $Q \in S(n+t, n+l)$ and that $\mathbf{N}$ is a subset of both P and Q. Consider the minor e_{PQ}. The rows in P of E are obtained by adding multiples of the first n rows of F to the rows in P of F. Thus $e_{PQ} = f_{PQ}$, since a minor is unchanged by adding multiples of some rows to others. Observe that the matrix E_{PQ} is block-lower-triangular, so that $e_{PQ} = | A | \cdot e_{P-\mathbf{N}, Q-\mathbf{N}}$. The identity (33) now follows by choosing $P = \mathbf{N} \cup (\{n\}+J)$, $Q = \mathbf{N} \cup (\{n\}+K)$. □

35 *Fact* Suppose $A \in \mathbf{F}^{n \times n}$, $B \in \mathbf{F}^{m \times n}$, $|A| \neq 0$, and let $G = BA^{-1}$. Suppose $J \in S(l,m)$, $K \in S(l,n)$. Then $|A| \cdot g_{JK}$ equals the determinant of the $n \times n$ matrix obtained from A by replacing the rows in K of A by the rows in J of B.

Remarks A simple example helps to illustrate the statement of the result. Suppose $A \in \mathbf{F}^{5 \times 5}$, $B \in \mathbf{F}^{3 \times 5}$, and let a^i, b^i denote the i-th rows of A and B respectively. Each of these is a 1×5 row vector. Suppose $J = (2,3)$, $K = (3,5)$. Then the above result states that

36 $$|A| \cdot g_{(2,3),(3,5)} = \begin{vmatrix} a^1 \\ a^2 \\ b^2 \\ a^4 \\ b^3 \end{vmatrix}.$$

Note that if B is a column vector then the above fact reduces to Cramer's rule for solving linear equations.

Proof Define

37 $$F = \begin{bmatrix} A & -I \\ B & 0 \end{bmatrix}.$$

Then G is the Schur complement of F with respect to A. By Fact (32),

38 $$|A| \cdot g_{JK} = f_{\mathbf{N} \cup (\{n\}+J), \mathbf{N} \cup (\{n\}+K)}$$

$$= \begin{vmatrix} A & -I_{.K} \\ B_{J.} & 0 \end{vmatrix}.$$

Note that each of the last l columns of the minor consists of all zeros except for a single 1. If we expand this minor about the last l columns using Laplace's expansion (Fact (16)), then the expansion consists of a single term, since the submatrix consisting of the last l columns has only nonzero $l \times l$ minor. Thus, if $K = \{k_1, \cdots, k_l\}$, then

39 $$|A| \cdot g_{JK} = (-1)^{\nu} (-1)^{l} \begin{vmatrix} A_{\mathbf{N}-K.} \\ B_{J.} \end{vmatrix},$$

where

40 $$\nu = \sum_{i=1}^{l} (n+i) + \sum_{i=1}^{l} k_i .$$

Now the minor on the right side of (39) is not quite in the form stated in Fact (35), since the rows of B occur *below* those of A, rather than *substituting* for them (as in Example (36) for instance). It is a matter of detail to verify that the parity factor $(-1)^{\nu+l}$ accounts for this difference. □

41 *Fact* Suppose $A \in \mathbf{F}^{n \times n}$, $B \in \mathbf{F}^{n \times m}$, $|A| \neq 0$, and let $G = A^{-1}B$. Suppose $J \in S(l,n)$, $K \in S(l,m)$. Then $|A| \cdot g_{JK}$ equals the determinant of the $n \times n$ matrix obtained from A by replacing the columns in J of A by the columns in K of B.

The proof is similar to that of Fact (36) and is left to the reader.

B.2 CANONICAL FORMS

Let $\mathbf{R}$ be a principal ideal domain, and let $\mathbf{F}$ be the field of fractions associated with $\mathbf{R}$. In this section, we study the sets $\mathbf{R}^{n \times m}$ and $\mathbf{F}^{n \times m}$, consisting of $n \times m$ matrices whose elements belong to $\mathbf{R}$ and $\mathbf{F}$, respectively. We prove the existence of two canonical forms, namely the *Smith form* on $\mathbf{R}^{n \times m}$ and the *Smith-McMillan form* on $\mathbf{F}^{n \times m}$.

A matrix $A \in \mathbf{R}^{n \times m}$ is a *left associate* of $B \in \mathbf{R}^{n \times m}$ (denoted by $A =_L B$) if there is a unimodular matrix $U \in \mathbf{R}^{n \times n}$ such that $A = UB$. A is a *right associate* of B (denoted by $A =_R B$) if there is a unimodular matrix $V \in \mathbf{R}^{m \times m}$ such that $A = BV$. A is *equivalent* to B (denoted by $A \sim B$) if there are unimodular matrices $U \in \mathbf{R}^{n \times n}$, $V \in \mathbf{R}^{m \times m}$ such that $A = UBV$. It is left to the reader to verify that $=_L$, $=_R$, $\sim$ are all equivalence relations.

We now commence our study of canonical forms.

1 *Lemma* Suppose $a_1, \cdots, a_n \in \mathbf{R}$, and let d_n be a g.c.d. of this set of elements. Then there exists a matrix $P_n \in \mathbf{R}^{n \times n}$ whose first row is $[a_1 \cdots a_n]$ and whose determinant is d_n.

Proof The proof is by induction on n. If $n = 2$, by Theorem (A.3.3) there exist $p_1, p_2 \in \mathbf{R}$ such that $p_1 a_1 + p_2 a_2 = d_2$. Now let

2 $$P_2 = \begin{bmatrix} a_1 & a_2 \\ -p_2 & p_1 \end{bmatrix}.$$

For larger values of n, let d_{n-1} be a g.c.d. of $a_1, \cdots, a_{n-1}$. By the inductive hypothesis, we can construct a matrix $P_{n-1} \in \mathbf{R}^{n-1 \times n-1}$ whose first row is $[a_1 \cdots a_{n-1}]$ and whose determinant is d_{n-1}. By Fact (B.1.11),

3 $$\sum_{j=1}^{n-1} a_j \, (-1)^{1+j} \, m_{1j}(P_{n-1}) = d_{n-1}.$$

so that

4 $$\sum_{j=1}^{n-1} \frac{a_j}{d_{n-1}} \, (-1)^{1+j} \, m_{1j}(P_{n-1}) = 1.$$

For convenience, let $z_j = -a_j / d_{n-1}$. Next, by Problem A.3.3, d_n is a g.c.d. of d_{n-1} and a_n. By Theorem (A.3.3.), there exist $x, y \in \mathbf{R}$ such that $x d_{n-1} + y a_n = d_n$. Now define

5 $$P_n = \begin{bmatrix} & & & a_n \\ & P_{n-1} & & 0 \\ & & & \vdots \\ & & & 0 \\ y z_1 & \cdots & y z_{n-1} & x \end{bmatrix}.$$

Expanding $|P_n|$ about the last column and using (4) gives $|P_n| = x d_{n-1} + y a_n = d_n$. $\square$

6 *Theorem* (Hermite Form) Every $A \in \mathbf{R}^{n \times n}$ is a left associate of a matrix B that is lower-triangular (i.e. $b_{ij} = 0$ for $j > i$).

Proof Let d_n be a g.c.d. of $\{a_{1n}, \cdots, a_{nn}\}$, i.e. the elements of the last column of A. By Theorem (A.3.9), there exist elements $p_1, \cdots, p_n \in \mathbf{R}$ such that

7 $$\sum_{i=1}^{n} p_i \, a_{in} = d_n.$$

By Problem A.3.4, 1 is a g.c.d. of the set $p_1, \cdots, p_n$. Hence, by a slight variation of Lemma (1), there exists a unimodular matrix $U \in \mathbf{R}^{n \times n}$ such that its *last* row is $[p_1 \cdots p_n]$. Now $(UA)_{nn} = d_n$. Also, for $i = 1, \cdots, n-1$, $(UA)_{in}$ belongs to the ideal generated by $a_{1n}, \cdots, a_{nn}$, and is thus a multiple of d_n. Let $z_i = (UA)_{in}/d_n$ for $i = 1, \cdots, n-1$, and define

8 $$U_n = \begin{bmatrix} 1 & 0 & \cdots & 0 & -z_1 \\ 0 & 1 & \cdots & 0 & -z_2 \\ \vdots & \vdots & & \vdots & \vdots \\ 0 & 0 & \cdots & 1 & -z_{n-1} \\ 0 & 0 & \cdots & 0 & 1 \end{bmatrix}.$$

Then U_n is unimodular since its determinant is 1. Moreover, $U_n UA$ is of the form

9 $$U_n UA = \begin{bmatrix} & 0 \\ A_{n-1} & \vdots \\ & 0 \\ a^{n-1} & d_n \end{bmatrix}.$$

In other words, the last column of A has been reduced to zero above the diagonal by means of left multiplication by an appropriate unimodular matrix. One can now repeat the procedure with the $n-1 \times n-1$ matrix A_{n-1}, and eventually arrive at a lower triangular matrix. For clarity, the next step of the algorithm is briefly outlined: Let d_{n-1} denote a g.c.d. of the elements of the last column of A_{n-1}. As above, there exists a unimodular matrix

$\bar{U}_{n-1} \in \mathbf{R}^{n-1 \times n-1}$ such that

10 $$\bar{U}_{n-1} A_{n-1} = \begin{bmatrix} A_{n-2} & \begin{matrix} 0 \\ \vdots \\ 0 \end{matrix} \\ a^{n-2} & d_{n-1} \end{bmatrix}.$$

Now define the $n \times n$ unimodular matrix

11 $$U_{n-1} = \text{Block Diag } \{\bar{U}_{n-1}, 1\}.$$

Then

12 $$U_{n-1} U_n U A = \begin{bmatrix} A_{n-2} & \begin{matrix} 0 \\ \vdots \\ 0 \end{matrix} & \begin{matrix} 0 \\ \vdots \\ 0 \end{matrix} \\ a^{n-2} & d_{n-1} & 0 \\ a^{n-1} & \cdot & d_n \end{bmatrix}.$$

The rest of the proof is now obvious. □

13 *Corollary* Every matrix $A \in \mathbf{R}^{n \times n}$ is a left associate of an upper triangular matrix $C \in \mathbf{R}^{n \times n}$.

Proof In the proof of Theorem (6), start with the first column of A instead of the last. □

14 *Corollary* Every matrix $A \in \mathbf{R}^{n \times n}$ is a right associate of a lower (resp. upper) triangular matrix.

Proof In the proof of Theorem (6), start with the last (resp. first) row of A. □

For rectangular matrices, the following result holds:

15 *Corollary* Suppose $A \in \mathbf{R}^{n \times m}$. Then A is a left associate of a matrix of the form

16 $$\begin{bmatrix} D \\ 0 \end{bmatrix} \text{ if } n > m, \; [D \quad E] \text{ if } n < m.$$

where D can be chosen to be either lower or upper triangular.

It is left to the reader to state and prove the result analogous to Corollary (15) concerning right associates of rectangular matrices.

Next, we start developing the Smith form. Recall that a matrix $A \in \mathbf{R}^{n \times m}$ is said to have *rank* l if (i) there is an $l \times l$ submatrix of A with nonzero determinant, and (ii) every $(l+1) \times (l+1)$ minor of A is zero. An *elementary row operation* on the matrix A consists of one of the following: (i) interchanging two rows of A, or (ii) adding a multiple of one row to another. An *elementary column operation* is similarly defined. It is easy to see that an elementary row (column) operation on a matrix A can be accomplished by multiplying A on the left (right) by a unimodular matrix. Thus a matrix obtained from A by elementary row and column operations is equivalent to A.

Now suppose $A \in \mathbf{R}^{n \times m}$ is a (left or right) multiple of $B \in \mathbf{R}^{n \times m}$, and let b_l denote a g.c.d. of all $l \times l$ minors of B. Then it follows from the Binet-Cauchy formula (Fact (B.1.19)) that b_l divides all $l \times l$ minors of A. Thus, if a_l denotes a g.c.d. of all $l \times l$ minors of A, we see that b_l divides a_l. >From this, it follows that if A and B are equivalent matrices, then a_l and b_l are associates, and A and B have the same rank.

17 *Theorem* (Smith Form) Suppose $A \in \mathbf{R}^{n \times m}$ has rank l. Then A is equivalent to a matrix $H \in \mathbf{R}^{n \times m}$ of the form

18 $$H = \begin{bmatrix} h_1 & 0 & \cdots & 0 & 0 \\ 0 & h_2 & \cdots & 0 & 0 \\ \vdots & \vdots & \vdots & \vdots & \vdots \\ 0 & 0 & 0 & h_l & 0 \\ & & 0 & & 0 \end{bmatrix},$$

where h_i divides h_{i+1} for $i = 1, \cdots, l-1$. Moreover, $h_1 \cdots h_i$ is a g.c.d. of all $i \times i$ minors of A, and the h_i's are unique to within multiplication by a unit.

Remarks $h_1, \cdots, h_n$ are called the *invariant factors* of A.

Proof Since A has rank l, it contains an $l \times l$ submatrix with nonzero determinant. By elementary row and column operations, this submatrix can be brought to the upper left-hand corner of A.

As in the proof of Theorem (6), there is a unimodular matrix $U \in \mathbf{R}^{n \times n}$ such that $(UA)_{11}$ is a g.c.d. of all elements in the first column of A. By elementary row operations, the first column of UA can be made to contain all zeros except in the (1,1)-position. Call the resulting matrix $\bar{A}$. If $\bar{a}_{11}$ divides all elements of the first row of $\bar{A}$, then all elements of the first row (except in the (1,1) position) of $\bar{A}V$ can be made to equal zero by a suitable choice of a unimodular matrix V of the form

19 $$V = \begin{bmatrix} 1 & -v_{12} & -v_{13} & \cdots & -v_{1m} \\ \cdot & 1 & 0 & \cdot & 0 \\ \cdot & \cdot & 1 & \cdot & 0 \\ \cdot & \cdot & \vdots & \vdots & \vdots \\ \cdot & \cdot & \cdot & \cdot & 1 \end{bmatrix}.$$

Moreover, the first *column* of $\bar{A}V$ will continue to have zeros except in the (1,1) position. On the other hand, if $\bar{a}_{11}$ does not divide all elements in the first row of $\bar{A}$, we can choose a unimodular matrix V so that $(\bar{A}V)_{11}$ is a g.c.d. of the elements of the first row of $\bar{A}V$. If this is done, the first *column* of $\bar{A}V$ may no longer contain zeros. In such a case, we repeat the above row and column operations. This process cannot continue indefinitely, because the original element $\bar{a}_{11}$ has only a finite number of prime factors (see Fact (A.3.13)). and each successive corner element is a proper (i.e. nonassociate) divisor of its predecessor. Thus, in a finite number of steps, we arrive at a matrix B equivalent to A such that b_{11} divides all elements of the first row as well as first column of B. By elementary row and column operations, all of the elements of the first row and column of B can be made equal to zero. Thus

20 $$A \sim B \sim \begin{bmatrix} b_{11} & 0 & \cdots & 0 \\ 0 & & & \\ \vdots & & B_1 & \\ 0 & & & \end{bmatrix}.$$

By proceeding to the first row and column of B_1 (i.e. the second row and column of B) and then repeating the procedure, we will eventually have

21 $$A \sim \begin{bmatrix} d_1 & \cdots & 0 & \\ \vdots & \vdots & \vdots & 0 \\ 0 & \cdots & d_l & \\ & 0 & & M \end{bmatrix}.$$

Now, $d_i \neq 0 \ \forall i$, since these are all divisors of the elements of the first l rows and columns of A; and none of these rows nor columns is identically zero. Next, M must be zero; otherwise A is equivalent to a matrix of rank at least $l+1$, which contradicts the fact that the rank of A is l.

Thus far we have shown that

22 $$A \sim \begin{bmatrix} d_1 & 0 & \cdots & 0 & \\ 0 & d_2 & \cdots & 0 & 0 \\ \vdots & \vdots & & \vdots & \\ 0 & 0 & \cdots & d_l & \\ & & 0 & & 0 \end{bmatrix}.$$

Let $D = \text{Diag}\,\{d_1, \cdots, d_l\}$. We will show that $D \sim \text{Diag}\,\{h_1, \cdots, h_l\}$, where h_i divides h_{i+1} for $i = 1, \cdots, l-1$. This is enough to prove the theorem. By adding columns 2 to l to the first column of D, it follows that

23 $$D \sim \begin{bmatrix} d_1 & 0 & \cdots & 0 \\ d_2 & d_2 & \cdots & 0 \\ \vdots & \vdots & \vdots & \vdots \\ d_l & 0 & \cdots & d_l \end{bmatrix}.$$

By multiplying the latter matrix on the left by an appropriate unimodular matrix, we get another matrix E whose (1,1)-element is a g.c.d. of $\{d_1, \cdots, d_l\}$ and whose first column is zero otherwise. Let $h_1 :=$ a g.c.d. of $\{d_1, \cdots, d_l\}$. Then

24 $$E = \begin{bmatrix} h_1 & \\ 0 & \overline{E} \\ \vdots & \\ 0 & \end{bmatrix},$$

where every element of $\overline{E}$ is in the ideal generated by $d_2, \cdots, d_l$, and is thus a multiple of h_1. Since the first row of $\overline{E}$ is a multiple of h_1, it follows that

25 $$E \sim \begin{bmatrix} h_1 & 0 & \cdots & 0 \\ 0 & & & \\ \vdots & & E_1 & \\ 0 & & & \end{bmatrix},$$

where every element of E_1 is a multiple of h_1. Now, by the preceding paragraph, $E_1 \sim \text{Diag}\ \{g_2, \cdots, g_l\}$ where h_1 divides g_i for all i. Now let h_2 be a g.c.d. of $g_2, \cdots, g_l$. Then clearly h_1 divides h_2. Moreover,

26 $$E_1 \sim \begin{bmatrix} h_2 & 0 \\ 0 & E_2 \end{bmatrix}.$$

Repeating this procedure, we finally get $D \sim \text{Diag}\ \{h_1, \cdots, h_l\}$ where h_i divides h_{i+1} for all i. This completes the proof of (18).

The divisibility conditions on the h_i imply that the product $h_1 \cdots h_i$ is a g.c.d. of all $i \times i$ minors of H. Since A and H are equivalent, the same product is also a g.c.d. of all $i \times i$ minors of A.

Now suppose A is also equivalent to another matrix

27 $$G = \begin{bmatrix} g_1 & 0 & \cdots & 0 & \\ 0 & g_2 & \cdots & 0 & 0 \\ \vdots & \vdots & \vdots & \vdots & \\ 0 & 0 & \cdots & g_l & \\ & & 0 & & 0 \end{bmatrix},$$

where g_i divides g_{i+1} for all i. Then the reasoning of the preceding paragraph shows that the product $h_1 \cdots h_i$ is an associate of the product $g_1 \cdots g_i$ for all i. Now, if $h_1 \sim g_1$ and $h_1 h_2 \sim g_1 g_2$, then $h_2 \sim g_2$. Reasoning inductively in this fashion, we conclude that $h_i \sim g_i$ for all i. □

28 *Corollary* Two matrices $A, B \in \mathbf{R}^{n \times m}$ are equivalent if and only if their invariant factors are associates.

Let $\mathbf{F}$ be the field of fractions associated with $\mathbf{R}$. We now begin a study of the set $\mathbf{F}^{n \times m}$ of matrices with elements from $\mathbf{F}$ and show the existence of a canonical form known as the Smith-McMillan form.

29 *Theorem* (Smith-McMillan Form) Suppose $F \in \mathbf{F}^{n \times m}$ has rank l. Then there exist unimodular matrices $U \in \mathbf{R}^{n \times n}$, $V \in \mathbf{R}^{m \times m}$ such that

30 $$UFV = \begin{bmatrix} a_1/b_1 & 0 & \cdots & 0 & 0 \\ 0 & a_2/b_2 & \cdots & 0 & 0 \\ \vdots & \vdots & \vdots & \vdots & \vdots \\ 0 & 0 & \cdots & a_l/b_l & \\ & & 0 & & 0 \end{bmatrix},$$

where a_i, b_i are coprime for all i; a_i divides a_{i+1} for $i = 1, \cdots, l-1$; b_{i+1} divides b_i for $i = 2, \cdots l$; and b_1 is a least common multiple of the denominators of all elements of F, expressed in reduced form.[3]

[3] See Fact (A.3.11) for the definition of a reduced form.

Proof Let y denote an l.c.m. of the denominators of all elements of F, expressed in reduced form. Then $yF \in \mathbf{R}^{n \times m}$, and the rank of yF is also l. By Theorem (17), there exist unimodular matrices $U \in \mathbf{R}^{n \times n}$, $V \in \mathbf{R}^{m \times m}$ such that

31 $$UyFV = \begin{bmatrix} h_1 & \cdots & 0 & \\ \vdots & \vdots & \vdots & 0 \\ 0 & \cdots & h_l & \vdots \\ & 0 & & 0 \end{bmatrix},$$

where h_i divides h_{i+1} for all i. Hence

32 $$UFV = \begin{bmatrix} h_1/y & \cdots & 0 & \\ \vdots & \vdots & \vdots & 0 \\ 0 & \cdots & h_l/y & \vdots \\ & 0 & & 0 \end{bmatrix}.$$

Let a_i/b_i be a reduced form for the fraction h_i/y, for $i = 1, \cdots, l$. Since $h_i \mid h_{i+1}$, let $h_{i+1} = h_i r_i$ where $r_i \in \mathbf{R}$. Let $[\cdot]$ denote the reduced form of a matrix. Then

33 $$\frac{a_{i+1}}{b_{i+1}} = [\frac{h_{i+1}}{y}] = [\frac{h_i r_i}{y}] = [\frac{a_i r_i}{b_i}] = a_i[\frac{r_i}{b_i}],$$

where in the last step we used the fact that a_i, b_i are coprime. Now (33) implies that $a_i \mid a_{i+1}$, $b_{i+1} \mid b_i$ for all i.

Finally, to show that $b_1 \sim y$, it is enough to show that h_1 and y are coprime, since a_1/b_1 is a reduced form of h_1/y. This is most easily done using prime factorizations. Let p_{ij}/q_{ij} be a reduced form of the element f_{ij}, for all i,j. Then y is an l.c.m. of all q_{ij}, and h_1 is a g.c.d. of all $yf_{ij} = yp_{ij}/q_{ij}$ for all i,j. Suppose t is a prime factor of y, of multiplicity of α. (By this we mean that t is a prime, t^α divides y but $t^{\alpha+1}$ does not) Then, from Fact (A.3.20), t must be a prime divisor of multiplicity α of *some* q_{ij}. The corresponding p_{ij} is not divisible by t, since p_{ij}, q_{ij} are coprime (see Corollary (A.3.18)). As a result, the corresponding term $yf_{ij} = yp_{ij}/q_{ij}$ is also not divisible by t (since t is a factor of multiplicity α of both yp_{ij} and q_{ij}). Thus h_1, being a g.c.d. of all yf_{ij}, is also not divisible by t (see Fact (A.3.15)). Since this is true of *every* prime factor of y, it follows from Corollary (A.3.18) that y and h_1 are coprime. □

Appendix C. TOPOLOGICAL PRELIMINARIES

In this appendix, a few basic concepts from topology are introduced. For greater detail, the reader is referred to [52] or [85].

C.1 TOPOLOGICAL SPACES

This section contains a brief introduction to topological spaces.

1 *Definition* Let $\mathbf{S}$ be a set. A collection T of subsets of $\mathbf{S}$ is a *topology* if[1]

(TOP1) Both $\mathbf{S}$ and $\emptyset$ (the empty set) belong to T.

(TOP2) A finite intersection of sets in T again belongs to T.

(TOP3) An artibrary union of sets in T again belongs to T.

The ordered pair $(\mathbf{S},T)$ is called a *topological space*, and subsets of $\mathbf{S}$ belonging to T are said to be *open*. A subset of $\mathbf{S}$ is *closed* if its complement in $\mathbf{S}$ is open.

2 *Examples* Let $\mathbf{S}$ be any set, and let T_1 consist of just the two sets $\mathbf{S}$ and $\emptyset$. Then $(\mathbf{S},T_1)$ is a topological space. T_1 is referred to as the *trivial* topology on $\mathbf{S}$. Let T_2 consist of all subsets of $\mathbf{S}$. Then $(\mathbf{S},T_2)$ is also a topological space. T_2 is referred to as the *discrete* topology on $\mathbf{S}$.

Suppose $\mathbf{S}$ is a set, and T_1, T_2 are topologies on $\mathbf{S}$. Then T_1 is *weaker* than T_2 (and T_2 is *stronger* than T_1) if T_1 is a subset of T_2, i.e. every set that is open in the topology T_1 is also open in the topology T_2. It is obvious that, for any set $\mathbf{S}$, the trivial topology and the discrete topology are respectively the weakest and strongest topologies that can be defined on $\mathbf{S}$.

---

[1] Throughout this section, upper case italic letters denote collections of sets, while bold face letters denote sets.

To give interesting and useful examples of topological spaces, the notion of base is introduced. To motivate this notion, recall the familiar definition of an open subset of the real line: A subset $\mathbf{U}$ of $\mathbb{R}$ is open if and only if, corresponding to every $x \in \mathbf{U}$, there is a number $\delta > 0$ such that the interval $(x-\delta, x+\delta)$ is also contained in $\mathbf{U}$. The concept of a base is an abstraction of this idea, whereby intervals of the form $(x-\delta, x+\delta)$ are replaced by more general sets satisfying appropriate axioms.

Let $\mathbf{S}$ be a set, and let B be a collection of subsets of $\mathbf{S}$ that satisfies two axioms:

(B1) The sets in B cover $\mathbf{S}$ (i.e. the union of the sets in B is $\mathbf{S}$).

(B2) Whenever $\mathbf{B}_1, \mathbf{B}_2$ are sets in B with a nonempty intersection and $x \in \mathbf{B}_1 \cap \mathbf{B}_2$, there exists a $\mathbf{B}(x)$ in B such that $x \in \mathbf{B}(x) \subseteq (\mathbf{B}_1 \cap \mathbf{B}_2)$.

Using this collection B, another collection T of subsets of $\mathbf{S}$ is defined as follows: A subset $\mathbf{U}$ of $\mathbf{S}$ belongs to T if and only if, for every $x \in \mathbf{U}$, there is a set $\mathbf{B}(x)$ such that $x \in \mathbf{B}(x) \subseteq \mathbf{U}$.

3 *Fact* A subset $\mathbf{U}$ of $\mathbf{S}$ belongs to the collection T if and only if it is a union of sets in B. T is a topology on $\mathbf{S}$. Moreover, T is the weakest topology on $\mathbf{S}$ containing all sets in B.

Remarks B is referred to as a *base* for the topology T, and T is the *topology generated* by the base B.

4 *Example* To clarify Fact (3), consider the set $\mathbb{R}^n$, consisting of n-tuples of real numbers. Let B denote the collection of "balls" $\mathbf{B}(x,\varepsilon)$ of the form

5 $$\mathbf{B}(x,\varepsilon) = \{y \in \mathbb{R}^n : \| x-y \| < \varepsilon\}$$

as x varies over $\mathbb{R}^n$ and ε varies over the positive numbers. Here $\|\cdot\|$ denotes the usual Euclidean (or any other) norm on $\mathbb{R}^n$. It is a straightforward matter to verify that this collection of balls satisfies axioms (B1) and (B2). Hence this collection forms a base for a topology on $\mathbb{R}^n$, in which a set

$\mathbf{U} \subseteq \mathbb{R}^n$ is open if and only if, for every $x \in \mathbf{U}$, there is a ball $\mathbf{B}(x,\varepsilon) \subseteq \mathbf{U}$. This coincides with the "usual" definition of open sets on $\mathbb{R}^n$. The one extra bit of information that comes out of Fact (3) is that a set is open if and only if it is a union of balls.

Proof of Fact (3) To prove the first sentence, suppose first that $\mathbf{U}$ is a union of sets in B; it is shown that $\mathbf{U}$ belongs to the collection T. Specifically, suppose $\mathbf{U} = \bigcup_{i \in I} \mathbf{B}_i$, where I is an index set and $\mathbf{B}_i$ is in the collection B for all $i \in I$. Let x be an arbitrary element of $\mathbf{U}$. Then $x \in \mathbf{B}_i$ for some $i \in I$, and for this i we have $x \in \mathbf{B}_i \subseteq \mathbf{U}$. Hence, by the definition of T, $\mathbf{U}$ is in T. Conversely, suppose $\mathbf{U}$ is in T. Then for every $x \in \mathbf{U}$ there is a $\mathbf{B}(x)$ in B such that $x \in \mathbf{B}(x) \subseteq \mathbf{U}$. It is now claimed that $\mathbf{U} = \bigcup_{x \in \mathbf{U}} \mathbf{B}(x)$, which would show that $\mathbf{U}$ is a union of sets from B. To prove the claim (and thereby complete the proof of the first sentence), note that $\mathbf{U} \supseteq \bigcup_{x \in \mathbf{U}} \mathbf{B}(x)$ since each $\mathbf{B}(x) \subseteq \mathbf{U}$; conversely, since every $x \in \mathbf{U}$ also belongs to the corresponding $\mathbf{B}(x)$, it follows that $\mathbf{U} \subseteq \bigcup_{x \in \mathbf{U}} \mathbf{B}(x)$. Hence the two sets are equal.

The second sentence is proved by verifying that T satisfies the three axioms (TOP1)-(TOP3). First, $\emptyset$ belongs to T since $\emptyset$ vacuously satisfies the defining condition for a set to belong to T. Next, since $\mathbf{S}$ is the union of all sets in B, it is also in T. Hence T satisfies (TOP1).

To establish (TOP2), it is enough to show that a nonempty intersection of two sets in T again belongs to T; it will then follow by induction that every *finite* intersection of sets in T again belongs to T. Accordingly, suppose $\mathbf{U}, \mathbf{V}$ are in T and let x be any element of $\mathbf{U} \cap \mathbf{V}$; we will show the existence of a $\mathbf{B}(x)$ in B such that $x \in \mathbf{B}(x) \subseteq (\mathbf{U} \cap \mathbf{V})$, which in turn will establish that the intersection $\mathbf{U} \cap \mathbf{V}$ is open. Since $x \in \mathbf{U}$ and $x \in \mathbf{V}$, there exist $\mathbf{B}_1(x), \mathbf{B}_2(x)$ in B such that $x \in \mathbf{B}_1(x) \subseteq \mathbf{U}$, $x \in \mathbf{B}_2(x) \subseteq \mathbf{V}$. Now, by axiom (B2), there is a $\mathbf{B}(x)$ in B such that $x \in \mathbf{B}(x) \subseteq (\mathbf{B}_1(x) \cap \mathbf{B}_2(x))$, which in turn is contained in $\mathbf{U} \cap \mathbf{V}$. Thus T satisfies (TOP2).

To establish (TOP3), suppose $\{\mathbf{U}_\alpha\}$ is a family of sets in T. Then each $\mathbf{U}_\alpha$ is a union of sets belonging to B, whence their union is also a union of sets belonging to B. Thus $\bigcup_\alpha \mathbf{U}_\alpha$ is also in T.

To prove the last sentence, let T_1 be another topology on $\mathbf{S}$ such that every set in B is contained in T_1. Since T_1 is a topology, arbitrary unions of sets in B again belong to T_1. Since every set in T can be expressed as a union of sets in B, it follows that every set in T is in T_1, i.e. T is weaker than T_1.

□

A topological space $(\mathbf{S},T)$ is *first-countable* if T has the following property: For every $x \in \mathbf{S}$, there is a *countable* collection of open sets $\{\mathbf{B}_i(x), i \in \mathbf{Z}_+\}$, each containing x, such that every open set containing x also contains some $\mathbf{B}_i(x)$. In view of Fact (3), the collection $\mathbf{B}_i(x)$, $x \in \mathbf{S}$ is a base for the topology T. Since the set $\mathbf{S}$ may be uncountable, the base for the topology may also be uncountable. However, in a first-countable topology, the collection of open sets *containing each particular point* has a countable base.

For example, consider the set $\mathbb{R}^n$ together with the topology of Example (4). For a fixed $x \in \mathbb{R}^n$, the collection of balls $\mathbf{B}(x,1/m)$, m an integer ≥ 1 is countable; moreover, every open set containing x also contains at least one of the balls $\mathbf{B}(x,1/m)$. Hence the topological space of Example (4) is first-countable.

A very general class of first-countable topological spaces is that of metric spaces, which are defined next.

6 *Definition* A *metric space* $(\mathbf{S},\rho)$ is a set $\mathbf{S}$, together with a function $\rho : \mathbf{S} \to \mathbb{R}$ satisfying the following axioms:

(M1) $\rho(y,x) = \rho(x,y) \; \forall x,y \in \mathbf{S}$.

(M2) $\rho(x,y) \geq 0 \; \forall x,y \in \mathbf{S}$; $\rho(x,y) = 0 \iff x = y$.

(M3) $\rho(x,z) \leq \rho(x,y) + \rho(y,z) \; \forall x,y,z \in \mathbf{S}$.

If $(\mathbf{S},\rho)$ is a metric space, then there is a natural topology that can be defined on $\mathbf{S}$. Let $\mathbf{B}(x,\varepsilon)$ denote the ball

7 $$\mathbf{B}(x,\varepsilon) = \{y \in \mathbf{S}: \rho(x,y) < \varepsilon\}.$$

Then the collection of sets $\mathbf{B}(x,\varepsilon)$ as x varies over $\mathbf{S}$ and ε varies over all positive numbers, is a base for a topology T on $\mathbf{S}$. T is referred to as the topology on $\mathbf{S}$ *induced* by the metric ρ. In this topology, a set $\mathbf{U}$ in $\mathbf{S}$ is open if and only if, for every $x \in \mathbf{U}$, there is a ball $\mathbf{B}(x,\varepsilon) \subseteq \mathbf{U}$.

Every metric space is first-countable as a topological space: For a fixed x, consider the *countable* collection of sets $\mathbf{B}(x,1/m)$, m an integer ≥ 1. Then every open set containing x also contains at least one of the sets $\mathbf{B}(x,1/m)$.

The question of convergence of sequences in topological spaces is examined next. Suppose $(\mathbf{S},T)$ is a topological space. A set $\mathbf{N}$ is said to be a *neighborhood* of $x \in \mathbf{S}$ if $x \in \mathbf{N}$, and $\mathbf{N}$ contains a set in T that contains x. In other words, a set $\mathbf{N}$ is a neighborhood of x if it contains an open set containing x. Note that a neighborhood itself need not be open. A sequence $\{x_i\}$ in $\mathbf{S}$ is said to *converge* to $x \in \mathbf{S}$ if every neighborhood of x contains all but a finite number of terms of the sequence $\{x_i\}$.

8 *Example* Consider the set $\mathbb{R}^n$ together with the topology defined in Example (4). Then a sequence $\{x_i\}$ converges to x if and only if, for every $\varepsilon > 0$, there is a number N such that $x_i \in \mathbf{B}(x,\varepsilon) \; \forall i \geq N$. This is the familiar notion of convergence in $\mathbb{R}^n$.

9 *Example* Consider a set $\mathbf{S}$ together with the discrete topology of Example (2). Let $\{x_i\}$ be a sequence in $\mathbf{S}$ converging to x. Since the singleton set $\{x\}$ is also a neighborhood of x, it follows that $x_i = x$ for all except a finite number of values of i.

10 *Example* Consider a set $\mathbf{S}$ together with the trivial topology of Example (2). Then $\mathbf{S}$ is the only neighborhood of any $x \in \mathbf{S}$. As a result, every sequence $\{x_i\}$ converges to every $x \in \mathbf{S}$.

The preceding two examples show that the meaning of convergence is very much dependent on the particular topology defined on a set. Moreover, in general. a sequence does not necessarily have a unique limit. To address the latter problem, we define Hausdorff topologies. A topology T on a set $\mathbf{S}$ is *Hausdorff* if distinct points have disjoint neighborhoods. It is easy to verify that if T is a Hausdorff topology on a set $\mathbf{S}$, then the limit of a sequence in $\mathbf{S}$ is unique if it exists. Also, if $(\mathbf{S},\rho)$ is a metric space, then the topology induced by the metric ρ is Hausdorff.

Let $\mathbf{S}$ be a set, and let T_1, T_2 be two topologies on $\mathbf{S}$, with T_1 weaker than T_2. Then, whenever a sequence $\{x_i\}$ in $\mathbf{S}$ converges to x in the topology T_2, it also converges to x in the topology T_1. However, it is possible that a sequence may converge in T_1 but not in T_2. This shows that convergence in T_1 is in general a weaker requirement than convergence in T_2 (and helps to explain the terminology).

Recall that a set in $\mathbf{S}$ is *closed* if its complement is open. One of the desirable features of first-countable topologies is that one can give an alternate, and very useful, characterization of closed sets. Suppose $\mathbf{U}$ is a set in $\mathbf{S}$. An element $x \in \mathbf{S}$ is said to be a *limit point* (or a cluster point, accumulation point) of $\mathbf{U}$ if every neighborhood of x contains an element of $\mathbf{U}$ other than x. It can be shown that a set is closed if and only if it contains all of its limit points. Now suppose $(\mathbf{S},T)$ is a first-countable topological space. Then, for each $x \in \mathbf{S}$, there is a *countable* collection of open sets $\{\mathbf{B}_i(x), i \in \mathbf{Z}\}$ such that every neighborhood of x contains at least one of the $\mathbf{B}_i(x)$. As a result, x is a limit point of a set $\mathbf{U}$ if and only if there exists a *sequence* $\{x_i\}$ in $\mathbf{U}$ converging to x such that $x_i \neq x \ \forall i$. Based on this, one can show that $\mathbf{U}$ is closed if and only if every convergent sequence $\{x_i\}$ in $\mathbf{U}$ has a limit in $\mathbf{U}$ (see Problem 5).

Now we come to the notion of continuity. Suppose $(\mathbf{S}_1,T_1)$, $(\mathbf{S}_2,T_2)$ are topological spaces, and f is a function mapping $\mathbf{S}_1$ into $\mathbf{S}_2$. Given any subset $\mathbf{U}$ of $\mathbf{S}_2$, its *preimage* under f is the subset of $\mathbf{S}_1$ denoted by $f^{-1}(\mathbf{U})$ and defined by

11 $f^{-1}(\mathbf{U}) = \{x \in \mathbf{S}_1: f(x) \in \mathbf{U}\}$.

The function f is *continuous at* $x \in \mathbf{S}_1$ if, whenever $\mathbf{U}$ is an open subset of $\mathbf{S}_2$ containing $f(x)$, its preimage $f^{-1}(\mathbf{U})$ is an open subset of $\mathbf{S}_1$ (containing x). f is *continuous* if it is continuous at all $x \in \mathbf{S}_1$. Clearly, whether or not f is continuous is very much dependent on the topologies T_1 and T_2. For example, if T_1 is the discrete topology, then *every* function $f:\mathbf{S}_1 \to \mathbf{S}_2$ is continuous. Also, one can easily verify the following: If f is continuous at $x \in \mathbf{S}_1$ and $\mathbf{N}$ is a neighborhood of $f(x)$, then $f^{-1}(\mathbf{N})$ is a neighborhood of x (see Problem 6). The converse is also true, but more difficult to prove: If $f^{-1}(\mathbf{N})$ is a neighborhood of x whenever $\mathbf{N}$ is a neighborhood of $f(x)$, then f is continuous at x. The proof can be found in [52].

12 *Fact* Suppose $(\mathbf{S}_1, T_1)$, $(\mathbf{S}_2, T_2)$ are topological spaces, and $f:\mathbf{S}_1 \to \mathbf{S}_2$ is continuous at $x \in \mathbf{S}_1$. Then, whenever $\{x_i\}$ is a sequence in $\mathbf{S}_1$ converging to x, the sequence $\{f(x_i)\}$ converges to $f(x)$.

Proof Let $\mathbf{N}$ be any neighborhood of $f(x)$. Since f is continuous at x, $f^{-1}(\mathbf{N})$ is a neighborhood of x. Since $\{x_i\}$ converges to x, $f^{-1}(\mathbf{N})$ contains all but a finite number of terms of the sequence $\{x_i\}$. This is the same as saying that $f(x_i) \in \mathbf{N}$ for all but a finite number of values of i. Since this is true for *every* neighborhood of $f(x)$, we conclude that $\{f(x_i)\}$ converges to $f(x)$. □

If the topologies T_1, T_2 are first-countable, then the converse of Fact (12) is also true: If $\{f(x_i)\}$ converges to $f(x)$ for all sequences $\{x_i\}$ converging to x, then f is continuous at x.

Problems

1. Let $(\mathbf{S}, \rho)$ be a metric space, and let $\mathbf{B}(x, \varepsilon)$ be defined by (7). Show that the collection of sets $\mathbf{B}(x, \varepsilon)$ is a base for a topology on $\mathbf{S}$ by verifying that axiom (B1)is satisfied.

2. Let $\mathbf{S}$ be any set, and define $\rho:\mathbf{S} \times \mathbf{S} \to \mathbb{R}$ by $\rho(x,y) = 0$ if $x = y$, 1 if $x \neq y$.
   (i) Verify that $(\mathbf{S}, \rho)$ is a metric space.

(ii) Show that the topology on $\mathbf{S}$ induced by ρ is the discrete topology.

3. Let $(\mathbf{S},T)$ be a first countable topological space, and let $\mathbf{U}$ be a subset of $\mathbf{S}$. Show that $\mathbf{U}$ is closed if and only if the following is true: $x_i \in \mathbf{U}$, $\{x_i\}$ converges to x implies that $x \in \mathbf{U}$ (hint: either $x = x_i$ for some i or else x is a limit point of $\mathbf{U}$).

4. Suppose $(\mathbf{S},\rho)$ is a metric space, and define

$$d(x,y) = \frac{\rho(x,y)}{1+\rho(x,y)}.$$

(i) Show that $(\mathbf{S},d)$ is also a metric space.

(ii) Show that ρ and d induce the same topology on $\mathbf{S}$.

5. Show that, if $(\mathbf{S},T)$ is a topological space and T is a Hausdorff topology, then a sequence in $\mathbf{S}$ can converge to at most one point.

6. Suppose $(\mathbf{S}_1,T_1)$, $(\mathbf{S}_2,T_2)$ are topological spaces, and $f:\mathbf{S}_1 \rightarrow \mathbf{S}_2$ is continuous at $x \in \mathbf{S}_1$. Show that, if $\mathbf{N}$ is a neighborhood of $f(x)$, then $f^{-1}(\mathbf{N})$ is a neighborhood of x.

C.2 TOPOLOGICAL RINGS AND NORMED ALGEBRAS

This section contains a brief introduction to the subject of topological rings. Roughly speaking, a topological ring is a ring, together with a topology, such that subtraction and multiplication are continuous operations with respect to the topology. Before talking about topological rings as such, it is necessary to define the product topology on the cartesian product of topological spaces.

Recall that the *cartesian product* $\mathbf{S}_1 \times \mathbf{S}_2$ of two sets $\mathbf{S}_1$ and $\mathbf{S}_2$ consists of all ordered pairs (x,y) where $x \in \mathbf{S}_1$ and $y \in \mathbf{S}_2$. Now suppose T_1, T_2 are topologies on $\mathbf{S}_1$, $\mathbf{S}_2$ respectively. The objective is to define a topology on the product set $\mathbf{S} = \mathbf{S}_1 \times \mathbf{S}_2$. This is done in the following way: Let B denote the collection of subsets of $\mathbf{S}$ of the form $\mathbf{U} \times \mathbf{V}$ where $\mathbf{U} \in T_1$, $\mathbf{V} \in T_2$. In other words, B is the collection of subsets of $\mathbf{S}$ formed by taking cartesian products of an open subset of $\mathbf{S}_1$ and an open subset of $\mathbf{S}_2$. It can be shown that B is a base for a topology on $\mathbf{S}$ (see Problem 1), which is referred to as the *product topology* on $\mathbf{S}$.

Now suppose $\mathbf{S}_i$, $i = 1,2,3$ are topological spaces, with the topologies T_i, $i = 1,2,3$. Let $\mathbf{S} = \mathbf{S}_1 \times \mathbf{S}_2$, and let T denote the product topology on $\mathbf{S}$. Suppose f is a function mapping $\mathbf{S}$ into $\mathbf{S}_3$. By definition, f is continuous if $f^{-1}(\mathbf{W})$ is an open subset of $\mathbf{S}$ whenever $\mathbf{W}$ is an open subset of $\mathbf{S}_3$. Now suppose f is continuous, that $(x,y) \in \mathbf{S}$ and let $f(x,y) =: z \in \mathbf{S}_3$. Since f is continuous, it follows that, whenever $\mathbf{N}$ is a neighborhood of z, $f^{-1}(\mathbf{N})$ is a neighborhood of (x,y). Recalling the definition of the product toplogy on the set $\mathbf{S}$, we see that, whenever $\mathbf{N}$ is a neighborhood of z, there exist neighborhoods $\mathbf{N}_1$ of x and $\mathbf{N}_2$ of y such that $f(x_1,y_1) \in \mathbf{N}$ $\forall x_1 \in \mathbf{N}_1$, $\forall y_1 \in \mathbf{N}_2$. In particular, $f(x,y_1) \in \mathbf{N}$ $\forall y_1 \in \mathbf{N}_2$, and $f(x_1,y) \in \mathbf{N}$ $\forall x_1 \in \mathbf{N}_1$. The conclusion is that if f is a continuous function, then for each $x \in \mathbf{S}_1$ the function $f(x,\cdot):\mathbf{S}_2 \rightarrow \mathbf{S}_3$ is continuous; similarly, for each $y \in \mathbf{S}_2$, the function $f(\cdot,y):\mathbf{S}_1 \rightarrow \mathbf{S}_3$ is continuous.

Now we come to topological rings.

1 *Definition* Suppose $\mathbf{R}$ is a ring and T is a topology on $\mathbf{R}$. The pair $(\mathbf{R},T)$ is a *topological ring* if the functions $(x,y) \rightarrow x - y$ and $(x,y) \rightarrow xy$ are continuous functions from $\mathbf{R} \times \mathbf{R}$ into $\mathbf{R}$, when $\mathbf{R} \times \mathbf{R}$ is given the product topology.

Several facts are immediate consequences of the above definition. First, if the function $(x,y) \rightarrow x - y$ is continuous, then *for each fixed* x, the function $y \rightarrow x - y$ is a continuous function from $\mathbf{R}$ into $\mathbf{R}$. In particular, taking $x = 0$, it follows that the function $y \rightarrow -y$ is continuous. Next, since compositions of continuous functions are again continuous, the function $(x,y) \rightarrow x + y = x - (-y)$ is also continuous. Similarly, the function $(x,y) \rightarrow y - x = -(x - y)$ is also continuous.

Given a subset $\mathbf{U} \subseteq \mathbf{R}$ and an element $x \in \mathbf{R}$, let $x + \mathbf{U}$ denote the set defined by

2 $$x + \mathbf{U} = \{x + y : y \in \mathbf{U}\}.$$

One can think of $x + \mathbf{U}$ as the set $\mathbf{U}$ "translated" by x. Now suppose $\mathbf{U}$ is an *open* subset of $\mathbf{R}$, and define the function $f_x:\mathbf{R}\rightarrow\mathbf{R}$ by $f_x(y) = y-x$. As seen earlier, f_x is continuous. Hence the set $f_x^{-1}(\mathbf{U})$ is open. But clearly this set equals $x+\mathbf{U}$. Thus we have shown that if $\mathbf{U}$ is open, so is $x+\mathbf{U}$ for all $x\in\mathbf{R}$. In other words, the topology on a topological ring is "translation-invariant," in the sense that translates of open sets are again open. A consequence of this is that, once we know all the open sets containing 0, we know all open sets.

We now speak briefly about *normed algebras*. Suppose $\mathbf{A}$ is a linear vector space over a field $\mathbf{F}$. This means that there is a concept of addition between members of $\mathbf{A}$, and scalar multiplication between an element of $\mathbf{F}$ and an element of $\mathbf{A}$. Now $\mathbf{A}$ is an algebra (over $\mathbf{F}$) if, in addition, one can define a product of two elements of $\mathbf{A}$ in such a way that $\mathbf{A}$ becomes a ring. Thus an algebra $\mathbf{A}$ has associated with it a field $\mathbf{F}$, and *three* operations: addition between elements of $\mathbf{A}$, multiplication between two elements of $\mathbf{A}$, and multiplication between an element of $\mathbf{F}$ and an element of $\mathbf{A}$. $\mathbf{A}$ is a *ring* with respect to addition and multiplication, and $\mathbf{A}$ is a *linear vector space* with respect to addition and scalar multiplication. For example, if $\mathbf{F}$ is any field, the set of matrices $\mathbf{F}^{n\times n}$ is an algebra over $\mathbf{F}$.

Now suppose $\mathbf{A}$ is an algebra over $\mathbb{R}$ (the real numbers) or C (the complex numbers). Then $(\mathbf{A}, \|\cdot\|)$ is a *normed algebra* if one can define a norm $\|\cdot\|$ on $\mathbf{A}$ such that

(NA1) $\|a\| \geq 0 \ \forall a\in\mathbf{A}, \ \|a\| = 0 \Longleftrightarrow a = 0,$

(NA2) $\|\alpha a\| = |\alpha|\cdot\|a\| \ \forall \alpha\in\mathbb{R}$ (or C), $\forall a\in\mathbf{A},$

(NA3) $\|a+b\| \leq \|a\| + \|b\| \ \forall a,b\in\mathbf{A},$

(NA4) $\|ab\| \leq \|a\|\cdot\|b\| \ \forall a,b\in\mathbf{A}.$

Axioms (NA1)-(NA3) are just the usual requirements for $(\mathbf{A}, \|\cdot\|)$ to be a normed space, and (NA4) is the additional requirement for it to be a normed algebra.

Let $(\mathbf{A}, \| \cdot \|)$ be a normed algebra, and let B denote the collection of balls

3 $$\mathbf{B}(x,\varepsilon) = \{y \in \mathbf{A} : \| x - y \| < \varepsilon\}.$$

Then B is a base for a topology T on $\mathbf{A}$. This is the same as the topology induced by the metric $\rho(x,y) = \| x - y \|$.

It is now shown that $(\mathbf{A}, T)$ is a topological ring. In order to do this, it is necessary to show that the functions $(x,y) \to x - y$ and $(x,y) \to xy$ are both continuous. Let us begin with subtraction. Suppose $\bar{x}, \bar{y} \in \mathbf{A}$, $\bar{z} = \bar{x} - \bar{y}$, and let $\mathbf{S} \subseteq \mathbf{A}$ be an open set containing $\bar{z}$. We must show that the set $\{(x,y) : x - y \in \mathbf{S}\} =: \mathbf{V}$ is an open subset of $\mathbf{A} \times \mathbf{A}$. This is done by showing that, whenever $(x,y) \in \mathbf{V}$, there are balls $\mathbf{B}(x,\delta), \mathbf{B}(y,\delta)$ such that $\mathbf{B} := \mathbf{B}(x,\delta) \times \mathbf{B}(y,\delta)$ is contained in $\mathbf{V}$. Since $\mathbf{S}$ is open and $x - y \in \mathbf{S}$, by definition there is an $\varepsilon > 0$ such that $\mathbf{B}(x - y, \varepsilon) \subseteq \mathbf{S}$. Now let $\delta = \varepsilon/2$, and suppose $x_1 \in \mathbf{B}(\bar{x},\delta), y_1 \in \mathbf{B}(\bar{y},\delta)$. Then

4 $$\| (x_1 - y_1) - (x - y) \| \leq \| x_1 - x \| + \| y_1 - y \| < 2\delta < \varepsilon.$$

Hence $(x_1, y_1) \in \mathbf{V}$. This shows that subtraction is continuous. Next, to show that multiplication is continuous, suppose $\bar{x}, \bar{y} \in \mathbf{A}$, let $\bar{z} = \bar{x}\bar{y}$, and let $\mathbf{S}$ be an open set in $\mathbf{A}$ containing $\bar{z}$. Define $\mathbf{V} = \{(x,y) : xy \in \mathbf{S}\}$. We show that $\mathbf{V}$ is open by showing that, whenever $(x,y) \in \mathbf{V}$, there exist balls $\mathbf{B}(x,\delta), \mathbf{B}(y,\delta)$ such that $\mathbf{B}(x,\delta) \times \mathbf{B}(y,\delta)$ is contained in $\mathbf{V}$. Since $\mathbf{V}$ is open, by definition there is an $\varepsilon > 0$ such that $\mathbf{B}(xy, \varepsilon) \subseteq \mathbf{S}$. Suppose $x_1 \in \mathbf{B}(x,\delta), y_1 \in \mathbf{B}(y,\delta)$ where δ is yet to be specified. Then

5 $$\begin{aligned} x_1 y_1 - xy &= [(x_1 - x) + x] \cdot [(y_1 - y) + y] - xy \\ &= (x_1 - x)(y_1 - y) + (x_1 - x)y + x(y_1 - y). \end{aligned}$$

6 $$\| x_1 y_1 - xy \| \leq \| x_1 - x \| \cdot \| y_1 - y \| + \| x_1 - x \| \cdot \| y \|$$
$$\| x \| \cdot \| y_1 - y \|$$

$$\leq \delta^2 + \delta(\| x \| + \| y \|).$$

Hence $(x_1, y_1) \in \mathbf{V}$ if δ is chosen such that

7 $$\delta^2 + \delta(\| x \| + \| y \|) \leq \varepsilon.$$

Such a δ can always be found since the left side of (7) is continuous in δ and equals zero when δ is zero. This completes the proof that multiplication is continuous.

8 *Fact* Suppose $(\mathbf{A}, \| \cdot \|)$ is a normed algebra, and let $\mathbf{U}$ denote the set of units of $\mathbf{A}$ (i.e. the set of elements of $\mathbf{A}$ that have a multiplicative inverse in $\mathbf{A}$). Then the function $f : u \rightarrow u^{-1}$ maps $\mathbf{U}$ into itself continuously.

Remark By the continuity of the "inversion" function f, we mean the following: Given any $u \in \mathbf{U}$ and any $\varepsilon > 0$, there exists a $\delta > 0$ such that

9 $$\| u^{-1} - v^{-1} \| < \varepsilon \text{ whenever } v \in \mathbf{U} \text{ and } \| u - v \| < \delta.$$

Proof Let $u \in \mathbf{U}$ and let $\varepsilon > 0$ be specified. Suppose $v \in \mathbf{U}$ and $v \in \mathbf{B}(u, \delta)$, where δ is not as yet specified. Then

10 $$\| u^{-1} - v^{-1} \| = \| u^{-1}(v - u)v^{-1} \|$$

$$\leq \| u^{-1} \| \cdot \| v - u \| \cdot \| v^{-1} \|$$

$$\leq \delta \| u^{-1} \| \cdot \| v^{-1} \|$$

$$\leq \delta \| u^{-1} \| \cdot (\| u^{-1} \| + \| u^{-1} - v^{-1} \|).$$

Solving for $\| u^{-1} - v^{-1} \|$ from (10) gives

11 $$\| u^{-1} - v^{-1} \| \leq \frac{\delta \| u^{-1} \|^2}{1 - \delta \| u^{-1} \|}.$$

Thus (9) is satisfied if δ is chosen such that the right side of (11) is less than ε. But such a δ can always be found, since the right side of (11) is continuous in δ and equals zero when δ is zero. This shows that the function f is continuous.

□

Problems

1. Suppose $(\mathbf{S}_1,T_1)$, $(\mathbf{S}_2,T_2)$ are topological spaces.
(i) Let T denote the collection of subsets of $\mathbf{S}_1 \times \mathbf{S}_2$ of the form $\mathbf{U} \times \mathbf{V}$ where $\mathbf{U} \in T_1$, $\mathbf{V} \in T_2$. Show that T is a base for a topology on the set $\mathbf{S}_1 \times \mathbf{S}_2$ (hint: Show that T satisfies the axiom (B1) of Section B.1).
(ii) Suppose B_1, B_3 are respectively bases for the topologies T_1, T_2. Let B denote the collection of subsets of $\mathbf{S}_1 \times \mathbf{S}_2$ of the form $\mathbf{B}_1 \times \mathbf{B}_2$ where $\mathbf{B}_1 \in B_1$, $\mathbf{B}_2 \in B_3$. Show that B is also a base for a topology on $\mathbf{S}_1 \times \mathbf{S}_2$.
(iii) Show that B and T generate the same topology on $\mathbf{S}_1 \times \mathbf{S}_2$ (hint: Show that every set in T is a union of sets in B).

2. Consider the ring $\mathbb{R}^{n \times n}$ of $n \times n$ matrices with real elements. Let $\| \cdot \|$ be any norm on the *vector space* $\mathbb{R}^n$, and define the norm of a *matrix* by

$$\| A \| = \sup_{x \neq 0} \frac{\| Ax \|}{\| x \|}.$$

Show that $(\mathbb{R}^{n \times n}, \| \cdot \|)$ is a normed algebra.

3. Consider the ring $\mathbb{R}[s]$ of polynomials in the indeterminate s with real coefficients. Define the norm of $a(s) = \sum_i a_i s^i$ by

$$\| a(\cdot) \| = \sum_i | a_i | .$$

Show that $(\mathbb{R}[s], \| \cdot \|)$ is a normed algebra.

REFERENCES

[1] V. Anantharam and C. A. Desoer, "On the stabilization of nonlinear systems," *IEEE Trans. on Auto. Control,* **AC-29**, pp. 569-572, June 1984.

[2] B. D. O. Anderson, "A note on the Youla-Bongiorno-Lu condition," *Automatica,* **12**, pp. 387-388, July 1976.

[3] B. D. O. Anderson and J. B. Moore, *Optimal Filtering,* Prentice-Hall, Englewood Cliffs, New Jersey, 1979.

[4] M. F. Atiyah and I. G. MacDonald, *Introduction to Commutative Algebra*, Addison-Wesley, Reading, MA., 1969.

[5] J. A. Ball and J. W. Helton, "A Beurling-Lax theorem for the Lie group $U(m,n)$ which contains most classical interpolation theory," *J. Operator Theory,* **9**, pp.107-142, 1983.

[6] C. I. Byrnes, M. W. Spong and T. J. Tarn, "A several complex variables approach to feedback stabilization of neutral delay-differential systems," *Math. Sys. Thy.,* **17**, pp. 97-134, May 1984.

[7] F. M. Callier and C. A. Desoer, "Open-loop unstable convolution feedback systems with dynamical feedback," *Automatica,* **13**, pp. 507-518, Dec. 1976.

[8] F. M. Callier and C. A. Desoer, "An algebra of transfer functions of distributed linear time-invariant systems," *IEEE Trans. Circ. and Sys.,* **CAS-25**, pp. 651-662, Sept. 1978.

[9] F. M. Callier and C. A. Desoer, "Simplifications and clarifications on the paper 'An algebra of transfer functions of distributed linear time-invariant systems,' " *IEEE Trans. Circ. and Sys.,* **CAS-27**, pp. 320-323, Apr. 1980.

[10] F. M. Callier and C. A. Desoer, "Stabilization, tracking and disturbance rejection in multivariable convolution systems," *Annales de la Societé Scientifique de Bruxelles,* **94**, pp. 7-51, 1980.

[11] F. M. Callier and C. A. Desoer, *Multivariable Feedback Systems,* Springer-Verlag, New York, 1982.

[12] B. C. Chang and J. B. Pearson, "Optimal Disturbance reduction in linear multivariable systems," *IEEE Trans. on Auto. Control,* **AC-29**, pp. 880-888, Oct. 1984.

[13] M. J. Chen and C. A. Desoer, "Necessary and sufficient conditions for robust stability of linear distributed feedback systems," *Int. J. Control,* **35,** pp. 255-267, 1982.

[14] M. J. Chen and C. A. Desoer, "Algebraic theory of robust stability of interconnected systems," *IEEE Trans. on Auto. Control,* **AC-29**, pp. 511-519, June 1984.

[15] J. H. Chow and P. V. Kokotovic, "Eigenvalue placement in two-time-scale systems," *Proc. IFAC Symp. on Large Scale Sys.*, Udine, Italy, pp. 321-326, June 1976.

[16] C. -C. Chu and J. C. Doyle, "On inner-outer and spectral factorizations," *Proc. IEEE Conf. on Decision and Control,* Las Vegas, Dec. 1984 (to appear).

[17] H. T. Colebrooke, *Algebra with Arithmetic and Mensuration,* John Murray, London, 1817, reprinted by Dr. Martin Sandig OHg, Wiesbaden, W. Germany, 1973.

[18] P. R. Delsarte, Y. Genin and Y. Kamp, "The Nevanlinna-Pick problem for matrix-valued functions," *SIAM J. Appl. Math,* **36,** pp. 47-61, Feb. 1979.

[19] C. A. Desoer and W. S. Chan, "The feedback interconnection of linear time-invariant systems," *J. Franklin Inst.,* **300**, pp. 335-351, 1975.

[20] C. A. Desoer and M. J. Chen, "Design of multivariable feedback systems with stable plant," *IEEE Trans. on Auto. Control,* **AC-26**, pp. 408-415, April 1981.

[21] C. A. Desoer and C. L. Gustafson, "Design of multivariable feedback systems with simple unstable plant," *IEEE Trans. on Auto. Control,* **AC-29**, pp. 901-908, Oct. 1984.

[22] C. A. Desoer and C. L. Gustafson, "Algebraic theory of linear multivariable feedback systems," *IEEE Trans. on Auto. Control,* **AC-29**, pp. 909-917, Oct. 1984.

[23] C. A. Desoer and C. A. Lin, "Two-step compensation of nonlinear systems," *Systems and Control Letters,* **3**, pp. 41-46, June 1983.

[24] C. A. Desoer and R. W. Liu, "Global parametrization of feedback systems with nonlinear plants," *Systems and Control Letters,* **1**, pp. 249-251, Jan. 1982.

[25] C. A. Desoer, R. W. Liu, J. Murray and R. Saeks, "Feedback system design: The fractional representation approach to analysis and synthesis," *IEEE Trans. Auto. Control,* **AC-25**, pp. 399-412, June 1980.

[26] C. A. Desoer and M. Vidyasagar, *Feedback Systems: Input-Output Properties,* Academic Press, New York, 1975.

[27] J. C. Doyle, "Robustness of multiloop linear feedback systems," *Proc. 17th Conf. Decision and Control*, Ft. Lauderdale, pp. 12-18, 1979.

[28] J. Doyle, "Analysis of feedback systems with structured uncertainties," *IEE Proc.,* Part D, **129,** pp. 242-250, Nov. 1982.

[29] J. C. Doyle, "Synthesis of robust controllers and filters," *Proc. IEEE Conference on Decision and Control,* pp. 109-114, 1983.

[30] J. C. Doyle and G. Stein, "Multivariable feedback design: Concepts for a classical/modern synthesis," *IEEE Trans. Auto. Control,* **AC-26**, pp. 4-16, Feb. 1981.

[31] J. C. Doyle, J. E. Wall and G. Stein, "Performance and robustness results for structured uncertainty," *Proc. IEEE Conf. on Decision and Control,* pp. 628-636, 1982.

[32] P. L. Duren, *The Theory of* H^p-spaces, Academic Press, New York, 1970.

[33] B. A. Francis, "The multivariable servomechanism problem from the input-output viewpoint," *IEEE Trans. Auto. Control,* **AC-22,** pp. 322-328, June 1977.

[34] B. A. Francis, "On the Wiener-Hopf approach to optimal feedback design," *Systems and Control Letters,* **2**, pp. 197-201, Dec. 1982.

[35] B. A. Francis, J. W. Helton and G. Zames, "H^∞-optimal feedback controllers for linear multivariable systems," *IEEE Trans. on Auto. Control,* **AC-29**, pp. 888-900, Oct. 1984.

[36] B. A. Francis and M. Vidyasagar, "Algebraic and topological aspects of the regulator problem for lumped linear systems," *Automatica,* **19**, pp. 87-90, Jan. 1983.

[37] B. A. Francis and G. Zames, "On optimal sensitivity theory for SISO feedback systems," *IEEE Trans. Auto. Control,* **AC-29,** 9-16, January 1984.

[38] F. R. Gantmacher, *Theory of Matrices*, Chelsea, New York.

[39] I. M. Gel'fand, D. Raikov and G. Shilov, *Commutative normed rings,* Chelsea, New York, 1964.

[40] B. K. Ghosh and C. I. Byrnes, "Simultaneous stabilization and simultaneous pole-placement by non-switching dynamic compensation," *IEEE Trans. Auto. Control,* **AC-28,** pp. 735-741, June 1983.

[41] K. Glover, "All optimal Hankel-norm approximations of linear multivariable systems and their L^∞-error bounds," *Int. J. Control,* **39**, pp. 1115-1193, June 1984.

[42] V. Guillemin and A. Pollack, *Differential Topology,* Prentice-Hall, Englewood Cliffs, N. J. 1979.

[43] C. L. Gustafson and C. A. Desoer, "Controller design for linear multivariable feedback systems with stable plant," *Int. J. Control,* **37**, pp. 881-907, 1983.

[44] E. Hille and R. S. Phillips, *Functional Analysis and Semigroups*, Amer. Math. Soc., Providence, R. I., 1957.

[45] K. Hoffman, *Banach Spaces of Analytic Functions*, Prentice-Hall, Englewood Cliffs, N. J., 1962.

[46] N. T. Hung and B. D. O. Anderson, "Triangularization technique for the design of multivariable control systems," *IEEE Trans. Auto. Control,* **AC-24**, pp. 455-460, June 1979.

[47] C. A. Jacobson, "Some aspects of the structure and stability of a class of linear distributed systems," Robotics and Automation Lab. Rept. No. 31, Renn. Poly. Inst., Dept. of ECSE, May 1984.

[48] N. Jacobson, *Lectures in Abstract Algebra,* Vol I, Van Nostrand, New York, 1953.

[49] T. Kailath, *Linear Systems*, Prentice-Hall, Englewood Cliffs, N. J., 1980.

[50] E. W. Kamen, P. P. Khargonekar and A. Tannenbaum, "A local theory of linear systems with noncommensurate time delays," submitted for publication.

[51] E. W. Kamen, P. P. Khargonekar and A. Tannenbaum, "Stabilization of time-delay systems using finite-dimensional compensators," *IEEE Trans. on Auto. Control,* Jan. 1985 (to appear).

[52] J. L. Kelley, *General Topology*, Van Nostrand, New York, 1955.

[53] P. P. Khargonekar and A. Tannenbaum, "Noneuclidean metrics and the robust stabilization of systems with parameter uncertainty," *IEEE Trans. Auto. Control,* (to appear).

[54] H. Kimura, "Robust Stabilizability for a class of transfer functions," *IEEE Trans. Auto. Control,* **AC-29**, pp. 788-793, Sept. 1984.

[55] P. Koosis, *The Theory of* H^p spaces, Cambridge University Press, Cambridge, 1980.

[56] V. Kučera, *Discrete Linear Control: The Polynomial Equation Approach,* Wiley, New York, 1979.

[57] T. Y. Lam, *Serre's Conjecture,* Lecture notes in Mathematics, No. 635, Springer-Verlag, Berlin, 1978.

[58] V. Ya. Lin, "Holomorphic fiberings and multivalued functions of elements of a Banach algebra," (English translation) *Funct. Anal. Appl.,* **37,** pp. 122-128, 1973.

[59] H. Lindel and W. Lutkebohmert, "Projektive modulen über polynomialen erweiterungen von potenzreihenalgebren," *Archiv der Math,* **28,** pp. 51-54, 1977.

[60] R. W. Liu and C. H. Sung, "Linear feedback system design," *Circ., Sys. and Sig. Proc.,* **2**, pp. 35-44, 1983.

[61] C. C. MacDuffee, *Theory of Matrices*, Chelsea, New York, 1946.

[62] H. Maeda and M. Vidyasagar, "Some results on simultaneous stabilization," *Systems and Control Letters,* (to appear).

[63] H. Maeda and M. Vidyasagar, "Infinite gain margin problem in multivariable feedback systems," submitted for publication.

[64] A. S. Morse, "System invariants under feedback and cascade control," *Proc. Int. Conf. on Math. Sys. Thy.,* Udine, Italy, 1976.

[65] B. Sz-Nagy and C. Foias, *Harmonic Analysis of Operators on Hilbert Space*, Elsevier, New York, 1970.

[66] C. N. Nett, "The fractional representation approach to robust linear feedback design: A self-contained exposition," Robotics and Automation Lab. Rept. No. 30, Renn. Poly. Inst., Dept. of ECSE, May 1984.

[67] C. N. Nett, C. A. Jacobson and M. J. Balas, "Fractional representation theory: Robustness with applications to finite-dimensional control of a class of linear distributed systems," *Proc. IEEE Conf. on Decision and Control*, pp. 268-280, 1983.

[68] C. N. Nett, C. A. Jacobson and M. J. Balas, "A connection between state-space and doubly coprime fractional representations," *IEEE Trans. Auto. Control*, **AC-29**, pp.831-832, Sept. 1984.

[69] G. Newton, L. Gould and J. F. Kaiser, *Analytic Design of Linear Feedback Controls*, Wiley, New York, 1957.

[70] L. Pernebo, "An algebraic theory for the design of controllers for linear multivariable systems," *IEEE Trans. Auto. Control*, **AC-26**, pp. 171-194, Feb. 1981.

[71] V. P. Potapov, "The Multiplicative structure of J-contractive matrix functions," *Translations of the Amer. Math. Soc.*, **15**, pp. 131-243, 1960.

[72] D. Quillen, "Projective modules over polynomial rings," *Invent. Math.*, **36**, pp. 167-171, 1976.

[73] V. R. Raman and R. W. Liu, "A necessary and sufficient condition for feedback stabilization in a factorial ring," *IEEE Trans. Auto. Control*, **AC-29**, pp. 941-942, Oct. 1984.

[74] H. H. Rosenbrock, *State Space and Multivariable Theory*, Nelson, London, 1970.

[75] J. J. Rotman, *An Introduction to Homological Algebra*, Academic Press, New York, 1979.

[76] W. Rudin, *Fourier Analysis on Groups*, John Wiley, New York, 1962.

[77] W. Rudin, *Real and Complex Analysis*, McGraw-Hill, New York, 1974.

[78] R. Saeks and J. Murray, "Feedback System Design: The tracking and disturbance rejection problems," *IEEE Trans. Auto. Control,* **AC-26**, pp. 203-218, Feb. 1981.

[79] R. Saeks and J. Murray, "Fractional representation, algebraic geometry and the simultaneous stabilization problem," *IEEE Trans. Auto. Control,* **AC-27**, pp. 895-903, Aug. 1982.

[80] R. Saeks, J. Murray, O. Chua, C. Karmokolias and A. Iyer, "Feedback system design: The single variate case - Part I," *Circ., Sys. and Sig. Proc.,* **1**, pp. 137-170, 1982.

[81] R. Saeks, J. Murray, O. Chua, C. Karmokolias and A. Iyer, "Feedback system design: The single variate case - Part II," *Circ., Sys. and Sig. Proc.,* **2**, pp. 3-34, 1983.

[82] D. Sarason, "Generalized interpolation in H^∞," *Trans. Amer. Math. Soc.,* **127**, pp. 179-203, May 1967.

[83] D. D. Siljak, "On reliability of control," *Proc. IEEE Conf. Decision and Control*, pp. 687-694, 1978.

[84] D. D. Siljak, "Dynamic reliability using multiple control systems," *Int. J. Control,* **31,** pp. 303-329, 1980.

[85] G. Simmons, *Introduction to Topology and Modern Analysis*, McGraw-Hill, New York, 1966.

[86] D. E. Smith, *History of Mathematics*. Vol. I, Dover, New York, 1958.

[87] A. A. Suslin, "Projective modules over a polynomial ring are free," *Soviet Math. Doklady,* **17,** pp. 1160-1164, 1976.

[88] M. Vidyasagar, "Input-output stability of a broad class of linear time-invariant multivariable feedback systems," *SIAM J. Control,* **10,** pp. 203-209, Feb. 1972.

[89] M. Vidyasagar, "Coprime factorizations and the stability of multivariable distributed feedback systems," *SIAM J. Control,* **13,** pp. 1144-1155, Nov. 1975.

[90] M. Vidyasagar, *Nonlinear Systems Analysis,* Prentice-Hall, Englewood Cliffs, N. J., 1978.

[91] M. Vidyasagar, "On the use of right-coprime factorizations in distributed feedback systems containing unstable subsystems," *IEEE Trans. Circ. and Sys.,* **CAS-25,** pp. 916-921, Nov. 1978.

[92] M. Vidyasagar, "The graph metric for unstable plants and robustness estimates for feedback stability," *IEEE Trans. Auto. Control,* **AC-29**, pp. 403-418, May 1984.

[93] M. Vidyasagar and N. K. Bose, "Input-output stability of linear systems defined over measure spaces," *Proc. Midwest Symp. Circ. and Sys.*, Montreal, Canada, pp. 394-397, Aug. 1975.

[94] M. Vidyasagar and K. R. Davidson, "A parametrization of all stable stabilizing compensators for single-input-output systems," (under preparation).

[95] M. Vidyasagar and H. Kimura, "Robust controllers for uncertain linear multivariable systems," *Automatica* (to appear).

[96] M. Vidyasagar and B. C. Levy, "On the genericity of simultaneous stabilizability," *Symp. Math. Thy. of Netw. and Sys.*, Stockholm, 1985.

[97] M. Vidyasagar, H. Schneider and B. A. Francis, "Algebraic and topological aspects of feedback stabilization," *IEEE Trans. Auto. Control,* **AC-27,** pp. 880-894, Aug. 1982.

[98] M. Vidyasagar and N. Viswanadham, "Algebraic design techniques for reliable stabilization," *IEEE Trans. Auto. Control,* **AC-27**, pp. 1085-1095, Oct. 1982.

[99] M. Vidyasagar and N. Viswanadham, "Reliable stabilization using a multi-controller configuration," *Proc. IEEE Conf. on Decision and Control,* pp. 856-859, 1983.

[100] B. L. van der Waerden, *Geometry and Algebra in Ancient Civilizations,* Springer Verlag, New York, 1984.

[101] J. L. Walsh, *Interpolation and Approximation by Rational Functions in the Complex Domain,* AMS Colloqium Publ., Providence, R. I., 1935.

[102] W. Wolovich, *Linear Multivariable Control*, Springer-Verlag, New York, 1974.

[103] W. M. Wonham, *Linear Multivariable Control: A Geometric Approach,* 2nd ed., Springer-Verlag, New York, 1979.

[104] D. C. Youla, J. J. Bongiorno, Jr. and C. N. Lu, "Single-loop feedback stablization of linear multivariable plants," *Automatica,* **10,** pp. 159-173, 1974.

[105] D. C. Youla, J. J. Bongiorno, Jr. and H. A. Jabr, "Modern Wiener-Hopf design of optimal controllers, Part I: The single-input case," *IEEE Trans. on Auto. Control,* **AC-21**, pp. 3-14, Feb. 1976.

[106] D. C. Youla and G. Gnavi, "Notes on n-dimensional system theory," *IEEE Trans. Circ. and Sys.,* **CAS-26**, pp. 105-111, Feb. 1979.

[107] D. C. Youla, H. A. Jabr and J. J. Bongiorno, Jr., "Modern Wiener-Hopf design of optimal controllers, Part II: The multivariable case," *IEEE Trans. Auto. Control,* **AC-21,** pp. 319-338, June 1976.

[108] D. C. Youla and R. Pickel, "The Quillen-Suslin theorem and the structure of n-dimensional elementary polynomial matrices," *IEEE Trans. Circ. and Sys.,* **CAS-31**, pp. 513-518, June 1984.

[109] G. Zames, "Feedback and optimal sensitivity: Model reference transformations, multiplicative seminorms and approximate inverses," *IEEE Trans. Auto. Control,* **AC-26,** pp. 301-320, April 1981.

[110] G. Zames and A. El-Sakkary, "Unstable systems and feedback: The gap metric," *Proc. Allerton Conf.,* pp. 380-385, 1980.

[111] G. Zames and B. A. Francis, "A new approach to classical frequency methods: feedback and minimax sensitivity," *IEEE Trans. Auto. Control,* **AC-28,** pp. 585-601, May 1983.

[112] O. Zariski and P. Samuel, *Commutative Algebra,* Vol. I, Van-Nostrand, New York, 1958.

Subject Index

U

W

Z